WHAT DOES A MARTIAN LOOK LIKE?

WHAT DOES A MARTIAN LOOK LIKE?

The Science of Extraterrestrial Life

JACK COHEN

and

IAN STEWART

John Wiley & Sons, Inc.

Published by John Wiley & Sons, Inc., Hoboken, New Jersey
Published simultaneously in Canada

First published in the United Kingdom by Ebury Press in 2002
under the title *Evolving the Alien*.

For general information on our other products and services, please contact
our Customer Care Department within the United States at (800) 762-2974,
outside the United States at (317) 572-3993 or fax (317) 572-4002.

Wiley also publishes its books in a variety of electronic formats. Some content that
appears in print may not be available in electronic books.

Library of Congress Cataloging-in-Publication Data:
Cohen, Jack.
What does a martian look like? : the science of extraterrestrial life /
Jack Cohen and Ian Stewart.
p.   cm.
Includes bibliographical references and index.
ISBN 0-471-26889-5 (cloth : alk. paper)
1. Life on other planets.   2. Exobiology.   I. Stewart, Ian, 1945–   II. Title.

QB54.C64 2002
576.8'39—dc21   2002068985

Printed in the United States of America

10 9 8 7 6 5 4 3 2 1

CONTENTS

The mind, that ocean where each kind
Does straight its own resemblance find;
Yet it creates, transcending these,
Far other worlds and other seas;
Annihilating all that's made
To a green thought in a green shade.

Andrew Marvell, The Garden, *1652*

Our biology, based on DNA and RNA and proteins, is an instantiation, a sufficient condition for Darwinian evolution. Could we now state all possible physical systems that might be capable of replication, heritable variation, and natural selection? I think not.

Stuart Kauffman, Investigations, *2000.*

PREFACE

There have been many books and television programmes about the origin of the universe, the early history of the solar system, and the origin(s) of life on Earth. They often mention the question of alien life, usually presented as 'life on other planets'. *What Does a Martian Look Like?* starts from the position established by such books and programmes: that there's a big universe out there with lots of planets more or less like Earth, and lots of other interesting things like neutron stars and solar atmospheres, Black Holes and immense dust clouds in which stars are forming. It addresses questions about life in that physical universe. How likely is it for life to appear on an aqueous planet, like Earth was four billion years ago? If we wait for evolution to proceed on such a planet, or look for one that's old enough already, will we find people? Klingons? ET? Unimaginable *somethings*? For that matter, what *is* life, anyway? Would we recognise alien life if we saw it? Given the rules that science has deduced about life here, can we work out what will happen, or *could* happen, elsewhere – and how it would be similar to, or different from, what has happened in our own little patch of universe? And what about other planets? Or places that aren't planets at all? Just how far away from our comfortable air-and-water world can we get, and still find life? Especially, intelligent life?

We will use serious science – orthodox biology, chemistry, astronomy, and physics – to offer some answers to these questions, but we don't think that being serious about an important subject forces us to wear a solemn face. Neither does it restrict us to sedate, difficult-to-understand scientific sources for our insights: we will also make use of ideas from the science fiction (henceforth SF) literature. We believe that the best SF writers, editors, and – occasionally – film producers have made some useful contributions to the *scientific* understanding of possibilities for alien lifeforms. Unfortunately, the area has also attracted, and accreted, great piles of rubbish, from the cutesy ET and 'alien' characters in midday TV series to the claims of UFO enthusiasts and the excesses of the 'aliens took my baby' crowd. Some of this material can be fun, but scientifically it's nonsense. That makes it a real challenge for both authors and readers to keep a clear head: to sort out the rational, exciting stories of what alien life might be doing, now, on other planets – and to distinguish them from wishful thinking about Paternal Aliens Come To Save Us From Ourselves, or other clichés like those of *Independence Day* and *Close Encounters of the Third Kind*. However enjoyable they might have been when you put your brain in neutral ('park' if you have automatic gears).

It's a difficult challenge: one that we all like to duck if we can. So we've used SF elements to lighten up some parts of the argument. You probably won't have read the same SF books as we have, though, so we've interspersed short synopses in boxes throughout the text. The aim of the boxes is to make it easy for you to relate to what we say about these books, even if you're not familiar with their contents. If you want to know more, you could always consider reading them. If you need an excuse, you could explain to your friends that you don't usually read rubbish like SF, but *this* is homework for a serious project. We have not given bibliographical details of these books because most of them are out of print, and all of them may reappear from new publishers at any time. Look on the internet. We've also started each chapter with a short piece of homespun SF: tales of two aliens, Cain and Abel, who are running tours to Earth's ecosystems. These vignettes are intended to help us see ourselves as others might see us, and put ourselves in an appropriate mood.

The genesis of this book goes back a long way, to an incident during Jack's first job. In 1958 he was a postdoc ('post-doctoral research assistant') at Birmingham University, in a Medical Research Council Unit run by Nigel Cruickshank. Jack was helping the British Association for the Advancement of Science to provide a Schools Lecture Service, involving local scientists and academics. Everyone taking part had to fill out a form describing their lectures, and Jack was busy writing titles like 'How does hair grow?' (his research topic then), 'Eggs, embryos, and adults', and 'Apes, angels, and ancestors' (he enjoyed arguing with creationists). Like all research scientists, he was reading the professional journals, but unlike nearly all research scientists, he was also reading SF, mostly in 'pulp' magazines like *Galaxy* and *Astounding Science Fiction*. There was a copy of *Astounding* on the lab bench next to the form, and Cruickshank noticed it. He picked up a pencil, and wrote, 'What does a Martian look like?' at the bottom of the lecture form. He and Jack looked at each other, suddenly seeing the possibilities, and Jack left the title in. In 1962, the first year of the Schools Lecture Service, this belated addition to the form received more requests than any other title except 'Science in the detection of crime'. In the second year, the physicist John Fremlin did a survey with the administrator, and called Jack into his office.

'Perhaps you could have a less . . . umm . . . populist . . . title? Give some of the other lecturers a chance? What's the lecture actually *about*?'

'It's about evolution, and what planets are like, and what we'd get if we evolved life in different circumstances . . . Actually, the circumstances wouldn't have to be very different to get different kinds of creatures, because evolution is as much accident as environment . . .'

The lecture was duly renamed 'The Possibility of Life on Other Planets'. In 1963 it beat the 'Science in the detection of crime' lecture as well, because all of the schools who'd been put off by the populist title asked for it under its new, respectable one. Between 1962 and 1986 Jack gave an evolving series of lectures with that title, and it became known to many organisations as POLOOP. He gave it to Astronomical Societies, Ladies' Sewing Groups, Biology Societies, an Evangelical Society, and a Jewish Society. He also delivered it as an adjunct to professional lectures when he visited the United States. It was given at least 360 times, and it changed as the scientific orthodoxy changed.

In 1968 Jack became entangled in SF fandom, because a technician in the Zoology Department had spotted the two shelves of SF books that he was using to expand the intellectual horizons of his zoology tutorial students. The technician alerted Peter Weston, an authoritative local fan with an almost professional interest in the author Robert A. Heinlein, who told Jack that there would be a con (SF convention) in Worcester in 1971, with all the big name authors in attendance. Indeed, in a week's time there would be a con in London with James Blish as Guest of Honour. That did it: Blish was Jack's most-admired author. So Jack went to the con, met Jim and Judy Blish, and started a close friendship that lasted till Jim's death in 1975. Jack's interest in alien design took a professional turn, and he has never looked back.

Previously, Jack had corresponded with and met the editor of *Astounding*, John Campbell Jr., several times between 1958 and 1964. During Campbell's editorship, *Astounding* was *the* premier SF magazine. In 1964 Jack also met and become friendly with the great Isaac Asimov, although Asimov was not primarily concerned with aliens – they hardly ever appear in his books. The main common interest was science, particularly its popular presentation. At the convention in Worcester, Jack met Anne McCaffrey, Brian Aldiss, Larry Niven, David Gerrold, Harry Harrison, Jim White and many other authors, and talked with them about credibility of aliens. POLOOP got its first feedback from SF fans, to mutual benefit. Then Jack spent nearly a week with Gerrold, setting up a credible, worked-out alien ecology whose creatures invade Earth. After some mishaps with publishers and long delays, this became the *Chtorr* series. The encounter with McCaffrey led to *Dragonsdawn* and other cooperations centred on the fictional world of Pern, which she had introduced in *Dragonflight* in 1968. Interactions with Harrison led to a highly credible biological background for the *West of Eden* trilogy. The relationship with Niven started slowly, when Niven found the McGuffin for *Legacy of Heorot* in some biology Jack was spreading around at the bar. (In SF parlance a McGuffin is a gadget or concept whose loss – or rumours of whose existence – causes the characters to search for it, fight for it, or flee from it, but which has no intrinsic meaning when the dust settles: Alfred Hitchcock originated the term in the context of thriller movies.) Niven then invited Jack to California to help design the ecology for the sequel, *Beowulf's Children* (*Dragons of Heorot* in the UK).

In 1995 Jack experienced a magical moment, of great significance for this book. He had been a poor boy in London at the end of the Second World War, and his father had been killed just after the war ended in 1945. So he earned money to go to university by collecting *Daphnia* (water-fleas) from local Emergency Water Supply ponds, and selling them to pet shops that dealt in tropical fish. He also collected balls of

Tubifex worms from sewage contamination along the banks of the river Thames, which gave a higher return on time and effort. Late one September afternoon in 1948, he had taken an SF magazine with a futuristic city on the front to the muddy, slimy riverbank, and tied his bicycle to a thick wooden beam projecting out of the mud. When he had collected a couple of jars full of worms he looked over at the bend in the Thames called the Isle of Dogs, just as the sun was illuminating its slums from behind his head. Inspired by the magazine cover, he vividly imagined the City of the Future that would rise there, with its impossibly tall buildings, helicopter pads, and elevated ramps for sleek, streamlined traffic. Nearly fifty years later, he was giving the POLOOP lecture to an SF audience at a con in Canary Wharf, a yuppie area of London on the site of the old Isle of Dogs, which looks remarkably like that SF City of the Future. He looked down from the futuristic hotel's twentieth floor at the bend in the Thames, and saw that very same wooden beam, still sticking out of the mud. He looked down into the eyes of his younger self. He told the audience what had happened – and realised that no editor would believe the story, just as the fifteen-year-old boy would never have believed his future.

This sounds like a prediction, but SF isn't really about predicting the future, although sometimes it manages to do just that *and* get it right. The true role of SF is more subtle: to explore the consequences of a 'what if?' assumption. But sometimes the answers have such a powerful effect on human minds that people *build* the future that the answer indicates. So it may have been with Canary Wharf, whose architects also grew up when the news-stands were full of pulp magazines with garish, futuristic covers. And so it may be with aliens: unless we have the imagination to understand their possibilities, we may never look for them in the right place, and so may never find them.

Ian's SF career started out much simpler than Jack's: from his early teens he collected and read thousands of SF books and magazines, and still does. He stayed away from cons and the other trappings of organised fandom. During a year in Connecticut in 1977–78 he met Ben Bova, the editor of *Analog* – the new name for the old *Astounding*. Bova convinced him to try his hand at writing SF short stories, and bought and published his first reasonable attempt, *And Master of One*. This was the first of several tales about a certain Billy the Joat (J.O.A.T.

– Jack of All Trades), a generalist of astonishing scientific and technological breadth with a well-honed contempt for bureaucrats.

Jack read the Joat stories, along with Ian's popular science book *Does God Play Dice?*, recognised a like mind, and got in touch. Together they produced two popular science books, *The Collapse of Chaos* and *Figments of Reality*. They also collaborated with Terry Pratchett on *The Science of Discworld* and its sequel *The Science of Discworld II: The Globe*; their SF novel *Wheelers* was published in 2000. When they started interacting and traded experiences, Ian was dragged into biological thinking much more happily than Jack had embraced mathematics, and became as frustrated as Jack was with sloppy thinking about aliens, particularly in the media. It therefore seemed natural, when publishers encouraged Jack to turn POLOOP into something more substantial, for Ian to come along for the ride: the collective entity that was Jack&Ian had already demonstrated how much bolder and brighter it was than either of us alone. So we decided to turn it loose on the question of alien life. As we expected, it didn't turn out entirely as we expected.

Jack&Ian
Coventry and Gorsley
in the auspicious year 2001

ACKNOWLEDGEMENTS

We inflicted the manuscript on many of our friends, and we are very grateful for all the comments. Ward Cooper, Jonathan Cowie, Dougal Dixon, Martyn Fogg, Henry Gee, Amanda Kear, Jackson Kirkman-Brown, Julian Moore, Spike Walker and Kotska Wallace gave very generously of their time. Greg Bear and Greg Benford, Russell Coope, Lionel Fanthorpe, Larry Niven and Jerry Pournelle, and Seth Shostak looked it over for us too. Thanks to all of you for your comments and suggestions.

Astrobiology and Xenoscience

*C*AIN AND ABEL *have walked and drifted in many strange places –* 'walked' *was not appropriate for many of them. Cain is a millipede four feet long, or alternatively one hundred and twenty-two feet along; Abel is – well, more of a gas cloud with a few solid bits, really. Our translator can give some flavour of their reactions to various Earth ecosystems, as they prepare a booklet for alien tourists. (The translator is necessarily imperfect; all alien concepts are replaced by the nearest human equivalent, however inappropriate.)*

They start, as they usually do, in a desert; it is a simple ecosystem with about 150 species. Abel sinks into the sand, while Cain turns over a flat rock. Under the rock is a small yellow scorpion, whose sting tries to penetrate Cain's armour and fails. Cain notes that the scorpion is nearly at the top of the ecosystem; he picks up a piece of its excreta and chews it thoughtfully. 'We should warn them about stings . . . and about this stuff,' *he says.* 'It is delicious, and could be addictive. This little chap ate two beetles last week. One was a fungus-feeder, one was a little carnivore; nematode worms, mostly.'

Abel lifts partly out of the sand. 'About three or four grams of algae every square metre,' *he says,* 'photosynthesising like mad. Lots of grazers on the algae, springtails, lots of kinds of mites . . . There's a fungus mat about ten centimetres down, catching all the excreta, shed skins, corpses; every couple of days it makes a set of spore-bodies at the surface, wafts spores around everywhere; only one in a million hatches, the others are eaten. Now, who's creaming off the top of this little system?'

Cain burrows feverishly in the surface layers, and reports: 'Down in that dip there's a frog who hasn't woken up for three years. It will be wet enough *for him this year, he'll eat a couple of beetles and go down again. And there's a dry stream bed with lots more life under its stones . . . a wind-scorpion who's just moulted: all it's had to eat is its shed cuticle; it might take a lizard some time in the next month.'*

'What is there for the tourists?'

'Um . . . quite a big gecko . . . and with some eggs it's guarding. And on the horizon, a maned wolf – they'll like that.'

'Enough here,' Abel declares, as he makes a few notes. 'More than enough to fill a prospectus.'

'Across the gulf of space, minds that are to our minds as ours are to those of the beasts that perish . . . regarded this Earth with envious eyes, and slowly and surely drew their plans against us.' In the opening paragraph of his 1897 story *The War of the Worlds*, the great H.G. Wells introduced his Victorian readers to the novel concept of an alien lifeform – a creature from another planet. (Anyone who has listened to the Jeff Wayne musical version narrated by Richard Burton will find 'minds immeasurably superior to ours' more familiar, which is less clumsy to the modern ear, but the above is what Wells actually wrote.) The wide Victorian readership of Wells's books and magazine stories found little difficulty in following his imagination out to the denizens of Mars, busily planning the invasion of Earth. The Victorians revelled in the harrowing tale of blood, death, and destruction as the Martians, disembodied brains enclosed in three-legged walking machines, laid waste to their home planet – or, at least, to England, which Wells pretty much identified as the home of humanity. And they were spellbound by the unexpected ending, in which Earth is saved not by Man, but by microbes. *The War of the Worlds* still has the power to thrill us, even in its mostly dreadful movie incarnations, and it has given us an icon for creatures from other planets that continues, subconsciously, to influence our thoughts about alien life.

Wells also gave us other memorable SF 'tropes' – generic story concepts – such as *The Time Machine* and *The Invisible Man*, concepts that most of his readers would otherwise not have entertained in all their lives. But there is a big difference between his Martians and those other concepts. Time machines and invisibility are simply fantastic – exercises of the imagination without any correspondence in reality, whereas Martians might really have existed earlier in that planet's history. In fact, as far as Wells knew, they might have been observing the Earth while he was writing his story. Envious eyes and all.

The War of the Worlds (H. G. Wells 1897)

Earth's observatories notice a brilliant light on Mars. A 'falling star' that lands near London turns out to be a cylindrical spaceship. A Martian emerges – 'a big, greyish, rounded bulk, about the size of a bear ... [it] glistened like wet leather ... There was a mouth under the eyes, the brim of which quivered and panted, and dropped saliva.' Others follow. The Martians construct giant tripodal fighting machines, whose terrible heat-rays terrorise the human inhabitants: Earth is being invaded by aliens. Panic-stricken crowds of refugees evacuate London and head for the coast, but few escape.

New machines like metallic spiders, with five legs and innumerable arms, appear. Inside them are Martians, who can now be seen to have the most grotesque appearance – living heads, little more than brains, with huge eyes but no nostrils, their mouths surrounded by sixteen tentacles arranged in two bunches of eight. Having no digestive organs, they take blood from living human victims and inject it into their own veins. They reproduce by budding, like the earthly micro-organism *Hydra*. Earth (or at least Southern England) is overrun by Martian vegetation, the Red Weed.

But there is a fatal flaw in the Martians' invasion plan. There are no bacteria on Mars, and 'directly those invaders arrived ... our microscopic allies began to work their overthrow ... they were irrevocably doomed.'

We now know that there are many planets out there in the galaxy, and we have good grounds for supposing that a number of these will have life. Not all scientists agree with these two statements, and we will discuss – and offer some answers to – their objections later; for now, assume that what we've said can be justified. If so, then life as we know it here on Earth is but one sample among many. And our biology, therefore, is but a small sample of all those extraterrestrial biologies. It is the only biology about which we know anything by direct observation, but we really do have a lot of information about its evolutionary history and its variety. Some two thousand years of study, more than half of it carried out in the last twenty years, has provided a large and reliable database against which to test our theories. We have a rapidly growing theoretical biology that sits squarely on our knowledge of life's history and function, its biochemistry and its adaptations, and to some extent its ecology, here on Earth. Presumably much of this theory can also be applied to life processes elsewhere: the process of natural selection, for example, applies to all simple lifeforms, though perhaps not to those that have organised their environment as human beings have.

Over the last decade or so, aliens have become scientifically respectable. Groups of scientists from widely scattered areas of expertise have become interested in extraterrestrial life. Evidence of possible life on Mars has hit the headlines. In 1996 a group of NASA scientists headed by David MacKay found tiny tube-like structures in the meteorite ALH84001, found in Antarctica in 1984. The tubes look rather like fossil bacteria, and there is good circumstantial evidence that the meteorite came to Earth from Mars, splashed off its surface by an impact. These 'nanobacteria' proved highly controversial; for a start, they were far smaller than Earthly bacteria. The controversy continues, but few scientists currently believe that the tubes have anything to do with Martian lifeforms. Nevertheless, the claim was taken seriously, and still is.

A new science of alien life has been emerging, but unfortunately in its most recent incarnation it has adopted the name *astrobiology*. It is – or, rather, claims to be – the serious science of alien life. The science is sound, often frontier work: planets encircling distant stars, the latest in genetics. As its name suggests, astrobiology is a fusion of modern astronomy and modern biology. So what's unfortunate about it?

Astronomy is about the stars, and other celestial bodies, and the key area here is the recent discovery of solid observational evidence that proves some stars other than our own *have* planets. Biology is also about a planet: this one. Of course the planetary aspects are in the background: biology is about living creatures that inhabit our planet. Biology is a vast and impressive subject, currently progressing at a remarkable rate; but in the context of alien life it suffers from one big drawback. It is restricted to those organisms that exist, or have existed, on Earth. Because we know so much about these organisms, it is all too easy to think that we therefore know a lot about life in general. However, there's no good reason to suppose that we do. The crucial question about alien life is: could it be *radically* different from life on Earth? If the answer is 'yes', then all that information about earthly lifeforms is actually misleading, and its quantity is irrelevant, along with most of its content. Of course the answer could be 'no', but our knowledge of how life works here doesn't imply that nothing else could do a similar job, and strongly suggests that there could be radically different alternatives. SETI, the Search for Extra-Terrestrial Intelligence, assumes that aliens will be much like us in all important respects, and in particular will develop technology and communicate by radio – but this assumption could be completely misguided. Absence of evidence is not evidence of absence, at least unless you've looked very hard for the evidence and not found any.

What this means is that astrobiology is the science of Earthlike planets supporting Earthlike life. It is, by and large, now. But many of the great pioneers like Carl Sagan, whose imagination was much wider, called themselves astrobiologists. And there are many mavericks now, esconced in that community but considering far-out possibilities. Today, then, astrobiology is an interesting topic, but it might be a very limited and narrow-minded one. And that's why the name, or more precisely, the attitude that it betrays, is unfortunate. Whether the current approach is too limited depends on whether the universe is like *Star Trek*, with humanoid aliens lurking on nearly every planet and few other kinds of alien anywhere (though *Star Trek* does also have an occasional very weird alien, such as a creature made solely of energy, to add a little variety without taxing the Effects Department too much). If we really live in a *Star Trek* universe, then astrobiology is entirely appropriate and we need nothing more; but we can't logically establish that if we start out by assuming it. It has to be a central part of the argument, not something that never even gets questioned. So, *whatever the answer,* astrobiology in its current form is by its nature too narrow-minded and too unimaginative to tackle the really big questions about aliens.

What else is there? Ever since the 1960s, the SF fraternity has been discussing *xenobiology*. This is a much better word. The Greek 'xenos' means 'strange', so xenobiology is, by definition, the biology of the strange. Of course very little xenobiology exists, though there is actually more than you might think, because many people have worked on theoretical alternatives to conventional life. Some references can be found in our technical reading list. But there's no useful body of *observational* xenobiology yet, and there won't be until (if ever) we come into contact with alien lifeforms.

Why, then, do today's scientists call the subject 'astrobiology'? Possibly because nobody learns Greek any more . . . but mainly because astrobiology was invented by astronomers who had a smattering of biology. If it had been invented by biologists who had a smattering of astronomy, it would have been called bioastronomy, but modern biologists are far too busy raising venture capital for biotechnology companies to worry about aliens. In fact, there is such a discipline. Jonathan Cowie lectures about it, telling his students as an example how coral layering in the Jurassic period can tell us that the Moon was nearer to the Earth then, with a shorter month. Either way, the name astrobiology itself is a warning: it betrays an unimaginative approach to a subject that absolutely cries out for imagination. Instead of opening up new worlds, new habitats, new types of lifelike organisation, astrobiology narrows everything down into two *existing* areas of science. One of which has its feet firmly set on Mother Earth, while the other is mostly looking for duplicates of Mother Earth.

There's a problem with the word 'xenobiology', too: it tacitly assumes that the way to make progress is to focus on the *biology* of aliens. In reality, the whole area has to be interdisciplinary. The biology is intimately entwined with the planetary science, and conversely. So as this book progresses, we will argue the case for a much wider kind of thinking – which, for ease of reference, we'll call 'xenoscience'. It's a pity that this is one of those graeco-latin hybrids, like 'television' and 'pentium', but 'xenology' doesn't sound like an interdisciplinary area, so we'll have to make the best of a bad job.

Our central theme will be the inadequacy of astrobiology and the need to replace it by xenoscience. Along the way we will point out some of the existing contributions to the foundations of this intriguing new science, and we will try to guess what xenoscience will eventually look like when some of the most glaring gaps are filled in. What we will *not* do is try to lay the foundations for xenoscience. That will need a lot of work by many people, and it can't be done in one popular science book. But we'll peer through the veil of the future, and try to see what might be built once those foundations have been made solid.

As things stand right now, xenoscience is a theoretical subject based on two kinds of information. Firstly, there is the one real example that we have to hand, which has generated an enormous database: Earth's biota. ('Biota' is a fancy word for 'lifeforms', the creatures that make up an ecosystem.) Secondly, and with far more authority than the bald data itself, we have the accumulated and tested knowledge of how Earth's living things began, and how they work, compete, die out, or change, through geological time.

Given this, it seems very strange that the most prolific, and apparently authoritative, science of extraterrestrial life has been written by astrophysicists. It is as if the chemistry of organic compounds had been written about by biologists, or the physics of stars by mathematicians. Of course, in appropriate circumstances it is entirely proper that this kind of discipline-hopping should occur: mathematicians, in particular, have made a speciality of extracting broad general concepts from a few examples, and then testing their universality very stringently. But it's not appropriate for a science of alien life. If we wish to apply the wisdom of biology to questions about aliens, then we must use the best biological knowledge that we have, both database and theory. Freshman Biology 1.01 is inadequate for the task. It is absurd to talk of the evolution of aliens, for example, using the 'folk' evolution models of the 1940s instead of today's better-informed models based in work on wild populations from the 1960s to the 1980s. Astrophysicists, by and large, seem not to have realised that there has been a revolution in evolutionary thinking during the 1980s, just as radical as the Newton/Einstein paradigm shift in physics.

One consequence of this reliance on folk biology is that astrobiologists are ruling out various scenarios for alien life, even though those scenarios already occur on this planet. Biology's view of life on Earth has undergone a revolution in recent years, with entirely viable lifeforms being discovered where previously they would have been considered impossible. The existence of 'extremophiles', bacteria-like organisms that live in boiling water or Antarctic cold, has overturned the conventional wisdom completely, and rewritten our theories of the origin of life. Other bacteria living high above the stratosphere, up where the Sun's ultraviolet radiation 'ought' to have killed them, make us wonder just how broad the range of habitats for Earthlike life might be. Computer experiments in 'artificial life' – mathematical systems designed to shed light on evolution – suggest that self-complicating systems arise with astonishing ease, and that complexity can grow out of simplicity of its own accord. It is therefore no longer clear that Earthly life, organised collections of big molecules based on carbon, is the only possible kind.

Against such a rapidly changing biological background, the contribution that astrophysicists *should* be making is the development of imaginative, front-line astrophysics. Instead of being surprised by the recent discovery that enormous numbers of planet-sized bodies exist in the void between the stars, they should have taken the possibility seriously long ago, if only on statistical grounds, and tried to work out whether such bodies could exist, and if so, how they might have formed. Now they're having to tear up their treasured belief that planets can be formed only as part of a star system, and play catch-up with observations.

Could an orphan planet like this support life? Surely not: with no nearby stars to warm it, the surface would freeze solid. Wouldn't it? Astrobiology's concept of habitable zones rules out such a world as a possible abode of life. But some xenoscientific thinking suggests that it might even harbour our kind of life. According to David Stevenson, radioactive elements like thorium-232 can provide a source of heat. And many of these planets will be covered in a thick blanket of molecular hydrogen from the protoplanetary nebula in which they formed. This blanket acts as an insulator, and could keep water liquid on such a world for ten billion years – twice the age of the Earth. So, even though the nearest star may be light years away, *Earthlike* life could still thrive on such a body.

Instead of looking for carbon copies of Earth, then, we ought to be theorising about and looking for the different kinds of planets, and other potential habitats for life, that exist out there in the wide universe. 'Exotic' habitats should be seen not as obstacles, but as opportunities;

instead of dismissing them with an airy wave of the hand and saying, 'Obviously life couldn't exist *there*', we ought to be asking, 'What would it have to be like if it did?'

We can get xenoscience off to a good start by recognising that its physics, and to some extent its chemistry, can be solidly rooted in observations. We know what stars are doing because we have observations of characteristic lines of emission and absorption in their spectra; we know what elements are there and approximately what quantities occur and in what layers of the star. Having found that about 2 per cent of stars have heavy planets – the only ones we can detect – we are confident that planetary systems are not rare, and may indeed be the rule. It would help if people stopped the silly practice of equating limitations of current observational techniques with limitations on the universe, though. We're getting rather tired of the claim that most solar systems contain weird, giant planets circling very close to their stars, when the real point is that these just happen to be the kinds of planets that are most easily detected with current methods. Again, absence of evidence is not evidence of absence.

At any rate, we know a lot about the properties of the planets in our solar system, especially the chemistry of their surfaces – but we also know that the biological possibilities of our own solar system, as seen through the eyes of physical scientists, gets them seriously wrong. Astrobiology was, and as we shall shortly see still is, very caught up on the concept of a 'habitable zone' around stars. By this is meant the region in which physical conditions – mostly the presence of liquid water – could lead to the production of human astrophysicists. A region running roughly from the orbit of Venus to the orbit of Mars was thought to be the spherical shell around our Sun that *alone* was favourable to the origin and maintenance of life. However, in the event much of that zone has disqualified itself. Venus seems to have suffered from a runaway greenhouse effect, making it hideously hot, dry, and full of sulphuric compounds. Mars seems to have been too small to hold on to its water, though the latest theory is that much of its atmosphere, and the water with it, was blown away by the solar wind when the planet's magnetic field died. This gives us the 'Goldilocks' view that Earth – like chair, bed, and porridge – is 'just right'. In consequence, the conventional astrobiological view is that the Sun's habitable zone is much narrower that we used to think, and that we are very fortunate indeed that the Earth has managed to scrape inside it.

This idea has led to very pessimistic views about the prevalence of life in our galaxy – if it's that difficult to pinpoint exactly the right distance from the Sun, how many extrasolar planets can have got it right? Very few, presumably . . . so much so that Earth might well be the only one! A typical example of this conservative way of thinking is *Rare Earth* by Peter Ward and Donald Brownlee, published in 2000, which is informative and rather too plausible for comfort, and we will be forced to defend our views against its line of thought on a number of occasions. Predictably, the authors are very uncritical about 'habitable zones', and trot them out at every opportunity. However, the belief in a very narrow habitable zone is a direct consequence of following a viewpoint that was mistaken in the first place. To clarify the position: the concept of 'habitable zone' involved here is that of a region round the star which, *without any modification or protection* (such as cooling by a cloudy atmosphere or reflective icecaps, or warming by the greenhouse effect) would be a comfortable temperature for human astrophysicists and their ancestors. Put a thermometer in empty space, and find how close it can get to the star, and how far away, while giving a reading that humans could tolerate. The habitable zone lies between two concentric spheres with those radii.

However, it has recently been discovered that Jupiter's satellite Europa has subsurface seas equal to or greater than Earth's in volume. This ocean lies beneath an ice layer tens of miles thick, and it exists because the satellite's hot core has melted the ice. Europa's ocean looks very promising as another local place to find life: it is entirely habitable (whether or not life does actually exist there) but is way outside the Sun's astrobiological 'habitable zone'. This is not the only example: Ganymede and Callisto, two other Jovian satellites, also probably have underground oceans, and Saturn's satellite Titan – even further outside the 'habitable zone' – is another place where some form of life might exist. So the astrobiological concept of a habitable zone is largely useless. Indeed, the Earth is probably outside it, and maintains a comfortable temperature only because its physical properties have been modified through co-evolution with its biology.

*

Habitable zones, 'just right' conditions, the kind of thinking that we associate with Goldilocks, are examples of a general kind of reasoning known as 'anthropic principles', made famous by John Barrow and Frank Tipler in their 1986 *The Anthropic Cosmological Principle*. The underlying idea has a compelling logic, and there's nothing much wrong with it, but its detailed elaboration has led to some seriously misguided thinking, as we will try to convince you now. The starting point is the question: 'Isn't it amazing that the universe is just the right sort of place to give rise to creatures like us?' Barrow and Tipler sensibly point out that this is a silly question. If the universe *wasn't* the right sort of place, then we wouldn't be here, so we wouldn't be asking the question. Yes, it *is* amazing that there *exists* a universe laden with such rich possibilities, such opportunities for the development of complex structures and patterns . . . but it's not amazing that (given we exist) we live in one. If the universe was unable to develop complex structures and patterns, then it wouldn't be suitable for a pattern as complex as us.

This argument, with its direct logic, constitutes the archetypal anthropic principle. The logic is a lot more subtle than it seems, and there are several standard mistakes that people often make when applying anthropic reasoning. They arise especially often in the area of astrobiology, where people are so impressed by the alleged 'fine tuning' of the solar system for human life that they conclude that no other kind of life is possible – and no other kind of solar system. Because if you changed *just one tiny thing*, lifeforms like ours could not exist. Change the balance of oxygen and nitrogen in the atmosphere *just a little,* make the oceans *just a bit* more or less salty, move the Earth *a few* million miles further in or out from the Sun . . . and we wouldn't be able to survive. How exquisitely finely tuned our existence is!

The same kind of argument turns up a lot in quasi-religious discussions of evolution and cosmology. The metabolic pathways of the cell are, if anything, even *more* finely tuned than the cosmological factors that influence Earth's ecology and climate. Change Planck's constant, which affects the laws of chemical bonding, by *just a little*, and carbon no longer forms long-chain organic molecules. Change the angles at which those bonds are placed by *the tiniest amount*, and the DNA double helix falls apart. Change the activation energy of ATP (it doesn't matter what any of that means, by the way) by *a few per cent*, and our cells wouldn't be able to get any energy.

Such statements make it sound as if life is incredibly fragile, as if we are perched on a knife-edge of existence with the abyss beneath us. They also lead people to claim that our universe, solar system, or cellular chemistry must be the *only* ones that can generate life. Obviously, these things have to be unique, if any single change, however tiny, makes them go wrong. Don't they?

No. We want to convince you that the kind of thinking involved is misleading, indeed wrong, even downright silly. The deduction of uniqueness is simply bad logic: the apparent fragility is an illusion for more subtle reasons. There is an element of misdirection in the way these anthropic arguments are always phrased, too, which is why we've italicised all the statements about how small the change can be and still make things go wrong. But small is not the point. Usually, big changes to those same quantities would also make everything go wrong (though not, it now transpires, when it comes to the hard core physics of the original Anthropic Principle, but we'll postpone that point until we've finished arguing this one). The misdirection is this: by making you think about the size of the change being contemplated, you are led not to think about the kind of change being contemplated. Which is: *only one thing at a time.*

We can use the same kind of reasoning to argue that your car cannot possibly work. If you made the tyres *the slightest amount smaller*, they wouldn't fit on the wheels. If there were *just a little bit* more water mixed in with the fuel, the engine wouldn't run. If the bolts had a thread *a few per cent different* in gauge from that in the nuts, all sorts of bits would fall off. And if the car were even a few millimetres further above the road, the wheels wouldn't touch the ground and it wouldn't be able to go anywhere.

All of the above statements are true. The last one is about how nature works rather than how cars are designed, but the others are part of a much longer statement of just how carefully a car has to be engineered before it runs at all, let alone reliably. However, it would be foolish to conclude from all this that there is only one design of car. On any road you see Fords, Volkswagens, Toyotas, Peugeots ... dozens of manufacturers, hundreds of models. And yes, if you try to replace the crankshaft of a Toyota with one from a Jaguar, it won't work. But that's because you do not get from a Toyota to a Jaguar by changing one tiny thing – or indeed by changing one thing. When, say, you make the crankshaft slightly shorter, the engine has to be modified so that the crankshaft still fits. And those changes alter the optimal settings for the valves, so those have to be changed too, along with the holes in which they are seated. Then the engine mountings must be redesigned to attach to the engine . . . and so on. You *can* get from a Toyota to a Jaguar, but only by making a coordinated suite of changes. No single change will work, though, and this is the logical fallacy in the deduction of uniqueness from anthropological arguments, because that's all that they ever address.

In *The Collapse of Chaos* we likened the situation to Conan Doyle's Sherlock Holmes stories. If you change *just one tiny thing* in a Sherlock Holmes story, it falls to pieces. *The Hound of the Baskervilles* works but *The Gerbil of the Baskervilles* does not. However, this does not mean that there is only one possible Sherlock Holmes story. In *The Red-Headed League*, a redhead is kept out of the way by persons of dubious intent, by the invention of a spurious society to which only redheads can belong. It wouldn't work if he was blond. But it *would* work if he was blond, *and* the perpetrators had invented *The Blond-Headed League*. Anything that works, in any kind of complex system – the universe, a cell, a car, a detective story – must be exquisitely 'fine-tuned'. That's what 'works' means. In the case of cars, it arises through the actions of a human designer; in Sherlock Holmes stories, through the action of a human author. But fine-tuning is not evidence of fragility, or 'difficulty', and it absolutely is not evidence of uniqueness.

Fine-tuning just reminds us that evolution and design home in on things that *work*.

That brings us to another, different point about fine-tuning, often confused or conflated with the uniqueness argument. Not only does fine tuning fail to guarantee uniqueness: it is not a valid argument for the existence of a designer. That was the whole point about Charles Darwin's theory of evolution. The clergyman William Paley pointed out that organisms were exquisite machines, like fine watches, and deduced from the existence of the 'watch' that there had to be a watchmaker. Darwin explained how organisms could become finely tuned to their environment without an organism-maker 'designing' them: natural selection would favour organisms that could survive for long enough to reproduce, and the better they fitted into their environment, the more chance they would have of doing that.

For similar reasons, alien organisms will not merely be Earth's organisms transferred to some other planet. They will not have to copy the finely tuned biochemistry that works here, even though it *is* finely tuned. Why not? Because evolution has fine-tuned it to work *here*, not on some other world. On another world, evolution will automatically fine-tune organisms to whatever the local environment may be – assuming that organisms can get going at all – and will thereby come up with a *different* evolutionary Sherlock Holmes story from ours.

The explanation of the apparent fine-tuning of the solar system – or the universe, or Earth, or Manhattan – for the existence of life is rather different: much closer to why car wheels do, in fact, touch the road. It is not that the universe was carefully set up to produce just the conditions that would suit *us*. It is that we evolved in a solar system that already existed, and evolution moulded us so that we fitted that solar system very well indeed.

Again, this is not a complete explanation of 'why we are here'. It makes it entirely clear that if we *are* here, then 'here' has to be the kind of place where we can survive. What it fails to explain is why the particular 'here' that exists is one in which anything interesting can happen. Why, for instance, does there exist a universe in which time passes, so that things can change? Yes, inside such a universe changes will occur – but why does such a universe exist at all? Some scientists think the answer is that in some sense all possible universes exist, so lifeforms appear in all those that are able to support them, and stupidly wonder why they're there. But this suggestion is rather speculative and not terribly helpful, because it fails to explain *why* all possible universes exist. And it offers no evidence that they do.

There is a more ambitious anthropic principle, which emerged from attempts to answer that much more difficult question. It is often called the Strong Anthropic Principle. It states that the reason why a universe can exist that is rich enough to give rise to things like us is because its *purpose* is to give rise to things like us. Planck's constant is what it is *in order for* carbon to be able to make big molecules, *in order for* humans to come into being. The main thing to appreciate about the Strong Anthropic Principle is that it does not have the same kind of compelling logic as the ordinary Anthropic Principle. Indeed, there is no clear logic to it at all. By analogy, think of someone who arrives at a dinner party having narrowly escaped a fatal accident on the highway. 'How amazing,' he says, 'that I am here to tell the tale.' But of course, if he hadn't avoided the accident, he wouldn't be here to tell the tale. So it's not amazing at all, and that's exactly the point of the ordinary Anthropic Principle. The Strong Anthropic Principle, in this analogy, would argue that the universe made sure that the accident would be avoided *in order for* the guest to appear at the party. The logic of the first situation is vivid, and it works on every such occasion. The logic of the second situation is non-existent, if only because it fails whenever someone does have a fatal accident on the way to a party.

As it happens, there is a more direct objection to the claim that universes like ours must be the only ones suitable for life. This deduction stems from the viewpoint of 'fundamental' physics. Physics describes the universe in terms of mathematical rules, 'laws of nature'. Those laws are elegant and concise, but they come with some added, unexplained baggage: various 'natural constants' like the speed of light, the charge on the electron, or Planck's constant in quantum theory. Most of the mathematics works whatever numerical values those constants have, but our universe uses specific numbers. The Anthropic Principle was an attempt to explain why it is those numbers and not others, and it did so by showing that if any one of the numbers is changed, then we wouldn't be around to object.

The classic example here is the 'carbon resonance' in red giant stars, the route whereby the universe made carbon, more technically known as the 'triple-alpha process'. We'll outline a few of the details, because they make it clear where the loopholes are, whereas the usual freewheeling description greatly exaggerates the numerological significance of this process.

In the triple-alpha process, three helium atoms collide and fuse to make a carbon atom. However, the odds on a triple collision occurring inside a star are very small. It is common enough for two helium atoms to collide, but it is highly unlikely that a third helium atom will crash into two others just as they are colliding. (Actually, in the hot plasma of a star the electrons of the atoms are stripped off, so we should say 'nuclei' instead of atoms. But 'atom' is a more familiar concept, so we'll ignore the missing electrons. By the time the carbon is found on a planet, the electrons have been replaced.)

Because the odds of a triple collision are so tiny, the synthesis of carbon must occur in a series of steps rather than all at once. First two helium atoms fuse; then a third helium atom fuses with the result. The first step leads to an isotope of beryllium, beryllium–8. Ordinary beryllium is beryllium–9, with an extra neutron, and is extremely stable. However, the lifetime of beryllium–8 is a mere 10^{-16} seconds, which gives that third helium atom a very small target to aim at. Calculations suggest that the universe hasn't been around long enough for all of the carbon that we observe to be made in this way. Unless – and this seemed to be the only alternative – the energy of carbon was very close to the combined energies of beryllium–8 and helium, so that if the third atom did arrive at the right time a stable carbon atom would very likely result.

A near-equality of energies like this is called a 'nuclear resonance'. In the 1950s Fred Hoyle predicted that there must exist a state of the carbon atom whose energy was about 7.6 million electron-volts above carbon's lowest-energy 'ground state'. Before the end of the decade, a state with energy 7.6549 million electron-volts was discovered. Now, the energies of beryllium–8 and helium add up to 7.3667 million electron-volts, so the resonant state of carbon has 4 per cent too much energy. Where is that extra energy to come from? It is supplied, with exquisite precision, by the temperature of the red giant star. It all seems very delicate. If the fundamental constants of the universe were different, so would that vital 7.6549 be. Different constants, no carbon. And without carbon, no *us*, indeed no life of the kind found on Earth. So our universe seems to be fine-tuned for the production of carbon, making it very special.

Impressive stuff, eh? Not really. For a start, the argument assumes that the temperature of the red giant and the 4 per cent energy discrepancy are independent. It assumes that you can change the fundamental constants of physics without affecting how a red giant works. That is plain silly. The structure of stars includes a built-in thermostat, which adjusts the temperature to make the reaction go at the correct rate. So the resonance in the triple-alpha process is about as amazing as the fact that the temperature in a fire is just right to burn coal, when in fact that temperature is *caused* by the chemical reaction that burns the coal. As soon as the fudge factor of the temperature of the star is permitted, the anthropic reasoning comes to pieces.

We've kept up our sleeve that there is a new theory, 2001 vintage, which casts doubt on the proposition that carbon production by red giants was ever important in the evolution of life on Earth. The theory does not concern carbon directly, but it changes our view of the production of all of the elements that are heavier than helium. The standard theory is that these elements were created many billions of years ago by nuclear reactions in large stars, which then exploded, scattering the heavy elements all over the galaxy. However, many meteorites contain elements that form from the radioactive decay of unusual isotopes of aluminium and calcium, and these must have formed at about the time the solar system originated. The orthodox picture of the early solar system is much too sedate to generate such elements, so it has been suggested that one of those large stars must have exploded somewhere close to the solar system, just as it was first forming. Such an event is highly unlikely; and if it really did happen, that adds weight to the idea that our solar system is very special and rare.

In 2001, however, a team of astronomers including Eric Feigelson discovered thirty-one young stars in the Orion nebula, all about the same size as our Sun. The big surprise was that they were extremely active, emitting flares of x-rays a hundred times as powerful as any that the Sun emits today, and a hundred times as frequently. The protons accompanying those flares must be energetic enough to create numerous heavy elements in any dust disc surrounding the star. If our own Sun was similarly active early in its history, those heavy elements that puzzle astrophysicists could have been produced right there in the dust disc. The important message here is not just about aluminium and calcium: it is that our entire picture of the production of elements heavier than helium probably needs revising. And that is likely to change our picture of where carbon comes from, and how special the universe has to be to produce it. It is dangerous to think that the universe must be special, merely because human scientists can't imagine alternatives to some improbable scenario. And especially when they haven't tried very hard.

Frankly, we find it rather disappointing that cosmologists can think of no more imaginative way to produce novel rules for a universe than to alter its fundamental constants. The SF literature beats them hands down, for instance with Philip José Farmer's 'pocket universes' in his World of Tiers series, or Colin Kapp's *The Dark Mind*. But in 2001 the physicist Anthony Aguirre discovered that even within this limited realm of changes, and without taking account of self-adjusting fudge factors like the temperature of a red giant, devotees of the Anthropic Principle had overstated their case. He found an entirely different set of numbers for the physical constants that would also fit the bill.

Aguirre restricted himself to the same kind of viewpoint espoused by Barrow and Tipler, taking the conditions for life to be things like the presence of carbon for long-chain molecules, heavy elements like calcium and iron, plus stars that remained stable for billions of years so that evolution could get going on their planets. In theoretical calculations, he found that the values of the physical constants in our universe are *not* the only ones that satisfy these conditions. Our universe started with a 'hot' Big Bang, and heavy elements formed only after the universe had cooled. Carbon formed in a very specific and delicate nuclear reaction in stars; heavy elements were mainly formed by supernovas (exploding stars). The alternative universe found by Aguirre starts with a 'cold' Big Bang, and carbon and the heavy elements all form during the first minute or so.

You can't get to it from our universe just by changing any particular constant, but you can get it by changing them all in a coordinated way. If our universe is the Toyota, Aguirre's is the Jaguar: they both work, but their parts are not interchangeable.

We're sure you will have got the message by now, and can readily spot this fallacy in reasoning. But to set the stage for further developments, we want to add to our argument a geometric image, the image of 'phase space'. This rather clever idea was invented by the mathematician Henri Poincaré about a century ago, and since then it has quietly pervaded every area of science, if not beyond. A phase space, for some phenomenon, is an imaginary 'space' whose points correspond to all the possible things that might occur. Not just those that *do* occur – those that, in some re-run of the universe, could reasonably occur instead. Phase spaces have a concept of 'closeness' or 'distance': things that differ from each other only a little are considered to be close together in their phase space; things that differ a lot are a long way apart. For example, DNA-space comprises all possible DNA sequences, and sequences that are very similar are close neighbours. Recall that an organism's DNA is a long series of four special molecules, called 'bases' or 'nucleotides', strung together: these molecules are adenine, thymine, guanine, and cytosine, usually identified by their initials A, T, G, and C. In this notation sequences AATGGCCA and ATTGGCCA are close together, differing in only one entry, the second; but TGGATTTG is a long way away from them both. Organism-space (more technically, phenotypic space) comprises all possible organisms, and organisms that are very similar are close neighbours – so a canary is closer to a seagull than either is to a hippopotamus. Sherlock-Holmes-space is the space of all Sherlock Holmes stories, including huge numbers of them not yet written, and stories that differ only by a few words here and there, or by routine changes to plot elements, are close neighbours. Car-space comprises all possible cars, and cars that are very similar are close neighbours – choose your own examples for distances – and so on.

When we ask, 'Why is our universe like it is?' we are implicitly contemplating that it might be different: we are thinking about other possible 'points' in universe-space. When we ask, 'Why are our metabolic pathways like they are?' we are implicitly thinking about metabolic-pathway-space. In this image, the problem with 'change just one tiny thing' is that when you change *one* thing, only, you move along one of the 'axes' of phase space. You go north, say. If you change some other thing, you go along another axis – say to the east. The point is that by starting at one particular point and changing only one factor at a time, you do not explore the whole phase space. You explore the axes – only. If you fail to appreciate that, then you will find yourself doing the scientific equivalent of claiming to have explored the whole of London, when all you've visited are Oxford Street and Charing

Cross Road. Or claiming to have explored the whole of New York, when all you've visited are Wall Street and Broadway.

While we're clearing up potential misunderstandings about ways of thinking, we'd better tackle another: the role and limitations of theories. Because humanity has not yet met any aliens, our discussion in this book must rely on theory. There is a widespread misconception that anything that is 'only a theory' can safely be ignored if you don't like its implications. Creationists, for example, make much of the folk belief that because the theory of evolution is 'only' a theory, people have a lot of choice about whether to believe it or not. On the contrary: the most tested, most reliable parts of science are the theories.

Strictly speaking, the dismissible 'only' opinions should be called *hypotheses*, not theories. A hypothesis is a suggestion put up in order to be shot down – if possible. Theories are coherent logical structures that encompass a broad and diverse range of observations; they have been up for disproof or replacement for a long time and have survived all of the alternative theories that have been proposed to explain the same data. Their survival qualifies them to be the best guess that is available, and you ignore them at your peril. Theories are what physicists, chemists and biologists use to understand their subjects as competent professionals. No scientific theory can ever be considered *true*, because the universe can always have some unexpected tricks up its sleeve; but it is foolish and ignorant to dismiss a theory merely because it *is* a theory. You have to assess the evidence, and if you don't like the theory you've got to come up with one that fits the observations *better*. And you have to do an honest job of looking for observations that it doesn't fit; you can't just pick and mix, keeping anything that seems consistent with the theory and ignoring anything that isn't.

To show you how far removed real evolutionary thinking is from the caricatures employed by creationists and their ilk, let's consider some real Earth-based evolutionary science, the evolution of swim bladders in modern bony fishes like goldfish, cod, and tuna. The jargon for such fishes is *teleosts*. The teleosts are the last, most recent, burst of evolutionary innovation among the vertebrates. Fishes have a very good fossil record – but the swim bladder doesn't fossilise. Nevertheless, we have a good idea of the early evolution of bony fishes, and we have some theories about their lungs and swim bladders. We can make good guesses about whether they had effective hydrostatic bladders, or just lungs, by determining whether their tails were symmetrical, which tells us that they were neutrally buoyant; they just sat there at any depth like modern teleosts.

We can also make comparisons with certain modern but archaic-looking fishes, such as lungfish and garpikes, for which there exist good fossil lineages. The pattern of blood vessels in the development of these modern forms, especially those archaic-looking 'primitive' forms, shows us how the gas glands, which teleosts use to secrete and absorb gases from the closed bladder, take the place of the open tube to the gut. We now have a good DNA story too, which confirms that these no-longer-missing links fit into the right place in the fish's family tree. This theoretical story is enormously convincing: the DNA, the blood vessels, the way modern representatives of halfway forms swim, the embryology, and the fossil series all fit together very nicely. That is to say, not perfectly, because we have doubtless misinterpreted some of the fossils, a few of the DNA sequences might be carelessly assembled, and the embryology is difficult: for instance, the eggs have lots of yolk, which makes development unusual in lungfish and garpikes. But with a couple of caveats here and there, the evolutionary tale is very reliable. As in most science, the overall story, the *theory*, is much stronger than any individual bit of evidence.

That was just a tiny bit of the theory of evolution, which is made up of thousands of similar stories, all well attested and well tested. It bears little resemblance to the naive versions criticised by creationists. There has also been recent work on the mechanisms by which species change, and here it seems that there are various rather different systems at work, in different circumstances and for different kinds of organisms. There are lots of mathematical models, some of which have had many validations from real life, while others have their being only in the most fantastic realms of the imagination. Later we will explain how these well-established theories of the origins and evolution of life on this planet can illuminate the situations we may reasonably suppose to have occurred, or to be occurring, on other Earthlike planets in our galaxy. And in places that aren't planets at all.

Lawrence Krauss's *The Physics of Star Trek* explained how the fantastic elements of that series could be achieved from science as we know it. We're not going to emulate that approach here, if only because most fictional aliens *can't* be achieved from science as we know it. We are, however, going to use well thought-out examples from the SF literature – not the film and TV stuff, by and large, because most of that is written by empty heads for empty heads – but novels that have extrapolated terrestrial biological theory to extraterrestrial locations sensibly, credibly, and often excitingly. Even where there are large gaps in the realisation of the extraterrestrial scenario – and there usually are, like the question of what *Dune*'s sandworms eat – we will use these examples to discuss exactly these questions of credibility. We have produced some of such fictional scenarios ourselves, and Jack has spent many years indulging in this almost-risk-free pleasure with many of the top authors, who have wanted just that extra bit of credibility – of make-believe, in the best sense – in their stories. The authors he works with do most of the writing, and they and the publishers take the risks: he suggests ideas for them to run with.

Most of these fictional extraterrestrial scenarios have been put together by novelists with little biological background. Because there has been much more publicity for the astrophysics, the authors have often bent the ear of a local astrophysicist or physicist to get a bit of help with orbits and things; occasionally they have asked chemists whether particular exotic atmospheres will 'work'. In contrast with these clever bits of 'real' science, they usually have such a strong 'feel' for the biology that they don't want to ask a biologist and complicate the situation. People, even novelists, think they 'know' about the biology. However, it seems that too many authors don't realise that it is possible to be sure and wrong; it feels just the same as being sure and right. So they just start in on the story once they've got the 'science research' – by which they mean the physics – right. They assemble a cast of more-or-less humanoid 'folk' aliens (six fingers, two hearts, copper-based blood), which are more or less vertebrate (reptilian ancestry, or altogether far too many felines) or robotic (but still with nose-above-mouth as if its ancestor was an Earth fish), and get on with the story. Reptile-based aliens have no finer feelings, feline aliens are carnivores at heart, robots can be relied on to do what only the most dastardly of Earthmen would consider . . .

Readers who have baulked at the astrophysics of a red star, which enters the Rukbat solar system for fifty years of 'Thread' menace every two hundred and fifty years, cheerfully accept alien fire lizards that are prototype dragons – but they make very cuddly toys, worn on shoulders at SF cons to identify the owner as a devotee of Anne McCaffrey's *Dragonflight* universe, centred on the planet Pern. We don't at all wish to be a wet blanket about this kind of fun with silly or incredible fictional aliens – we enjoy Keith Laumer's *Retief* stories about alien diplomats too, even though the scenarios involved are scientifically ludicrous. We just want SF fans to widen their experience to include the possible as well as the entertaining; non-fans can ignore the entertainment part. And we want scientists to appreciate the 'what if?' value of good, scientifically credible SF scenarios, and to learn to distinguish them from entertaining 'sci-fi' trash.

Now we come to a crunch, about SF tropes in general. Why, you may ask, do we worry about the biology of *Alien*, or *Dune*, when we don't worry about faster-than-light travel, or telepathy, or cloning identical people, or all the other practical impossibilities put in to make other SF stories work? There are two answers to this question, and the first concerns the spectrum of possible scenarios. There are some impossibilities that stretch credulity too far. SF is about 'suspension of disbelief', and disbelief cannot be suspended if it hangs by too thin a thread. If the heroine has breakfast in London, then gets on her bicycle and has lunch in New York, we want a good explanation. The question is obvious to everybody, and that bicycle had better be pretty spectacular (or does New York, in the future, have outliers in London?). A colleague of ours lives at the other end of that spectrum. He loves reading westerns – cowboy novels – and in Zane Grey paperbacks he can maintain his suspension of disbelief about how many shots can be fired, and how accurately, and so on. But he can't watch western movies, because he is a geologist (actually a palaeontologist). He *knows* that when the bad guys gallop up that rise and look straight down into the valley, the valley must be at least ten miles from the rise. He has special knowledge, which upsets his appreciation of western films.

We have special knowledge too, and it upsets our appreciation of the biology of *Star Trek* aliens. In this book, we share it with you in pursuit of a further goal, which is based in science and rationality: what can we sensibly say about the appearance of real aliens? In this connection, we *don't* go for suspension of disbelief. In scientific thinking, critical disbelief is important, and it should not be suspended. Many of the arguments we discuss here were originally put together to answer critical points made by our scientific colleagues. When we go and see *Star Wars* movies, on the other hand, we do suspend our disbelief, as far as we can . . . but we *can't* believe that aliens will look like people.

There is a second answer to why we accept faster-than-light travel, but cavil at *Dune* sandworms. It is a matter of how much work has been put in by the author or film director. Gadgetry like faster-than-light travel is usually 'justified' by a phrase, a bit of scientific gobbledegook, that shows they know there is a scientific problem – that the question has occurred to them. With a little more effort, they could do the same for any dodgy biological gimmicks . . . but they seldom do.

A glaring example of this discourtesy on the part of an author upset Jack when he was a boy, just starting to read SF. It's a problem with physics, not biology. In Jules Verne's *Twenty Thousand Leagues Under the Sea* the submarine *Nautilus* was very impressive. Verne had a reputation for accurate science, though in retrospect, we have a feeling that this rested on the presence of technical details, not the logic behind them, and that it was more smoke-and-mirrors. Anyway, he took care to give the submarine's dimensions, and the thickness of the steel hulls, and so on. And then Verne told his readers that 'anti-gravity rays' were stored in the fore and aft tanks, so that she would be buoyant in the water. Now Verne, and his readers, knew that steel ships can float without gravity rays; after all, such ships were a familiar part of their lives. If Verne had done the sums with his own figures, he could have worked out (as young Jack did) that making *Nautilus* just ten feet longer would have rendered it neutrally buoyant. That Verne had not bothered to think such a basic piece of engineering through was discourteous to his readership. Giving a silly answer showed that he had not taken the question seriously, even though steel ships are part of the story. So we would like to know that the makers of *Alien*, and Frank Herbert, had sat at a restaurant table disfiguring tablecloths and napkins with worries about the biology . . . but there's no evidence of that.

*

Many extraterrestrial stories do have the benefit of some biological credibility, however; either because, like David Brin, Harry Harrison and Hal Clement, the authors have some genuine feel for the subject, or because they have sought expert advice. Sometimes the advice doesn't 'take': many authors seem to believe that 'putting in some science' is the way to make any story credible to SF readers, and they just pick up a couple of 'good ideas' and weave them into a beautifully-crafted but ludicrous tale. Ian Watson has been guilty of several examples, particularly the River that separates the Goodies and the Baddies and which only his lead characters can cross. Sometimes the problem is the other way round: the author gets a 'good idea' for the story's central theme, and then after two years' hard work he or she realises that there's a logical inconsistency at the heart of the plot. Too many stories have an all-powerful alien race that comes in three-quarters of the way through the book to work some magic: 'We are so far in advance of you Earthmen that you will think this is magic – perhaps you had better treat us as your gods.' This frequent trick allows the author to persuade the reader that transcendent science has rescued the story from an otherwise incredible climax – think of *2001: A Space Odyssey* without the special effects. Sometimes the author has been a biologist and a good novelist (Thomas Bass's *Half Past Human* and *The Godwhale* spring to mind) or the biologists have been around at the initial crafting, as in the *Epona* series of Wolf Read, published in *Analog*. Epona is an invention of the internet-based 'Contact' group of SF fans, whose view of the SETI approach is thinly disguised contempt. They came up with a highly original scenario, and the stories are set within it. On the planet Epona, the land is temporary; the idea is to make the evolution of land life very difficult, and then see how it could happen anyway. The dominant species is the dracowolf, with flying intelligent forms. There are mobile trees. Species swap between plant and animal forms repeatedly as evolution pursues its path. The details have all been worked out with care and sound xenoscientific insight.

Jack, as will become clear, has contributed to the biological credibility of many of the more dramatic alien ecologies, but some of the best and most enjoyable alien stories – by authors like Clifford Simak, David Brin, Larry Niven, Jerry Pournelle, Jack Chalker, Harlan Ellison, Philip K. Dick, Greg Bear, and James White – have been produced without such advice. These examples have contributed to the genesis of this book far more than the bad ones. They have taken our imaginations off among the stars, and that's what we intend to do here, too. Not a *Physics of Star Trek*, but an imaginative discussion of what the biology of real aliens is about, with illustrations from many of the great works that have excited us.

Mission of Gravity (Hal Clement 1954)

Mission of Gravity is in many ways the archetype of alien planet settings, and the description is much more detailed than Wells's in *The War of the Worlds*. The story is set on the planet Mesklin, supposedly like two spinning dinner-plates with their edges opposed, with nearly 700g of gravity at the poles, but only 3g at the equator because of the mitigating effects of centrifugal force. (Normal Earth gravity is 1g.) It is the story of an odyssey undertaken by centipede-like Mesklinite sailors on a set of rafts, under the direction of an Earth-manned Galactic research vessel. They start from the North Pole and make their way to the South Pole to rescue a 'gravity-probe', sent there by humans who are in orbit around the planet and in radio contact with the Mesklinites.

Barlennan, the captain, and his First Mate Dondragmer experience numerous adventures on the journey. They meet lower-gravity cultures (with canoes, and even bows-and-arrows near the equator). The physics is dramatically illustrated by their adventures, especially its counter-intuitive aspects. For example, anywhere in the high-gravity regions, one seems to be in a bowl, because gravity affects the density of the atmosphere, which refracts light to give the 'bowl' illusion. So the aliens' maps are relief maps in bowls. There is a moderately successful attempt to portray certain ecosystems, especially the marine life, some of which is enormous compared to the tiny, high-gravity sailors.

A successor, *Star Light*, showed the Mesklinites investigating another high-gravity planet, and Clement has depicted other alien physics (and not-very-detailed biology) in many other stories. *Cycle of Fire* is set on a planet with alternating hot and cold phases with different lifeforms. *Close to Critical* involves a world with an atmosphere of sulphur dioxide, sulphur trioxide, water, and sulphuric acid, close to the 'critical point' at which all three can coexist. There are raindrops fifteen feet across, which temporarily suffocate the inhabitants.

In pursuit of that aim, we have not restricted our fanciful examples to those that have seen print before, by other hands or by ours. Hal Clement (the pseudonym of Harry Clement Stubbs) put the matter very clearly. When you buy a slim book of one of his novels, he explains, you also buy an invisible – much thicker – book of all the background thinking that went into making the context of the novel workable. His classic *Mission of Gravity* is probably the most widely recognised example of a story that depends on such an invisible book.

To take a more trivial example, if the author wants to have the spaceman run up the spaceship ladder with a large lump of gold – say a six-inch cube – in each hand, then the least of the problems will be that he's got to run up it no-hands. Gold is very heavy indeed, even though most of the readers won't have had experience of large lumps of it. Sometimes the author gets it wrong instructively, and then author and some of the more critical readers can wade into a long and informative discussion. There was a lot of fannish correspondence about a scene in Harrison's *Deathworld Two,* where the heroes lower a rope over a cliff – on a planet whose gravity is twice that of Earth's – and then swing over their enemies. Readers reckoned that the rope would not swing as a simple pendulum, but would have two or three waves down it, making the scene much more difficult to write, or indeed envisage. There was more than a year's correspondence about this in *Analog* magazine. And Niven was assailed at a worldcon by engineering students from MIT bearing banners saying 'The Ringworld is unstable'. They'd done the sums, for the fun of it.

We mention these examples to show that there is a large coterie of SF fans who enjoy reading the invisible book too. For them, mostly, we have invented some alien scenarios, to make some of the points about what aliens could actually look like and how they could behave, as distinct from being able to show that they won't look like Earth creatures, but they might behave like them. We have chosen to do this from several viewpoints, and to keep the framework in, as it were, so you can see some of the more explicit steps that go into working out alien ecosystems for various frivolous or serious purposes. We have not done this so that you can lift the ready-made alien ecologies off our pages to set your stories in, as Ellison and others did with the *Medea* series, set on a planet with a commonly-agreed astrophysics and a rather nice but far too terrestrial biota, but of course you can if you want to.

We do it to show the fun we get from this kind of multi-dimensional jigsaw puzzle, and to illustrate that the professional biology is somewhat different from the 'folk' biology so readily disbursed by astrobiologists.

There have been few, if any, fannish discussions about the biology of the various stories: just the physics. Perhaps the fans feel that biology isn't hard science, able to be argued about with figures on the restaurant napkin – or menu in more serious discussions. The physical sciences are rigid, governed by precise laws. Biology is flexible. There is considerable commitment in media science programmes, the SF fan community, and possibly the whole 'lay' appreciation of science, to the proposition that biology isn't a 'real' science. It's like psychology: mostly a matter of opinion. Actually, psychology has now earned its stripes by inventing techniques that let us watch the brain thinking. So far both the spatial and temporal resolution leave a lot to be desired, but watch this space. Equally, biology has developed many new ways to investigate life. Biology's flexibility implies that its rules are subtle, not that they don't exist. But the media have stuck to showing the techniques of biology and some of its data, so that's all that SF readers have been exposed to. The theoretical side of biology has not settled down to a series of sound bites yet, so it has not been disseminated by the usual Public Understanding of Science routes. But there is a vast and growing body of modern theoretical biology, and we will have to look at some of that before we can set up our alien biologies. We'll also have to clear away the fuzz that has made thinking about the real science of alien life so very difficult, from TV images and from conservative astrobiology colleagues who will doubtless be outraged by our purveying their subjects so naively. The fun bit – actually creating credible alien landscapes – can come only after a couple of chapters to set up the necessary context, but it's worth waiting for.

Not a bit like astrobiology.

2

THE INVISIBLE BOOK

C AIN AND ABEL *have been in the desert about a quarter of an hour, and feel that so much time spent has given them a good idea of what is going on in that ecosystem. Their jeep transforms back from the unremarkable rock that has been its most recent disguise, unfurls to let them in, and becomes invisible. They screech away in a cloud of sand, spoiling the illusion.*

Their next stop, about an hour later, is on a tidal beach. Rich pickings. These boundary ecosystems will give the tourists a chance to stretch their tentacles — or anything else stretchable — some of them down in the water, some of them up on the rocks under the seaweeds. Some could catch a quick meal, too. And no doubt will.

'I think you bashed a fender on the jeep,' Cain says accusingly.

'It will heal while we're exploring this beach. Wow, look at all these lifeforms! How many species do you think are here?'

'About half a million,' says Cain. Abel is a thin spray of mist lying along a fresh-water stream outlined in bright green algae as it runs into the sand. 'Half of them are bacteria and their viruses, and the grazers on the algae on all the wet sand grains; there must be thousands of kinds of grazers, none of them big enough for the tourists to see without technical assistance. It's the strand-line they all get caught on, with the sand-hoppers as the top of that system — the seagulls cream them off. Most of our tourists come from planets with small moons, not a monster like this one.'

Abel has found some more complex lifeforms. 'Yes, and look at all the seaweed on the rocks, graded in kinds according to how long they're exposed at each low tide. Barnacles and mussels and fan-worms filtering the water, limpets and winkles eating the weeds. Crabs and fishes, that little octopus waiting for the water to come in and flood its little cave. Fishes of twenty kinds, seals and porpoises further out, with luck we'll get some dolphins in for the tourists. Maybe a few humans, replaying that crucial time in their evolution . . .'

The emerging discipline of xenoscience is one of the few places where science and science fiction come together in a completely natural way. As we've just seen, the kind of thinking about potential alien lifeforms that can provide a sound foundation for xenoscience is very similar to the kind of thinking that ought to go into the planning of any work of 'hard SF' – science fiction that makes a serious attempt to get its science right. Aside from that little matter of faster-than-light travel or matter transmission that injects an element of speculation into the story, that is.

Many of you will have read the kind of SF that we have in mind, but some of you may not have done, and in any case we can't assume you've all read the *same* SF. However, it will be useful to have some common ground on which to base various discussions: an SF story that we can dissect and criticise, and which we can be sure you've read (or, at least, have been offered an opportunity to read). The only way we could think of to achieve that was to include such a story here – which in turn implied that it should be short, suitable, and preferably without copyright problems. So we're going to use one of our own SF stories – which has the added advantage that we're not going to annoy some innocent author by tearing his or her masterpiece to bits in the services of improving the foundations of xenoscience. The worst we can do is annoy each other, and we do that fairly often anyway. Of course, having broken the ice by criticising our own writing, we will also feel more comfortable about doing the same to other people's.

As it happens, the story we're going to use made its debut in an impeccable place, the scientific journal *Nature*. What was *Nature* doing publishing an SF story? It is one of the world's leading science journals – for instance, it published Francis Crick and James Watson's discovery of the double-helix structure of DNA – but it also contains editorial material, news, opinion, and occasional quirky pieces to keep its readers awake. What (especially to a readership of professional scientists) could be more quirky, yet more appropriate, than an SF story with a scientific sting in its tail? In fact, the journal published a whole series of SF stories: one every week for over a year. The series was called 'Futures', and it was partly *Nature*'s way of celebrating the millennium and partly the work of a faction within the journal that wanted to insert a little bit of SF between its covers. In all, more than fifty SF authors contributed. Each was given the same brief. Write about something that in your view typifies a possible science-related development over the next thousand years, keep the length to about a thousand words, and put a twist in the ending. Preferably, think about the story for a few days or weeks, and then write it all in one go. So that's what we did, and the editor kindly printed our tale last of all, to round off the series.

We'll present the story, without comment. Then we'll try to unwrap some of the thinking behind it, the allusions, Clement's 'invisible book' that lies behind the scenes in all hard SF. Possibly that discussion will convince some of our more sceptical readers that there is a little more to science fiction writing than they'd appreciated. Finally, having shown you what an amazing job we did in such a limited space, we'll analyse what we did wrong, by applying the philosophy of xenoscience to the scenario in which our tale is set.

First, the story . . .

'Monolith' (with apologies to Arthur C. Clarke)

At my age, you get nervous. Most of your friends have already discovered their role in life – male, female, caretaker. If you are a late developer, like me, though, you can't help worrying. So much depends on chance . . . There you lie, half-submerged in your comfortable bed of oxyhydride slush, wondering which role fate will select for you. Will you risk the danger of the skies in the hope of chancing upon an unencumbered female, to sink joyously into the huge mounds of her soft, yielding flesh, arresting your development at malehood? Will you wander for your entire adolescence, alone and unloved, until increment by tiny increment you grow into a fat, gravid female? Or will you squander your heredity in the comfort of the slush-marshes, until one day you discover a clutch of eggs, settle down, and become a caretaker?

One part of me once wanted to fly free, to risk the terrors of the open skies – even to make the perilous ascent to the ceiling of warm rock that roofs the world, keeping it safe from the fires above. In past ages belief in the overhead fires was little more than superstition, but modern science has shown that the fires really do exist: we have even observed hot, molten rock dripping from cracks in the cosmic ceiling, to cool into grotesque abstract stalactites as it encounters the dense fluid of the upper sky. Ours is a sophisticated, informed cosmology – the immense, finite-but-unbounded biolayer, sandwiched between the eternal fires of the rock ceiling and the indescribable cold of the undermarsh icefields. There are even savants who claim to know the total volume of the universe.

To fly . . . to become sexed . . . once, I thought to make the attempt. With tremulous strokes of my tailflukes I cast off from my familiar semifluid homelands, abandoning the safety of the slush-tower forest for the dark domain of the liquid skies. I say dark, but that is a conventional exaggeration, for everywhere the skies are lit by the glow of living creatures in myriad colours – an ever-changing panorama of subtle communication in languages as yet not understood. And that was my undoing. My confidence, I thought, was secure; somewhere in the distant heights was the rocky roof of the universe, and I would accept no lesser goal. Until I saw the gigantic, glowing mass of an engulfer hesitate, sniff the currents, and then – I still dissolve at the very recollection – turn towards me, and commence its hunting gyrations.

I was terrified. I fled. It followed. In a moment of unparalleled horror, I felt it engulf me . . . and pass by, suddenly revealed as a sham. I had fled in ignominy from nothing more fearsome than a shoal of nimmows, mimicking the bioluminescent spectrum of the most feared predator of our skies. Chastened, glowing a virulent green with embarrassment, I recognised that the path of parenthood was not for one as timid as me, and I allowed myself to settle back into the homely contortions of slush-towers and polylith-patches at the Bottom of the World. Then, finally, I knew myself to be a nonentity, doomed to become a caretaker; and thereafter my thoughts revolved solely around the size and proportions of the egg-clutch upon which I would eventually imprint. My greatest ambition was to assist in the hatching of the most numerous and most perfectly formed squablets that the universe has ever seen.

Although I continued to delude myself into the belief that I might yet regain the courage to become sexed, subconsciously I knew that my time was running out. Many times I passed the glowing lure of an egg-mass . . . until one day, inevitably, I blundered upon one that smelt of my family. At once I was entranced, ensnared, ensorcelled – lost. In that instant I became a caretaker. I tended my clutch with single-minded devotion. I flew endlessly around my eggs, fanning them with my dorsal fins to enhance the flow of nutrients. I chased off sundry predators, my fears now utterly subjugated to the hormonal imperatives of my caretaker role . . .

I was happy.

Then – tragedy struck. I still cannot understand it. It is a thing utterly beyond my comprehension. I do not believe it can be of this universe, yet this universe, by definition, constitutes all that there is. The Monolith is a transcendent mystery, which would be wonderful if it had involved anybody's brood other than my own.

This is what happened. Over many tedious circadian cycles I had assembled a most marvellous bed of decaying organics, ready for the implantation phase of my squablets' nurture. I painstakingly scraped away the patches of rampant polyliths that disfigured the serenity of the locale. Though I say it myself (and I am not one for self-aggrandisement) it was the most perfect carrion-midden ever prepared by a doting caretaker in the history of the universe. And then – barely a cycle before implantation was due – disaster struck. I was rearranging fronds of putrescent geloids to bring my masterwork even closer to the pinnacle of protectiveness when – without warning, with a reality that was total and brutal – something indescribably awful caused the entire slush-marsh to shudder. Then some ghastly *thing* poked up from the marsh, right through the middle of my carrion-midden, ruining my life's work. At first, I thought it some rapidly growing species of polylith. But as the sediment settled and the sheer weirdness of the thing began to impress itself on my vision, I saw that it was all of one piece – not poly, but mono.

The Monolith is still there. Occasionally, parts of it move. What it is, I do not know. Perhaps its most inexplicable feature is its markings, which scholars braver than I have since delineated by the light of their own protuberances. I record them here for your enlightenment, in the hope that someone cleverer than I might decipher their meaning:

NASA EUROPA EXPEDITION

SF *aficionados* will understand immediately what was going on in this sad but instructive story. Anyone who has not grown up with the genre may find it all a bit puzzling, and at least some of our friends and acquaintances have confessed that they didn't have the foggiest idea what it was about. That's understandable, and not our fault: if you had no idea what a six-gun or a rustler was, you'd have trouble understanding a cowboy story.

It is the last three words that give the game away. Up to that point, even a hardened SF fan would still be puzzled by the strange setting, an unbounded but finite world sandwiched between molten rock above and ice below. It would, perhaps, remind them of one of Farmer's pocket universes. But NASA EUROPA EXPEDITION reveals all. Why is it upside down? We're coming to that.

Of course, it only gives the game away if you have the right item in your personal scientific database. Hard SF is constructed on the common ground of such items, and its devotees file them away in their heads in the same way that a fan of westerns files away Colt 45, branding-iron, and saloon. This particular database item concerns Europa, one of Jupiter's four main satellites. That giant world has at least 38 others, at the most recent count, but when Galileo was gazing through his telescope he could see only four: Io, Europa, Ganymede, and Callisto. When NASA's Voyager probes passed through the Jovian system in 1979, humanity was treated to some of the most spectacular images of distant worlds that it had ever seen. Io was a planet of sulphur, with dozens of active volcanoes, looking – especially in colour-enhanced photographs – like a huge pizza. Europa was an enigmatic pale blue ball upon which the deity had doodled millions of red-brown lines, some thick, some thin. Ganymede was dominated by the concentric rings of a gigantic, ancient impact. And Callisto was dark and brooding, with craters everywhere, like our own Moon yet different.

Europa, it quickly became clear, has a surface made of ice. Ordinary ice, frozen water, not something more exotic like dry ice, frozen carbon dioxide. And it looks like it's relatively thin ice, which has cracked and shifted over the aeons. It may be only a few miles thick, it may be 60 miles (100km) thick, but compared to Europa's diameter it's pretty thin. The brownish-red doodles are cracks in the ice where material of a different colour has welled up. Ice is exciting because it signals the presence of water, which (the natural 'default' assumption of most scientists) is essential for life. On the other hand, it occurs here in frozen form, which (again, according to conventional wisdom) is not at all suitable for life. And ice was what everyone expected, because previous Earth-based observations made it likely, and because ice is a pretty common substance. Chemically, all you need to make ice is hydrogen – which is by far the most common element in the universe, amounting to about 94 per cent of all matter – and oxygen, which makes up about 0.1 per cent. Physically, what you need is *cold*, and there's plenty of that going if you don't get too close to stars. So Europa, Ganymede, and Callisto seemed to be ordinary, typical satellites, with very little going for them in terms of astrobiological interest.

Since 1995, though, when the Galileo spacecraft visited the Jovian system, there has been a dramatic shift in opinion about Europa as a potential abode for life. The possibility was first raised about 1983, but the 'habitable zone' enthusiasts pooh-poohed it. This was, perhaps, the first specific scientific discovery to reveal the ever-widening gap between astrobiology and xenoscience. The discovery that changed everything was the realisation, already mentioned in the preface, that Europa has an *ocean*. You can't see it on the surface, though, because it's not there: the surface, as it should be, is nothing but solid ice and the odd rock or two from impacting asteroids.

Europa's ocean is underground.

Most of Europa is a rocky core, and the ice forms a thick layer on top of the rock. It is now believed, on the basis of detailed observational evidence, that the core is hot enough to melt the lower layers of ice. Calculations suggest that the heat is probably not residual heat left over from the time when the satellite was molten, although more recent calculations related to another satellite, Callisto, suggest that the insulating effect of the top layer of ice may have been underestimated. Some of the heat may come from the decay of radioactive elements, but most of it is probably caused by Jupiter's strong gravitational field. Jupiter is so massive that nearby Io, Europa, and Ganymede are locked in a fierce mathematical embrace, a '1:2:4 resonance' in which Io revolves four times round the planet for every two revolutions of Europa and every one revolution of Ganymede. Callisto just fails to make it a 1:2:4:8 resonance; its orbital period is about 9.4 times that of Io, instead of a neat and tidy eight times. Moreover, all four satellites always present the same face to the parent planet, just as the Moon does to the Earth – and the cause is the same: 'tidal' friction.

As Jupiter's four main satellites perform their act of synchronised spaceflight, they are *squeezed* by Jupiter's gravitational field – a bit like those toy balls you can buy to relieve tension by squeezing them in your fist. A rocky satellite doesn't distort as obviously as a rubber ball when you squeeze it, but it does distort nevertheless; and the more it resists the distortion, the more friction there is, which generates heat. This friction is the cause of Io's prodigious volcanoes, and it is greatest on Io because that's the closest of the four satellites to Jupiter. Europa comes next, and although it's not as active as Io, it now seems clear that Europa's core is warm enough to melt much of its ice coating. Most likely, Europa's surface consists of about 1–10 miles (2–16km) of solid ice, with a thicker layer of slush beneath that, and all of it floating on a liquid ocean that may be 50–100 miles (80–170km) deep. How can we tell? The biggest clue is Europa's magnetic field, which does not behave like it would if the satellite were solid. Water with dissolved salts conducts electricity, and electricity produces magnetism: Europa's magnetic 'signature' looks like that of a world with a salty ocean. There also seem to be places where liquid water has welled up from beneath the surface, and various other collateral evidence.

If there really is an ocean beneath the Europan ice, then it has been there for a long time – probably over a billion years. And there's a good energy source, the hot rocks below. This conjunction of energy, water, and time makes Europa's ocean an ideal place for life to evolve.

By now you'll have got the point: 'Monolith' is actually set in Europa's ocean. 'Oxyhydride slush' is plain old ice slush – the usual posh scientific name for water is 'hydrogen monoxide' (H_2O, remember?), but the molecule could equally well be described as oxygen dihydride, so that's a small clue for anyone who can recall some very basic chemistry. However, Europa's ocean has ice on top, hot rocks below – so why does our creature think that the ice is below and the rocks are above? This is definitely not just a linguistic point: in SF, when you 'translate' an alien concept into English, you try to do so honestly. The point behind this is that if you're a slightly buoyant creature floating in an ocean, you will tend to move *upwards* – away from the satellite's centre. So you will feel as if 'gravity' is pulling you that way. In short, the Europan concept that naturally translates into the Terran 'down' happens, confusingly, to be what Terrans would usually term 'up'. ('Terran' is SF for 'Earthman'.) This is why we put NASA EUROPA EXPEDITION upside down at the story's climax. *Not* because it would necessarily appear that way to the creature – NASA can write on its equipment in any orientation it wants – but because psychologically, seeing the words upside down makes the reader more likely to realise that the whole *world* is upside down.

Finally, you need to know that at the time we wrote the story, NASA was planning an expedition to Europa to penetrate the ice layer and explore the ocean beneath. It won't get off the drawing board for a long, long time – if ever – but *Nature* offered us a thousand-year window. So what the story is about is this: a Europan creature suddenly coming into contact with an alien artefact, a NASA probe that has burrowed or melted its way through the ice and finally penetrates to the slushy zone where ice turns into water. We, the readers, then understand that the creature's view of its world is an inversion of our own view – but of course the poor beast has no idea what's happened.

That's the bare bones of it. Unless you know about Europa's ocean and the potential NASA visit, you won't understand what the story is about.

However, there's plenty more going on behind the scenes. The title is a second clue, and to decipher it you need to have seen the movie *2001: A Space Odyssey*, or to have read the book. The author was Arthur C. Clarke, one of the giants of SF, and the director of the movie was Stanley Kubrick. The part of the plot that is relevant here is that intervention by distant, unknowable, incredibly advanced aliens sets the Earth's apes on the evolutionary path to spaceflight. The immediate agent of this development is a strange monolith, a rectangular black slab. The apes dance around the slab, and it *does* things to their minds . . .

Our story, then, is a parody of this part of *2001*. But now the 'incredibly advanced aliens' are *us*, and the role of the apes is played by the Europan caretaker. The joke is that *we* have no idea what we're doing, we don't even know the Europan is down there. What it sees as a *2001*-style monolith is actually an accident.

2001: A Space Odyssey (Arthur C. Clarke 1968)

Three million years ago, in what will one day be called Africa, the ape Moon-Watcher comes face to face with a New Rock, a monolithic rectangular slab that has appeared from nowhere – though it *had* been preceded by a new bright light in the sky . . . The monolith manipulates his brain, and those of his fellows, and leads them to discover that a bone club can kill a marauding leopard. Cut to a spaceplane heading for the Moon: the apes have absorbed the lesson 'tools' and gone well beyond the Bone Age. In particular, they have absorbed the generic concept 'weapon' so well that there are now thirty-eight nuclear powers, with bombs in orbit.

On the Moon a similar monolith is discovered, buried in the crater Tycho. When it is disturbed, it suddenly transmits a signal beamed towards Saturn. Frank Poole and Dave Bowman go out in the *Discovery* to find out what's receiving the signal, but Poole is killed by the ship's rogue artificial intelligence HAL before they even get to Jupiter. Bowman shuts HAL down and continues the mission.

In orbit around Saturn's moon Japetus he finds a gigantic monolith. 'Call it the Star Gate. For three million years it had circled Saturn, waiting for a destiny that might never come . . . Now the long wait was ending. On yet another world, intelligence had been born and was escaping from its planetary cradle. An ancient experiment was about to reach its climax.' Bowman tries to land on the monolith, only to discover that 'The thing's hollow – it goes on forever – and – oh my God – *it's full of stars!*' He falls through the Star Gate, via the galaxy's equivalent of Grand Central Station, to a faraway alien stellar system, a huge, crimson sun orbited by a White Dwarf star. In the surface of the red star he glimpses travelling strings of bright beads, organised patches of hot gas hundreds of miles across . . . are they alive? He never finds out. He comes to rest in the most unlikely environment he could conceive of: 'an elegant, anonymous hotel suite that might have been in any large city on Earth.' The aliens – Clarke does not describe them, but we infer that they have powers beyond human conception – have prepared a familiar welcome for him.

When he sleeps, the hotel dissolves back into the mind of its creator. Bowman's mind is run in reverse, extracting all of his memories for safe keeping. At the same time, a replacement Bowman comes into being: a newborn baby. This 'Star-Child' now has those same inconceivable alien powers, and transports himself back to a thousand miles above the Earth, where he watches orbiting nuclear bombs becoming active. He detonates the bombs harmlessly, and then settles down, not quite sure what to do next with the planet of which he is now sole master . . .

'But he would think of something.'

So far, the only overt science that's been mentioned in relation to our tale is Europa's ocean. However, we also built in some speculations about the biology and culture of the aliens. Let's pull those speculations out, and see what science lies behind them.

For a start, our Europans have three 'sexes' – or, as we refer to them, roles: male, female, caretaker. Terran bees have a similar arrangement: drone, queen, worker. There is an interesting population genetics question here: what's 'in it' for worker genes when they don't get to reproduce? The narrative role of the three sexes is mainly to add local colour and show readers that this beast is genuinely alien. It also – unlike bees – has a choice of sex. That probably sounds really way out, but something very similar happens on Earth. This is an example of the gap between the biology known to most astronomers, which seems to be the basis for most astrobiology, and genuine Earth biology. Our own planet is far weirder than anything permitted elsewhere by astrobiological orthodoxy.

The Terran creature upon which our main Europan character is based is the deep-sea anglerfish. That's right, the funny-looking thing like a football with a light on a stick attached to its snout. The football-like beast is a female anglerfish: the male is very small, about finger-sized, and spends its life *attached to a female*. Moreover, a juvenile anglerfish's sex is not determined by its genetics: like many fish, it has the potential to assume either sex. The juvenile anglerfish swims around in the ocean depths looking for a female, and if it finds one, it attaches itself to her and develops into an adult male. If, after a reasonable time, it fails to find one, it continues growing and becomes a female. The trigger that determines the sex is not genetic: it is some chemical associated with females, probably a hormone, which causes the juvenile to develop into a male.

Our Europan is similar, but with the added ingredient of a third, neuter 'sex': caretaker. The hormonal trigger of the anglerfish is replaced by something more complex. Probably – we never really worked this out in detail because it's irrelevant to the narrative, the 'invisible book' has many missing pages and paragraphs – it is the warmth of the 'upper' ocean, near the hot rocks, that triggers development into male or female. The trigger for 'caretaker' is closer to home: encountering an egg-mass laid by close relatives. The signal here would be something like 'smell' – that is, biochemical 'identifiers'. The evolutionary undercurrent is interesting: only creatures that have the courage to brave the long, dangerous journey to the 'upper' ocean get to choose a sex and produce offspring. The timid ones wander around near the slush-layer, where the eggs are (by inference) laid, and eventually encounter egg-masses laid by their relatives and become caretakers. In a sense, there is a fourth 'sex': Not Yet Determined. Creatures too timid to choose a sex, and too insensitive to pick up the odours of their relatives' egg-masses, are rightly eliminated from any aspect of the breeding cycle.

Jurassic Park (Michael Crichton 1991)

The book is considerably more intelligent than the movie, which is a thin story held together by brilliant special effects, somewhat better animated than BBC television's *Walking with Dinosaurs*. This programme was known at the BBC Natural History Unit as 'Making it up as you go along With Dinosaurs'. Crichton writes technothrillers rather than SF: the distinction is obvious to any SF fan, but invisible to the outside world. You almost certainly know the movie plot, but we'll summarise its scientific basis, which is also the basis for the book version.

Mosquitoes that are 65 million years old and preserved in amber, contain small quantities of dinosaur blood, hence dinosaur DNA. By sequencing this DNA, *Jurassic Park*'s 'mad scientist' – he's not completely mad but even in the book he fits the standard movie icon – brings dinosaurs back to life. For safety, only female dinosaurs are created, so that if they escape they can't breed. Ian Malcolm, a visiting mathematician who knows about chaos theory and is generally rather cool, predicts that it is too complicated not to go horribly wrong, and of course he's proved right. Frog DNA has been used to fill gaps in the dinosaur sequence, which makes the female dinosaurs behave in some respects like frogs. Crucially frogs can *change sex* . . .

The scientific errors, unfortunately, are gross. In order to make a dinosaur, even nature needs more than its DNA sequence. In particular, it needs a female dinosaur to set the developing egg off on the correct trajectory. This 'dinosaur-and-egg' problem implies that you can't make a dinosaur unless you've already got one, in which case (give or take a male as well) you don't need to sequence ancient DNA in amber. As for the sex change: DNA 'instructions' depend on context. In the context of a frog, certain bits of frog DNA can help trigger a sex change. In the context of a dinosaur, those same bits of DNA probably do something very different: most likely they kill the developing embryo. If they do anything at all, they could just as plausibly make its skin slightly rougher, or change its eyes' sensitivity to light. It all depends on what other bits of DNA have done already, and on how the embryo has developed at the stage when the frog DNA is 'expressed', that is, acted upon.

It also turns out that eukaryote DNA more than a few tens of thousands of years old falls to pieces and can't be sequenced at all: claims of success in the scientific literature were due to modern contamination, not to ancient DNA. But that wasn't known for sure when Crichton wrote the book.

In this bit of the story we're riding one of our scientific hobbyhorses: the enormous scope for *non*-genetic factors in biological development. Since the 1950s the media, aided and abetted by a lot of professional geneticists who ought to know better, have created a myth of genetic determinism. The enormous prominence given to the magical powers of DNA has led a lot of people to believe that once you have listed an organism's DNA code – its *genome* – then you've pretty much understood that organism. After all, the genome is the information needed to make the organism, isn't it? And surely, once you know the information, you know everything about the organism? This is why thousands of private investors, and national governments, have poured tens of billions of dollars into biotechnology companies and academic research. They have been promised that once we know the DNA sequence of a human being, we will be able to cure hundreds of new diseases. Those promises have been made not just by the media, but by thousands of professional biologists who really should know better. In this myth, the *only* important thing you need to know about an organism is its genome. This is the belief that lies at the core of *Jurassic Park*: that the genetic 'blueprint' for the organism determines everything about it. However, organisms are far more complicated than just a 'message' written in DNA.

We don't want to rehearse the biological arguments at any length, because we've already discussed them in *The Collapse of Chaos*, *Figments of Reality,* and *The Science of Discworld* – but the list of things that affect how an organism develops, but aren't DNA, is enormous. Parents often supply 'privilege' – extra food, such as yolk in an egg or milk. The example of the anglerfish, already mentioned, is instructive: here the fish's DNA *does not even tell it which sex it should be.* Anyway, by making our Europan turn into a caretaker when it smells a familial egg-mass, we are hinting at this particular prejudice of ours.

Let's take this point one step further. All vertebrates start much the same: the embryos are very similar at what is called the *phylotypic stage* (stage typical of the phylum): all annelid worms look similar at their phy-lotypic stage, as do all gastropod molluscs (snails). Very different eggs (think of chicken egg, mammal egg, frog-spawn, caviar among verte-brates) all converge on to the phylotypic stage, and the strange thing is that they don't need their genetic instructions to do it. The egg architecture, plus a suite of clever molecules called informosomes copied from *mother's* genetics, guides the pre-phyletic embryo to its phylum-typical shape, with its nuclei in lots of different kinds of cells. These different cells call up a different developmental programme from each of their nuclei, so that dif-ferent genes are expressed in liver, kidney, nervous system and skin. In a very real sense, vertebrates are vertebrate because their mothers were: they made eggs that developed into vertebrate phylotypic embryos, which then read out their genes in the characteristic vertebrate way.

Because of this two-step development, it is much easier to have a well-controlled passage of information: think of the egg as the tape-player, the chromosomal DNA information as the tape, which has on it instructions for making the ovary that makes eggs – tape-players – of the appropriate kind. That rather destroys the background 'science' of *Jurassic Park*, because it suggests that the tape player evolves as well as the DNA tape. Ostrich eggs, and a bit of frog DNA to fill in the gaps in the tape, just won't work. Nature agrees: rat nuclei (complete good-condition nuclei, not DNA that's been through a mosquito gut and then been bombarded by cosmic rays for 70 million years . . .) won't develop properly in mouse eggs, and vice versa. And rat and mouse are a lot closer together than ostriches and tyrannosaurs, or frogs and velociraptors.

The future may have special tricks . . . but the *Jurassic Park* 'science' was based on 'folk' DNA genetics: the belief that the organism is only its DNA writ large. The film is an entertainment, not an educational tool, and it achieved its purpose. But we should bear in mind at least 'tape-player and tape', and not simple media-talk about the DNA being the blueprint for the organism.

We feel vindicated in this view by the recent 'triumph' of the Human Genome Project, whose most significant result from the first 'whole draft' is that humans have about 34,000 genes instead of the expected 100,000. On a 'blueprint' model, we need a lot more genes than other, less complex, organisms, in order to make us so amazingly complicated. Actually, humans have less than twice as many genes as a nematode worm's 19,099. So: stuff the blueprint model, along with all other thinking that imagines an organism to be a coded list of proteins. The human body contains far more proteins than genes, possibly ten times as many, certainly three times as many. So much for the standard assumption of molecular biology, that one gene makes one protein. This means that the human genome project was excellent basic science, because the first thing it turned up was strong evidence that its entire theoretical basis was wrong. This is how science *learns*: it makes mistakes and then corrects them. Disproofs are always an advance.

However, the same discovery knocks a big hole in the promises made to governments, investors, and the general public about all of those diseases that will be curable once we know the complete human genome. In effect, there aren't enough genes to cause all those fascinating diseases, so fixing the genes can't fix the diseases. Of course, humans are very complex, and a lot of that complexity may well be inherent in our genome(s) – but if so, it's certainly not there in any form that you can 'read off' from the genetic codebook. The geneticists, having belatedly been forced to admit all this, though not very candidly, are now busily making the same mistake all over again by enthusing about the 'proteome' – a complete list of every protein in the human body, along with 'the' function (assumed to be unique) of that protein. And a whole slew of other 'omes', such as the ugly and linguistically ignorant 'cellome'.

Our SF story exploits other sciences than biology, too. The Europan tales of molten rock coming 'down' from the roof of the world is modelled on the recently discovered 'black smokers' on the Earth's ocean floors, in the mid-ocean ridges where new material is welling up between continental plates that are spreading sideways. Nowadays there is a popular theory that this habitat, which to us seems bizarre, may have been the 'cradle of life', the place where *all* earthly life began. Not up at the ocean's surface, in the sunlight, but down in the murky dark, where the energy comes in the form of superheated water and sulphuric acid. We now know that bacterial-grade organisms (technically, we no longer call these particular ones bacteria) can survive – indeed thrive – in environments where the temperature is hotter than boiling water. We call such organisms 'extremophiles' – lovers of extremes. The very name betrays our inability to get away from our parochial prejudices: yes, such an environment seems extreme *to us* – but we're not the organisms that have to live there. To the extremophiles themselves, temperatures of 100°C are nice and comfortable: *we* would be the ones living in an extreme environment, namely, an extremely cold one. There is an inversion of this in the SF literature: in E.E. Smith's *Spacehounds of IPC*, an alien species is held spellbound by the sight of an Earthman working, unprotected, with *molten ice*. Not only that: he takes his shirt off and allows the terrible heat to play against his uncovered skin.

Short though 'Monolith' is, it incorporates a whole series of biological 'universals' – general tricks that may be expected to occur in numerous alien environments, but in ways that cannot be predicted in detail because they would depend on local conditions. The fictional Europan ocean contains predators – 'engulfers' – who hunt the species to which our protagonist belongs. Hunting prey is a very general trick with evident evolutionary advantages, so we would expect predation to evolve quite soon after something evolves that can be made to play the role of prey. An equally generic trick for evading the attentions of predators, or to gain other evolutionary advantages, is mimicry. Creatures of one kind evolve the ability to resemble creatures of another kind. Here a shoal of 'nimmows' mimics the form and behaviour of the fearsome engulfer. The name, by the way, is an in-joke: see 'samlon' later. In this case the mimicry is a collective feature: it takes a large number of tiny creatures to produce the credible semblance of an engulfer. Shoals of fish in Earth's oceans may to some extent be playing a similar trick when they move 'as a whole' in response to unseen currents and swirls of water.

We have also endowed the Europan ecology with the property of bioluminescence. Some chemicals emit light, and many of Earth's creatures have latched on to this useful evolutionary trick. Well-known examples include glow-worms and fireflies, but the most spectacular examples occur in the deep ocean trenches, where bioluminescence is the only source of light. And we have equipped our creatures with the ability to produce, and sense, pheromones – special, highly characteristic identifying chemicals. Our protagonist is ensnared in the non-reproducing gender of a caretaker by an egg-mass that smells of its family – that is, it emits 'familiar' pheromones. Each of these chemical tricks is arguably generic, so it is legitimate to employ them in the Europan environment. What would be illegitimate would be to get too specific, and use the *same* chemicals that occur on Earth. This would be as bad as giving the Europan a terrestrial nose to 'smell' with.

You should now be aware that whatever the literary merits and demerits of 'Monolith', there is a lot more to the story than first meets the eye. The same goes for most 'hard SF': the aforementioned invisible book, which is often considerably longer than the story itself, and which informs the narrative to keep it scientifically honest. The intelligent way to read SF is mentally to reconstruct the invisible book for yourself, as you go along – to hone your xenoscientific reflexes.

What did we get *wrong*? Here are three xenoscientific criticisms. First, the story offers no hint about how Europan life might have evolved. At the very least, the invisible book should have worked this out – and if it had done so, we would have told a few parts of the story in a subtly different way. Maybe some references to Europan analogues of extremophile bacteria, say. Another problem is that having three sexes is a Terran parochial (modelled on 'queen, worker, drone' for bees). But our worst sin was to give our Europan several human-like emotions, such as embarrassment, fear, and pride. We did that for narrative ease and simplicity – in a story, readers must be able to identify with the drives and needs of the characters, or nothing makes sense. Aliens may well have emotions . . . but they are unlikely to correspond neatly with our own, which evolved to suit our particular conditions.

To end this chapter, let's close the circle by revisiting *The War of the Worlds* and subjecting it to the same kind of xenoscientific critical analysis. Wells's story is a powerful one, but its invisible book is not terribly well thought out, despite his reputation for getting his science right. By now you won't be surprised that he gets most of his *physical* science right, but falls down on the biology. His over-evolved, hi-tech Martians are well realised and as plausible in themselves as we could hope, especially for the period in which Wells wrote. But it is not at all plausible that they would be susceptible to Earthly bacteria, because Earth's pathogenic bacteria have co-evolved along with their hosts and are finely adapted to the host species. And it is even less likely that they could feed on human blood, or any kind of terrestrial blood, because Earth's biochemistry would not suit their alien metabolism. So two of the book's most dramatic scenes, including its surprise ending, don't actually work.

3

ALIENS, ALIENS AND ALIENS

*N*OW TO THE *forest. Grand beech trees reaching fifty metres into the sky, and a clearing where two had fallen and been chopped up by humans. Cain is stretched across one of the tree sections, counting the rings with his feet, while Abel has sunk right into the leaf-mould, leaving just a breath of vapour at the surface.*

'I love these economical systems,' says Abel. 'The fungi on these tree roots have stretched a hundred metres down and across, and have found magnesium- and zinc-rich rocks, and the minerals are being pumped back up into the tree. The fungi found the minerals only two years ago, the top two layers of leaf-mould are really beautifully nutritious – hang on, this system is running a bit nitrogen-short ... Aha, I've found it, there's a fungus here killing nematodes and springtails and sucking them dry, returning their nitrogen to the tree, that's why the carnivore levels are running a bit hungry.'

'About six hundred species in the mould,' says Abel, 'not counting the viruses.'

'There are birds here, little deer, rabbits and rats and mice and squirrels,' replies Cain, 'moths and butterflies and lots of little wasps whose babies grow up in caterpillars, hundreds of species of sap-feeders on each tree, greenfly and scales and mites, and all their parasitic wasps too. Centipedes and millipedes and earwigs. About another three hundred common kinds.'

'There's a badger sett over there,' Cain points out excitedly. 'We must warn the tourists not to get the badgers telling stories, they never stop.'

'I've got an idea,' responds Abel. 'The understory is mostly brambles and nettles, nothing to eat them down, about twenty ferns and mosses, some

lichens. The tourists will like this: they count the species they find, anyone with over five thousand gets a free trip next century.'

Our story 'Monolith' is fiction, but now we can start to turn the thinking behind it into a new science – xenoscience, the science of alien life. 'Alien' is a very potent word, with many deep-seated associations. Tens of thousands of people definitely believe, and millions are alleged to believe, that they have been involved in 'contact' with alien creatures, of a variety of (mostly humanlike) kinds. There is a huge literature about Unidentified Flying Objects (UFOs) and their purported extraterrestrial origins. The notorious incident in which parts of a strange 'craft' landed in a field near Roswell, New Mexico, somewhere during the first week of July 1947, played a major role in starting the UFO craze. It has spawned at least ten television spectaculars, some of them dressed up as 'investigations'. Roswell is close to a huge US Air Force base. 'Mac' Brazel, a New Mexican rancher, rode out with his neighbour's son and discovered some strange debris. A few days later he reported his find to Sheriff George Wilcox, who passed the information on to Major Jesse Marcel of the 509 Bomb Group. On 8 July the Commander of 509 Bomb Group issued a press release to the effect that wreckage of a crashed disk had been recovered. A few hours later a new press release contradicted the first, saying that a weather balloon had mistakenly been identified as wreckage of a flying saucer.

This 'Roswell incident' gave rise to the theory that the US government was secretly in contact with aliens, and that alien corpses had been recovered from the wreck and taken to the US Air Force's secret Area 51 installation at Groom Lake, Nevada – where, for example, the U–2 spy-plane and the F–117 stealth fighter were tested during development. Later, a grainy, out-of-focus film purporting to show the dissection of the aliens became public. Dave Thomas, writing in the January-February 1995 issue of the *Skeptical Inquirer*, gives strong evidence for a more prosaic theory. There was indeed a government cover-up, and the 'weather balloon' story was untrue, but it wasn't anything to do with aliens. The wreckage was actually that of a balloon flight forming part of Project Mogul, a top secret project to monitor Russian nuclear weapons using high-altitude microphones. One such train of balloons, flight number 4, was launched on 4 July 1947, and components of the balloon-train account for much of what seems to have been found on the ground. This debris is still believed by millions to have been the wreckage of an alien spacecraft, despite all the evidence to the contrary.

Roswell is a prime example of the cultural belief in 'aliens' muddying the waters of rationality and history. The success of *The X-files* shows just how influential such beliefs are.

The movie *Alien* and its sequels have been seen by more than a billion people. The biology involved in *Alien* is impossible, but it obviously gave most of its viewers what they wanted – a frisson of terror. Films of the supernatural, tied to ancient fears of the dark – 'ghoulies and ghosties and long-leggety beasties', in the words of the old Scottish prayer – have been very successful, even when tied to apparently scientific backgrounds, as in the classic *Frankenstein* movies. Or there may be no pretence at science and rationality at all, as in *Nightmare on Elm Street* and *The Witches of Eastwick*, where frightening magic is shown convincingly on screen.

We have a strong impression, however, that the *Alien* films had a much wider audience; they appealed to a broader public than the classical horror film or the modern special-effects-magic film. The horrific element of *Alien* touches something special. The larva bursting out of a human chest is an image that has taken on a mythic status in most places where the film has been shown. In contrast, the later mother alien is portrayed as a rather fanciful dragon, chasing and being chased in a saga that has not changed much since the medieval tale of *Sir Gawain and the Green Knight*. The third *Alien* film was intended to be creepy, and again had a scientifically unbelievable basis. If a culture can afford a 'prison planet', then it can turn the lights up a bit and not used chipped china mugs – unless you think they were cleverly crafted art pieces. But it went down well at the box office, because everyone knew that there would be an alien larva bursting out of somebody's chest again – it was too good a filmic image not to be squeezed for all the juice it had. Even killing off Sigourney Weaver in a bath of molten metal didn't stop her being resurrected for yet *another* sequel, either.

In much of the literature of SF and fantasy, as well as in film and TV versions, the alien makes its appearance in the role of *icon*. It is a symbol of all that we fear: it doesn't need to *work* on a rational level. Very few well-known SF series have left out aliens – Asimov's *Foundation* and *Robots* series come to mind – and nearly all of the best-known SF stories and films, like *The Day of the Triffids*, *Gremlins*, *ET*, *War of the Worlds*, and *Childhood's End*, are known by the aliens that play central roles.

How, and why, has the concept of the 'space alien' become such a potent image in our society? Both *ET* and *Star Trek* are enormously successful cultural icons, and this is clearly related to their attachment to 'space' and 'alien'. What does this phenomenon relate to? What potent mythic images in other societies does it resemble? Trolls in Norwegian myth? Leprechauns? Witches in medieval England or eighteenth-century New England? Sea serpents, or the 'Here Be Dragons' logos on the edges of the explored areas of old maps? Does our attraction to the alien resemble the mythic priesthood in the Judaeo-Christian Bible, bringing rules for living and hope for mankind from on high? Think of *Close Encounters of the Third Kind*, or the Martian-instructed Michael Smith in Robert Heinlein's *Stranger in a Strange Land*. We hope that you thought, 'Well, it's a bit like each of those' as you read that list. We all have lots of mythic and scientific links with the word 'alien'.

In this book we want to deal professionally with only one of these links. We will examine the likelihood of other lifeforms and other evolutions than those that have arisen in our own story here on Earth, and we will apply solid biological principles to those alien evolutionary stories. Many of the principles that have been derived from humanity's studies of terrestrial biology are applicable much more widely, and not only to our own evolutionary history on this particular planet. Some of those principles, such as the Darwin–Wallace principle of natural selection, should be applicable to all natural biologies everywhere. The reasoning is straightforward. We know that the physical and chemical principles discovered by human scientists apply widely in the universe, because we can see the absorption lines of familiar elements in the spectra of distant stars, and what we see there behaves just as it does on Earth. We also expect that general principles like natural selection will apply to lifeforms on planets around those distant stars, and even to lifeforms that aren't on planets at all.

*

The word 'alien' has other connotations, though, and we need to clear them away from our discussion of alien lifeforms, including some that have been presented as a factual addition to that discussion, like little green men in flying saucers and the 'Greys' – small humanoids with bulging, teardrop-shaped heads and big, dark eyes. We have to talk about the various usages of 'alien' first, and some of the myths associated with those usages, because the unexamined assumptions that they involve would otherwise cloud the issue of the real biology of real aliens.

In its regular cultural usage – for instance, in the Customs Department at an airport – the word 'alien' means 'strange', especially in the sense of coming from somewhere culturally different. 'Undesirable Aliens' are people who smuggle themselves into your country by nefarious means, and we reserve the phrase for people we expect to be really strange. (Mind you, many fictional aliens *are* undesirable: the 'Affronters' in Iain M. Banks's *Excession* are a glorious example.) Familiar or neighbouring cultures don't produce these 'aliens': the people that you call 'alien' are the odd guys from various parts of the world that you can't even place on a map with any confidence, who you don't think will understand how to live properly in your country. This judgment may be wrong, or it may not, but either way that's the attitude.

Excession (Iain M. Banks 1996)

Iain Banks writes mainstream novels and SF alternately, adding an 'M' to his name for the latter. Most of his SF novels are set in the Culture, a futuristic civilisation that combines individual freedom with a strong sense of galactic responsibility. In particular, the denizens of the Culture assume the role of galactic police, taking action – usually imaginative and often nasty – to prevent wars.

Excession introduces the Affronters, shaped like a slightly flattened ball six feet (two metres) across, suspended from a veined gas-sac that allows it to float. The main eyes and ears are on two stalks above the fore beak; a rear beak protects the creature's reproductive organs. Between six and eleven tentacles are attached to the central mass, of which at least four end in flat paddles. The number of limbs possessed by an adult Affronter largely depends on how many fights it has had, and who won.

Colonel Alien-Befriender (first class) Fivetide Humidyear VII of the Winterhunter tribe, an Affronter, is having a diplomatic dinner with the human Byr Genar-Hofoen in the regimental mess. A formal dinner with Affronters is an interesting affair, held around a collection of huge circular tables, all of which have a good view of the bait-pit where animal fights take place. Much of the food is alive and must be stabbed with a fork as it rushes past. The most interesting item of cutlery is a small harpoon, which is used to purloin food from the plates of fellow diners. The trick is to *flick* the line in such a manner as to avoid depositing the harpooned meat in the pit, where the scratchounds will get it.

Two thousand years ago an early Culture General Contact Unit came upon a dead star that appeared to be a trillion years old, and nearby they found a gigantic sphere emitting perfect black-body radiation. Three years later, when a follow-up expedition arrived, both had vanished. Now a very similar sphere has turned up, and the Culture faces an OCP – an Outside Context Problem. Most civilisations encounter such a problem only once, in much the manner that a sentence encounters its final full stop.

To complicate the politics, diplomacy breaks down and the Affront declares war on the Culture. The Affront is also very interested in the sphere, and the Culture desperately needs to understand it before the Affronters do.

In mundane fiction of the early twentieth century ('mundane' is a term that the SF world employs for everything from Shakespeare to Agatha Christie), it was common to have a Hungarian, a Mongolian or an unidentified Mittel-European play the role of Stranger. Chinese were useful, because they also looked recognisably different from westerners: Fu Manchu was obviously 'alien' to Sax Rohmer's western readers. If somewhere had to host vampires, then Romania had some nice wooded mountains and Transylvania was a very romantic name. Dracula couldn't have been Welsh or Canadian, though. Russians, suitably bearded – or particularly Mongols, with their associations to half-forgotten schoolbook mentions of Genghis Khan – could be brought into the storyline to do melodramatic things that your average Englishman or Yankee would have difficulty with, like chopping up young ladies. The beards were alien-looking, too, on naturally less hairy people. Chinese villains were icons: they didn't have to *do* anything except be evil, and wander about in depressing dock scenes. Nobody wondered about their mothers or sisters saying, 'Poor Ching has gone off to those Triads; I wonder if he'll remember to put on clean underwear every day?' Ching's function in the plot, like that of imperial stormtroopers in *Star Wars*, was to be killed routinely.

These culturally different alien people could also be supposed to instigate those amazing scenarios where the heroine is roped to railway lines in the path of a speeding locomotive, or the hero is left in a cellar with the fuse lit on a keg of gunpowder. No one that the reader or moviegoer actually *knew* would do anything so silly, but they could believe it of culturally different, *alien* people with strange beards who hailed from the unknown reaches of Transylvania or Manchuria. They could believe that Chinese iconic villains could devote all their (short) lives to pursuing vengeance for their Celestial Boss all over the world's docklands, only needing the odd bowl of rice.

That technique has been adopted in many so-called SF dramas, like *Star Trek* in its various incarnations, in order to bring in characters who have no cultural connections with the main plot. They don't use our money, and they aren't interested in our sex games. They are there to convey information to the reader or viewer about the assumed background, or to add a dash of local colour. Kear says that *Star Trek* aliens are all 'one small step' aliens – just one change from middle-class Californians: Vulcans are middle-class Californians who talk about logic a lot; Bajorans are middle-class Californians who talk about spirituality a lot; Klingons are middle-class Californians who talk about honour and war a lot. Sometimes they play the part of the Chorus, which goes back to the ancient Greek dramatists, explaining what's happening to the science-fictionally illiterate reader or viewer: 'Ah, Doctor Millefeuil, I see that your people have discovered the Gravity Transductor!' Mostly, however, as in the old Victorian melodramas, they provide unlikely plot elements that would be too unbelievable if the actions were performed by more familiar characters.

These 'aliens with unspeakable habits', with which *Star Trek* is riddled, throw no light on the biology of creatures living on the planets of other stars. However, not least because of *Star Trek* and *Babylon 5*, where humanoid creatures of various persuasions are involved in simple-minded plots with strong resemblances to old melodramas, it is very difficult to get that usage out of our minds when we think about real biologically-credible aliens. Klingons, ET, Greys, or alleged bodies being dissected after the Roswell incident (the associated alien-dissection movie was recently confirmed as a fake by a relative of one of the hoaxers) will not help us to think rationally about real biology that is wider than we've met on Earth. Nor, come to that, will the odd people supposed to be aliens in daytime TV shows, where someone (usually Robin Williams) is telepathic, can magic goods out of thin air, or can do other tricks, which save the writer from having to invent plausible plots. These zero-dimensional alien characters clutter people's minds with fantastic associations, and for the purpose of this book they hinder rather than help. They are like the Flintstone family on TV, which hinders our thinking about human evolution: cavemen and dinosaurs failed to coincide by more than sixty million years. Similarly, though, they can be fun to watch – which is why they have taken up such deep residence in our psyches.

Another associated word is 'alienation'. This is essentially a psychological usage. Someone who feels excluded from social acceptance and behaves in ways different from the social norm (Norman Bates in *Psycho*, for example, or a pacifist among the indelibly warlike Yanamomo tribe of Venezuela) may be said to be alienated, especially if it is added that some occurrence during that individual's development makes it plausible that their behaviour is reasonable *to them*. Our interest in this usage will concern cultures that intelligent aliens might have developed, and the extent to which we might understand those cultures. There are some very intelligent stories based on this scenario, where it is shown that in order to understand – or even to interact usefully with – a very different culture, we might need to become alienated ourselves. An example occurs in Blish's *A Case of Conscience*, in which a human biologist, who is also a Jesuit, attempts to save alien souls. His actions leads to tragedy because he and his church don't understand the nature of the aliens and their society. There are excellent mundane stories with this theme too, including *Captain Corelli's Mandolin* and *One Flew over the Cuckoo's Nest*.

Authors set up these alienated people, and the fictional problem is to convince the audience that, despite the alienation portrayed, we can understand the (usually tragic) alienated character. How much more difficult will it be to communicate with *really* alien lifeforms? Even fictionally? Most SF stories simply apply the Dr Dolittle model: this guy can understand the aliens because I, omniscient author, say he can. We did this ourselves with our speaker-to-aliens character Moses in *Wheelers*: guilty as charged. Sometimes, as in Mary Doria Russell's *The Sparrow*, the aliens are anatomically humanoid, and their terrifyingly cruel behaviour (in this case, remodelling people's hands without anaesthetic) is seen to be entirely proper, indeed charitable from their viewpoint.

All too often, though, this approach is a cheat, both in SF and in mundane fiction. The novelist's resonance is with alienated people, who might do these things for their own reasons, and the novelist relies on the reader's imagination to expand *human* alienation to the alien(ated) portrayal in the novel. We, the audience, think we can expand our imagination to do this, because we have read a lot of novels, have seen a lot of films and even TV soaps, in which our feeling that we understand the alienated character has apparently been validated. We can come away from such an experience with the strong feeling that if we had been that person, in that circumstance, with those constraints, and that strange desire to eat human flesh . . . or molest little boys . . . or kill a bull elephant . . . then we would have emerged from the experience in just that way. Sometimes the author puts in a twist, showing that we were nearly right, but because Amanda was having an affair with Horace, or because the African hunter tripped at just that moment . . . and then we come away feeling that we were very close to sorting out what the people really would have done in the circumstances.

What nonsense!

All that we have done is to come to the same conclusion as the author – or sometimes the producer, who changed the author's original ending because he found it unbelievable. There is no evidence – and the best psychologists agree – that we can take ourselves even a tiny bit out of our individual grooves. We simply cannot imagine what it is like to be a criminal, or a five-year-old girl, or – especially – someone who is genuinely alienated.

Neither can we imagine what it is like to be an animal. Some people are convinced that they know what their cat or their dog is thinking – and we're prepared to believe that they're right, much of the time. But this is only like following the author's line of thought in the resolution of a novel. We can't know what the cat or the dog *feels like* when it has that thought, that's the problem. A famous essay by the philosopher Thomas Nagel, called 'What is it like to be a bat?' has caused many people to reflect on these issues. The closest any human can come to feeling like a bat is to feel like a person who is trying very hard to feel like they imagine a bat must feel. We can't even imagine what animals that are closely related to us, chimps and gorillas in zoos, feel like. Jack used to visit the London Zoo's mad gorilla 'Guy' in the 1940s and 50s, and was very conscious that Guy was not simply 'a person in a bad way'. Animals that seem very close to humans, like bonobos ('pygmy chimps') give us (Jack and Ian) the same feeling that we get with small children: much of what they're doing can be sympathised or empathised with, but occasionally they do something quite mad – probably just for the hell of it – and we are bewildered.

If we can't really put ourselves inside the head of another mammal, or one of our own children, then how can we possibly grasp the truly alien? Yet we will see that a few authors can persuade their readers that they might. We are sceptical, but we get the same pleasure that you do by second-guessing authors or producers, and some of these stories are far enough away from ordinary human concerns for us to feel that our psychological reflexes have been severely tested.

There is an easy way to invent aliens that appeal to our cultural reflexes, and *seem* to be convincing and comprehensible. This is to base them on terrestrial animals. So we find cat-people and dog-people and lizard-people and bird-people. There is a general issue here, and we need to appreciate it so that we don't fall into this trap. In 1978 Paul Shepard wrote a book called *Thinking Animals*. Its message was that humans become thinking animals by thinking (about) animals. All human cultures use animal images to educate their children. Even the most urban of families has animal toys, even though they are often iconic animals like teddy bears, wombles, or Disney characters. The nursery stories that all humans hear when they are toddlers are about iconic animals of various kinds, and this is still true even though most urban children will never see a chicken except as a frozen corpse in the freezer, and certainly not a wolf – not even in a zoo. TV natural history programmes are extremely popular, but again what we are shown are iconic animals: not real smelly parasitised brutes going about their everyday business, but episodes of a cheetah Being A Carnivore, or a lobster Being Reproductive.

Iconic animals are culturally useful, because in all human languages many character traits are tied to nursery animal icons. The commonality of these stories enables everyone to understand Athene's Wise Owl, why the Wolf wants the Little Pigs, and why Chicken Licken thinks The Sky is Falling In.

Iconic animals are dangerous, for the same reason. It is customary, in science-based books such as this one, to pretend that readers are totally rational creatures. The readers collude with this custom, of course. They sit there reading with a serious face, saying 'Mary, it is a very good point this chap makes; I've never thought of it just that way before,' and pretending that they're changing their minds as the author tells them new things. It is customary, too, in this kind of book with scientific pretensions, to avow that the authors are rational souls, just like their readers. However, in this book we will not do this. *We know your secret.* You were not programmed rationally, like your computer – and neither were we. When you were born you already had a lot of specialised nerve-cell hardware in place, none of it designed to be rational (effective, yes, but that's quite a different matter). You had a whole lot of nervous connections that enabled you to suck, to eliminate excesses, to make vocal noises, to sleep. Very soon after, you could recognise a smile. That is to say, you could respond to an upwardly curved line anywhere on your retina by sending signals down just those nerves that made you yourself smile – and you didn't use a mirror to check you'd got it right: blind children smile too, in response to a pleasant stimulus. You were pre-wired . . . but not rationally.

Then your first real programming started to change that wiring. You were subjected to all kinds of different data inputs: the sounds you heard and the sights presented to you became your experiences, the accidental and purposeful touching and handling of objects chose what data you got. Those experiences rewired your nervous circuits. Babies are different from each other, mostly because of differences in genetics – but also because some mothers smoke and some don't, some listen to Mozart and some prefer the Beatles. However, two-year-olds differ from each other much more than babies do, because their brains are wired up differently. Your early learning experiences involved baby-language and cuddling from your mother, then strange nursery stories, songs and rhymes about a universe of talking animals . . . full of iconic princes, fairies, witches . . . You learned what 'sly' means, because it's what the Fox did in your stories.

Icons are symbols, not realities. Real foxes are not sly – they can't be, they don't think like humans, and 'sly' is an attribute of *people*. If you had been an Inuit (Eskimo) child, the Fox would have been your icon for 'brave' and 'fast'. A real fox can be fast, because that's not an emotional attribute, but it can't be brave in the human sense because bravery is a high-level human emotion. In order to be brave, you have to proceed with some course of action despite having an idea of what horrible consequences it might entail, and there's no evidence that the minds of foxes function on that level. A fox can be terrified, though, because that's a much more basic emotion. Even so, we cannot meaningfully compare what it feels like to be a terrified fox with what it feels like to be a terrified human – except that it's evidently a nasty experience for both.

Fox is the hero of Eskimo stories, but the villain in British or American ones. Neither of these is the actual biological fox, of course. Instead, it's the fox-in-the-mind that affects the way you think. Most modern westerners, partly at least because the universalising effect of Disney has replaced the old stories, share cunning foxes, wise owls, brave lions – and cute Mickey-style mice. We know what dinosaurs looked like, because the older ones among us learned it from *Fantasia* and the younger ones learned it from *Jurassic Park*. Our mental picture of the Stone Age is irrevocably contaminated by the Flintstones and Raquel Welch. So let's not pretend that we can simply and easily be rational beings about animals and evolution, about early human cultures, about all these things that moulded our minds. If we are to approach rational thinking about these emotive subjects, about which we all carry so much fantasy, we must work at it. And getting to know how iconic animals in the nursery, and aliens in the media, have affected us, is part of the challenge. By rising to that challenge, we can have a more credible claim to rationality when we think about the biology of extraterrestrial lifeforms.

Childhood's End (Arthur C. Clarke 1953)

The story opens on an Earth that has been shocked into submission following the arrival of the alien Overlords, who banished war and other evils from the planet, without ever leaving their spacecraft, which are suspended above Earth's principal cities. Under their benevolent dictatorship, the Earth becomes a scientific and industrial Utopia, with food and housing freely available. Humanity's ancient devils of aggression, prejudice, and greed are swept away. So much so that when the Overlords finally reveal themselves, the impact is minimal – even though 'there could have been few . . . anywhere in the world, who did not feel the ancient terror brush for one awful instant against their minds before reason banished it forever.' For the Overlords are, literally, devils: they have leathery wings, horns, a barbed tail . . . and they are as black as ebony.

They come, we are told, from a low-gravity world (hence the wings). They always wear a belt with many complex gadgets, to adjust their weight in Earth's uncomfortably high gravity. They dislike strong sunlight and wear dark glasses. They can breathe terrestrial air, but carry small canisters of gas from which they occasionally refresh themselves.

Humans live contentedly with their new masters for another fifty years. But then the Overlords' true intentions start to emerge. A child is born with strange powers, a sign for which the Overlords have been patiently waiting. More such children are born. Humanity is beginning to evolve into a new, and to the aliens more desirable, form. '*The stars are not for Man*,' say the aliens: unreconstructed humans are not to be trusted, and cannot join the Overmind, the totality of galactic races. Having succeeded in their aim of evolving humans into a more acceptable species, the Overlords depart, taking the children with them. Human civilisation disintegrates, an event witnessed by one man, who stows away on one of the Overlords' spacecraft.

Eighty years later he returns, and understands why the Overlords so closely resemble a human myth: the classic 'devil' is a racial memory of the Overlords, but one from the future, when humanity is falling apart. In the final pages, we learn that the Overlords, too, are barred from greatness by the forces of the Overmind – they too must adjust to the cosmic reality.

Religion, too, has permeated all cultures, and in the West the primary influence has been the Judaeo–Christian versions. The interaction of pagan threads with this dominant theology has provided many images

– witches, devils, Father Christmas – that have provided the basis for another set of myths and stories. In different countries, different images have been paramount, at least proverbially: banshees and leprechauns in Scotland and Ireland, trolls in Norway, werewolves and perhaps vampires in middle-Europe. With the arrival of a new spirit of enquiry at the end of the nineteenth century – one that pretends to rationalism and to a healthy disbelief in these older superstitions – there may be a lacuna, a hole needing to be filled. Perhaps people need 'the unknown' as a goad to those little thrills we get as we walk home on dark nights. It is very difficult, in our experience, to forget the irrational, the 'supernatural' contributions to our pictures of what the natural world is like. We all have elf-shaped circuits in our minds, and God-shaped ones, and devilish ones (a trait exploited to the full in Arthur C. Clarke's *Childhood's End*).

Many of the best SF and fantasy books exploit these peculiarities of how culture and evolution have wired our minds. J.R.R. Tolkien's *Lord of the Rings* trilogy is an excellent example of the deliberate use, by an intelligent, myth-conscious author, of a great kaleidoscope of these irrational images. We can all pick up a mind-picture of a dwarf, a hobbit, an orc, or an ent. Tolkien's friend C.S. Lewis, in his anti-science trilogy *Out of the Silent Planet*, *Voyage to Venus*, and *That Hideous Strength*, exploited these western cultural myths to associate science with devilish images. In his insidiously Christian-evangelical *The Lion, the Witch and the Wardrobe* he associated Christianity with the pure and powerful (remember Aslan the lion?), the innocent lamb and the Nature-bound fish.

In many places the space alien has supplanted the older ghosties and creepies as the Watcher in the Night, who will get you if you don't carry garlic, or mumble The Lord's Prayer, or – in parts of middle America, according to recent films about 'alien abductions' – carry a gun. Some questionnaire results have suggested that one in three Middle Americans claims to have undergone some kind of 'Alien Experience', though this goes down to one in ten in New York, Seattle or Los Angeles. Are these figures urban myths? Are ordinary Americans *really* that naive? A Roper poll of 1994 put the figure at four million, about one in fifty. This is still very high. In a *Nature* book review, Sue Blackmore summarised the typical abduction experience:

The stories are remarkably consistent as well as outrageous. People are woken in the dead of night or, less commonly, taken from their car or workplace, and confronted by large-headed, small-bodied, huge-eyed grey aliens who transport them magically into a spacecraft. Here they are taken down curved corridors, laid on a flat table, and subjected to humiliating or terrifying mental, medical and gynaecological procedures. Eventually they find themselves back in bed but with two or three hours 'missing'.

Such experiences are probably 'waking dreams', the result of sleep paralysis. While we are asleep, our brains ensure that (mostly) we don't move our muscles like we would do when awake. This protection fails for sleepwalkers. But sometimes, as we are waking up, the sleep paralysis continues. This is rather scary, and makes us think that the dream we'd been having was true; it also makes it easier for us to *remember* the dream. But why is the dream so consistent in these accounts? One reason is that so many abduction stories have been told that we all know the plot, and this could influence the dream. In the 1990s Michael Persinger, a neuroscientist, advanced the theory that the experiences are the result of magnetic fields affecting the firing of neurons in the temporal lobes. The abduction stories may be similar because our brains are similar.

We have kept up our sleeves one devastating feature of the Roper report, in which 2 per cent of Americans reported being abducted by aliens. As Joel Best pointed out in his 2001 book *Damned Lies and Statistics*, they did nothing of the kind. The researchers were worried that a direct question about alien abductions would put people off, so they came up with five symptoms of abduction instead. Anyone who scored four or more out of five on those questions was *deemed* to have undergone an abduction experience. The questions were things like 'Have you ever woken up paralysed with the sense of some strange presence in the room?' So really it was a survey about sleep paralysis, and only the researchers thought that it had anything to do with alien abduction.

Cattle mutilations, crop circles, and unexplained electrical failures have also been attributed to alien activity. In the past such events would doubtless have been blamed on 'supernatural forces': demons, witches, ghostly manifestations, or the Little People. This kind of explanation is not so far away from the minds of even the most rational, and this fragile rationality illuminates the world of those who believe in alien abductions.

Here's an example of how fragile our rationality can be. Jack's Great-Aunt Becky came out of Poland when she was about ten: her sister Yetta had come out earlier, when their family was killed in a pogrom. No one ever found out how Great-Aunt Becky escaped, but she was very pragmatic indeed about nearly all of life; she would be called streetwise now. The family had pretty high standards of street-wisdom, even for those times: Becky and Yetta found themselves in a family where Yetta's husband had inherited a market stall and a horse from his father, and a place in street-markets up and down the east coast of England. He graduated to a grocer's shop in Norwich Market, and later bought a little flat in London for his sister-in-law. The pride of her life was a radio. (Actually she called it a 'wireless', not a radio, for historical verisimilitude.) She knew – and told people – that the Little People who were in the radio were 'on our side' in the Second World War. So she liked listening to the radio, though she was confused about whether the voices were those of people you could actually meet in the street. One of her acquaintances claimed to have met Tommy Handley, the star of a popular radio comedy, in the doctor's waiting room. Great-Aunt Becky never spoke to her again – not from malice, but from doubt. When Jack and his friend Hugh said that they believed Tommy Handley was real, and was quite likely to need a doctor, she tried very hard to get them to distinguish the world of the radio from the Real World. She was also very unsure – even frightened – of television. The Little People on television were much more 'real', more potent, than the disembodied voices on the radio. She would run from a room in which the television was turned on. She simply *knew* that this was Little People trouble, which sensible people didn't get involved in. Cars, on the other hand, weren't frightening magic, because they often went wrong and even ordinary people, like Jack's grandfather, could repair them.

Jack has met many people who believe themselves to have been abducted by aliens, or undergone other alien-contact experiences, and talking to them seems to him to be just like talking to his Great-Aunt

Becky. She was not at all *nutty*: she had no patience with friends who believed in mediums who communicated with the spirits of dead people (there isn't anybody there to talk to or listen to, she said). Similarly, alien nuts are usually rational enough about other things, and it takes a particular kind of event to get them out of their delusion. Jack was interrogating a woman on a radio programme, who had been going around the media circuit for six months saying how she had been transported up to the Alien Ship ('. . . but Mum, you were sitting there by the fire and you just woke up sudden!' said her daughter) and they had taken out her baby from her womb – she had the scar to prove it. 'How pregnant were you?' asked reproductive biologist Jack – unfortunately only after the (live) programme had started. There was a sudden silence, and you could see that no one had thought to ask her that indelicate but crucial question before. She said, 'Oh, I wasn't pregnant. Can I say on the radio that I had my period the week before? They're magic, you know . . .' and then dried up completely. Afterwards, she was completely convinced that her 'abduction' had actually been a vivid dream – as her son-in-law and daughter had been trying to persuade her for half a year. We wish we could persuade some of the more persistent saucer-watchers that, whatever UFOs are or are not, they have only the most gratuitous connection with space aliens. Like radio voices with the Little People.

What have we learned from all this? That there is a busy media market in 'alien' manifestations, a popular culture of 'alien' characters on television and film, a general belief that many people have been interrupted in their lives by 'alien' intervention, and an association of some fringe 'science', especially 'space science', with imminent alien contact. This market contaminates any serious thought about real aliens. So, having pointed out the danger, we are not going to say anything more about this kind of alien. They are irrelevant to the real aliens that are now living their ordinary lives on the planets of other stars, or elsewhere in the universe.

4

NOT-QUITE LIFE

*T*HE HUGE TUBEWORMS *tower above the aliens' tiny vessel, which passes effortlessly into, out of, and through the sulphur-loaded hot water and the reef-like masses of living creatures growing on the corpses of last week's growth. Cain is here only in effigy, immersed in Abel, who really enjoys the high pressure and the company of multiminds like his own; simple as they are, he is apt to be nostalgic about them. Cain is still not used to Earth's evolutionary symbioses: every cell of the tubeworms, clams, shrimps and other creatures in the peculiar lightless world of the great deeps is itself a little community. Some of them are regular Earth life. But there are strange guests here, in most organisms of the smoker community. All the cells of the tubeworms, for example, play host to sulphur-metabolising bacteria and archaeans . . .*

'These landladies don't need to make any other living,' says Cain. 'The tubeworms have no guts of their own; they just live off, and on, their new guests.'

'They're everywhere,' says Abel. 'I keep getting caught up with them, it's like having bits of me go stupid. It's very nostalgic, though: my original ancestors lived in the deeps, just like this; got into – well, you might as well say "radio" contact with each other, and took over their world. I keep getting little blips from these little intracellular bugs that they have here.'

'Call this "deeps"?' comments Cain. 'We're less than a thousandth of the way to the planet's core. Not like Europa, that'll be exciting; the ocean's much *deeper there.*'

'Yes, but we haven't got permission from the dominant lifeforms there yet.'

'True. I guess Europa will have to wait.'

'How many species here, Abel?'

'Dunno – but that's a plus: we'll get the tourists to count levels and management structures, and a few of them will work out that it's the sulphur-bugs in charge, running the whole shebang from the bottom-up and the top-down. And their viruses. Symbiosis and parasitism and predator/prey and diseases all overlap here, like that crazy symbiotic planet over in Lyra. Plenty of toys here for our guests.'

Is life a remarkable, highly unlikely accident? If so, we may live on the only inhabited planet in the universe. If not, alien life is certain. Our assessment of the odds depends on what we mean by 'life'. Whatever life is, it has to be complex enough to be able to reproduce and to respond to its surroundings. In his book *Investigations*, complexity guru Stuart Kauffman has in effect defined a lifeform to be any 'autonomous agent', which is any entity that can reproduce and can carry out at least one thermodynamic work cycle. That is, it can redirect energy so that it can 'pump up' processes round to their starting point again, and pass that ability on to copies of itself that it can make. With this definition, life could in principle be made from anything, up to and including a vacuum. The quantum-mechanical vacuum is a seething mass of particles and antiparticles, being created and annihilated in amazingly complicated ways: a vacuum has more than enough complexity to organise itself into an autonomous agent. Could this actually happen? We have no idea, but at least we're aware there's a question.

Naive astrobiologists take a rigid line and act as if the only possible kind of life is carbon-based with DNA heredity. Xenoscientists wonder how rich the alternatives might be. In this chapter and the next we make the case, based on what we know about life on our own planet, that life is a generic property of complex systems of matter or energy. Life is 'downhill' to complex chemistry, or to complex anything else – we should not be as obsessed by molecules as the molecular biologists are on this planet – which means that it is *easy* for life to arise. If that's correct, then there have to be aliens out there in the galaxy, including many of kinds that no human being has ever contemplated. But, and it's a big but: conditions have to be suitable, and there has to be enough time for inevitable but rare key events to occur.

We now have a fairly good understanding of when and how life originated on Earth. As long as we concentrate on general features of this process, it can inform our understanding of how alien life might get started on another world. The mistake is to get too attached to particular elements of detail, things that seem to have been very important here but might be irrelevant in an alien context. For instance, on this planet life is very closely tied to, and could not work without, DNA. Would a Europan caretaker have DNA? We don't know, but it would be silly to assume that it *must* have DNA, just because that's what is used here. In this chapter and the next we take a look at life on our own planet, and examine which useful lessons we can draw from it, and which silly ones we should avoid. This chapter is mostly about the kinds of chemistry that could lead to 'pre-life' – complex systems of molecules that have some resemblance to living systems, but don't really qualify as life proper. The next chapter is about life itself, on this planet, because we want to use our understanding of Earthly life as a springboard to possible alien lifestyles.

First, though, we need to clear away one possible distraction, the alleged conflict between evolution and theology. The history of the Earth has been contentious since the Biblical Genesis story, or the 'God's spittle crystallising out of the void' theory, or other creation narratives. All of these special creation stories seem in various ways to be implausible, don't agree with the evidence, or raise tricky philosophical issues – so people ask awkward questions. Our priestly ancestors were bright enough to realise that these questions might cause doubts, and that trying to answer them could easily make things worse because the answers might sound unconvincing, or because the mere act of taking the question seriously would give it too much importance. So they took a very direct way out of the impasse, and made the questioning itself heretical. A particularly informative example is the 'omphaloidean heresy' of late medieval Christianity, and we describe it here because it will have an important bearing on how to think about aliens.

The omphaloidean heresy is to ask whether Adam had a navel. Not to have an opinion about the answer: *to ask the question*. Because, if you have a fundamentalist turn of mind, that question *alone* is explosive. Why? The navel is the external, visible sign of the umbilical cord that links mother to child in the uterus. Adam, if made by an act of special creation, didn't have a mother, didn't grow in a uterus, and didn't need to be created with a navel. But if he was an actual historical figure, not an allegory, then he either had a navel or he didn't – and both answers spell trouble.

If Adam had a navel, then God was creating things *as if* they had a history that pre–dated the creation itself. If Adam had a navel, then presumably there were also tree-rings in the trees in the Garden of Eden, documenting seasons that had never been, but allowing the trees to be held up by the same process that would hold up all later trees. The soil in the Garden of Eden would have looked as though it had been formed from the decay of previous leaves that had grown on previous trees, even though there weren't any. There would be fossils in the sandstones and limestones under the Garden, because it is in the nature of sedimentary (sedimentary-seeming, in this case) rocks to have fossils, just as it is in the nature of trees to have tree-rings. The Sun would have been created as if it had been going for about ten billion years, the Earth would seem *in all respects* to be a bit less than about five billion years old, and the Moon would have craters apparently made by early bombardment, because a Moon like that would have been bombarded if it had existed at that time. So, if Adam had a navel, the world picture would be just like it is because the all-powerful God made it so – and you won't catch Him out.

So what's the problem for the priests? Since the universe was created in a form that *in every respect* resembles one that wasn't, the rest of us might as well go along with the *untrue* story that the creator wrote into the structure of our Universe. We should assume that the Earth is five billion years old, because it really does look like that, perfectly, complete in every detail, no matter what new instruments we invent or what discoveries we make. Creationists who point to inconsistencies in the scientific data about the ages of fossils believe that they are criticising the science. But if the creation story is true, as they believe, then actually they are finding mistakes in the consistency of God's creation – which is probably sinful in a person of faith.

What of the alternative: that Adam didn't have a navel? The Christian theologians, being caught on Aristotelian logic, thought that this was the *only* alternative. It implies that God made the father of us all differently: Adam belonged to a different theological species, so his sin with Eve could not be transmitted as original sin to the whole of mankind. The cause of the imperfect world that we inhabit could not logically be ascribed to a justly set up Fall of Man. Since God set it up imperfectly to begin with, He can't blame *us*. This was a much bigger problem for medieval theologians than it is for today's, even if they are fundamentalist Christians or Jews. At any rate, you see why even *thinking* about Adam's navel had to be heretical.

Christian people usually believe in the perfection of God's creation *and* the existence of original sin. Only literally-minded fundamentalists need *worry* about the omphaloidean heresy – but, paradoxically, they must worry about it by not even thinking it . . . At any rate, from either a theological or a scientific standpoint, it is necessary to accept that the universe has a consistent story to tell, however it was created or appeared. It is this story that science tries to unravel, and it is this story that aliens will have to fit. So, even if you believe in special creation, and whether or not you think that human beings are the only conscious creatures in the universe, you still need to pay attention to what science has to say about the prospects for alien life.

A Case of Conscience (James Blish 1958)

This is a very unusual item: an SF novel rooted in a theological and ethical dilemma, with an intelligent and sympathetic attitude to the theology. It began as a short story, and was expanded, and significantly changed, in a subsequent novel, described here.

The year is 2049, and the scene is the planet Lithia, which orbits the star Alpha Arietis in the constellation of Aries. A Terran scientific commission is studying Lithia on behalf of the United Nations. The planet is a biologist's paradise, with forests that drone to the sound of insects, lungfish that bark, and flying squid. And it has intelligent aliens. The Lithians have technology – fast jet aircraft, but no telephones. They know nothing of magnets, but have an amazing ability to control static electricity.

Lithians are tall and slender, with low voices like the soughing of the wind. Their eyes have nictitating membranes, like Earth cats, and they have heat-sensitive pits behind their noses. The commission's biologist, the Jesuit Father Ramon Ruiz-Sanchez, encounters a theological problem. The Lithians have no concept of hatred, greed, or envy – but they also have no concept of God. How can an alien Eden, devoid of sin, exist without God? Who, if not God, created the Lithians? Ruiz-Sanchez sees his research setting him firmly on the road to heresy.

Worse is to come. The mindless, barking lungfish turn out to be the Lithians' children. Lithian females lay eggs in abdominal pouches, then go to seek males to fertilise them. On Migration Day, when all the eggs are fertilised and ready to hatch, the women wade into the sea, and the children are born. Then they are left to fend for themselves. In the climax, Ruiz-Sanchez is instructed by his Church to exorcise the whole planet of Lithia – with unexpected and disastrous consequences.

Science, being human, is imperfect. It gets things wrong. Unusually, it *admits* it gets things wrong, and tries to put them right. Being carried out by humans, it does this in a flawed way: elderly scientists go to extreme lengths to suppress new ideas that contradict the theories on which their careers have been built. But *new ideas come in anyway*. Understanding the universe is an ongoing process of improvement and modification, not an ever-growing pile of sterile 'facts'. It is the way of thinking that matters most, not the conclusions that are drawn at any given moment.

For instance, the 'scientific' estimate of the age of the Earth has changed many times. This doesn't mean that scientists don't know what they're doing. On the contrary, it means that they are willing to revise their opinions when new evidence becomes available and theories have to change. Science is a way of thinking, not a body of timeless 'facts'. In the nineteenth century, the only heat-making processes that western rational scientists knew about were combustion and frictional heating, so they had to assume that the Sun had once been hotter, and was now cooling down – like a coal fire going out. Earth's history therefore had to be limited to the Sun's lifetime to that stage, which could be calculated in several ways. Forty million years seemed an outside estimate, and ten million looked more reasonable.

However, Charles Lyell's *Principles of Geology* of 1830, which was used by Darwin to estimate the time that had been available for species to change, needed the Earth to be at least twenty times as old in order to make sense of parts of the Earth's topography – such as the Cliffs of Moher on Ireland's west coast, which show an unbroken sequence of several million years of sedimentation. Lyell's Principle of Uniformity assumed that processes that can still be observed today, like sedimentation, must have been going on in much the same way throughout history, and a short lifetime for the Sun was not in accord with this. Philip Henry Gosse, in *Omphalos*, made much of this apparent discrepancy, realising that even if Adam did have a navel and the Earth therefore should look very old, it should look *consistently* old. With the discovery of radioactivity in the early twentieth century, and the invention of plausible energy-producing atomic-fusion processes, there was a sigh of relief all round: the Sun could be much older than forty million years, and so, therefore could the Earth. That particular problem was relegated to history.

We now have many convergent kinds of evidence putting the age of the Sun at some ten billion years, with about forty billion more to go. These ages, like the astronomical distances and evolutionary times that we will consider later, simply cannot be understood as part of the regular things we think about in ordinary life. All of the 'Think of the solar system as the size of a football field/baseball pitch . . . then the Sun is . . .' stories don't really work for us, and we don't think they work for anyone, really. The universe is *old* in a way that we cannot sensibly reduce to anything in daily life, and the most important thing to understand is that it is *not* so reducible. That doesn't mean we can't learn to think about it – for instance, we can do calculations with big numbers even though we have no feel for their sizes in ordinary terms. What it does is stop us dragging unexamined everyday assumptions into our thinking.

Fifty years ago there were several theories about the early atmosphere: whether there had ever been a high-temperature phase in Earth's aggregation into the planet we know, how the Moon formed, whether our solar system was typical or a rare happenstance. From the early twentieth century, however, there has been a growing scientific consensus that the Earth has something close to a five-billion-year history. The details have not remained the same throughout the last hundred years, though: there have been many radical suggestions about the gases that make up the atmosphere, the movement of continents, and so on. But the picture has now crystallised, so that we can use Earth history as a model for our thinking about life's history on this planet –

and, indeed, about the chemical and geological history of other planets.

We now have general agreement that the Earth began to aggregate soon after the sun 'ignited' at the centre of our solar Nebula, about six billion years ago. By five billion years ago the Earth was about two thirds its present size, and it was still being subjected to massive impacts from bodies in the solar system that had not yet 'settled down' into regular orbits, or into the various regions – asteroid belt, Kuiper belt, Oort Cloud – where there are still many small bodies. This was the 'bombardment phase' of the solar system's history, and it left its mark on the Moon, Mars, Mercury, and many other bodies. On Earth, erosion long ago removed all traces, but the Earth, too, must have been bombarded by innumerable lumps of rock, some of them gigantic. One of these impacts, perhaps involving a body as big as Mars, dislodged a large lump of the early Earth, which became the Moon, probably formed as a chain of molten spheroids, which aggregated into our familiar satellite. The Moon, without an atmosphere, has craters that document that early bombardment. So, it turns out, has Earth, but they have been eroded by wind and rain, hidden by topography, and smeared by continental movements.

Fifty years ago we thought that the Earth's early atmosphere was 'reducing', basically methane, carbon monoxide, carbon dioxide, and water vapour, together with ammonia, cyanogen, cyanides, and formaldehyde, plus a trace of sulphur as hydrogen sulphide. These last gases, from ammonia on, are poisonous to most of today's animals, including us. Back in the 1950s Stanley Miller realised that nonetheless, the word 'poisonous' did not, in this context, mean 'inimical to life'. Instead, it meant 'involved in living processes to the extent that it upsets them'. He realised that nearly all possible gases don't get involved in life at all: they might suffocate you by keeping oxygen away, but they don't poison you. So the list of atmospheric gases was very significant. He started scientists thinking about the biogenic (life-forming) potential of the basic ammonia/water/carbon dioxide mixture by using sparks to simulate ancient lightning, and he found that this process led to the formation of many interesting chemicals. Miller and his successors showed that this basic atmosphere would generate a great variety of interesting organic compounds under the action of ultraviolet light, lightning, and various catalysts that are produced during volcanic eruptions. These compounds include most of the twenty amino acids (building blocks for proteins), nitrogenous bases (building blocks for nucleic acids), sugars (energy currency of life), and fats and oils (partitions which life uses to keep different chemistries separate). Essentially, this work established that the early Earth, if indeed it had that kind of atmosphere, would be an organic-chemistry Earth. The building blocks for life would not have to be synthesised: they would be there for the taking.

It was a wonderful series of experiments, with just the sort of result that would appeal to scientists. If science was a belief system, Miller's experiments would have been an indelible part of the catechism. In fact, though, this comfortable little story of the origins of living chemistry has been discarded. Why change such a winner? The reason is straightforward: even though the story appealed to all the scientists' prejudices and was very repeatable and demonstrable, *it didn't fit the historical evidence*. So they threw it out.

We don't now believe that the early Earth's atmosphere was like that. It was probably mostly hydrogen and nitrogen, with lots of water, some methane and carbon dioxide and monoxide, a little ammonia, and some sulphur compounds; it is not nearly so easy to persuade this mixture to produce the feedstock for life. But we have also discovered that this kind of synthetic organic chemistry goes on even in space, so that comets and asteroids would deliver quantities of these basic organics, comparable with those to be attained by Miller processes in the other 'wrong' atmosphere. Further, scenarios for the origin of living processes have moved on from the 'primordial soup' or 'pre-biotic broth' models of the 1930s to 1970s, in which these organic building blocks were thought to be aggregated by concentration (were tidal pools necessary?). Then we thought that these Miller chemicals might self-assemble – in just the way that a pile of stones might assemble itself into a building . . .

Something's wrong here.

Fifty years ago we dealt with this little difficulty by a lot of arm-waving and some fancy calculations. Typical of that style was an evening that Jack spent with Eleazar Dawidowicz and Asimov in about 1966, trying to work out how likely was the assembly of a nucleic acid triplet. Such a molecule consists of three DNA 'code letters', or bases, and it specifies one amino acid, a component of a protein, according to the famous 'genetic code'. A triplet looked like the smallest chemical unit that could qualify as 'interesting' in the context of the origin of life. Asimov had done his PhD on nucleic acid assembly, catalysed by oil films or by clays, and had kept up his interest; Dawidowicz was a biophysicist. Asimov had worked out what would happen on the basis of molecular-collision chemistry if we just filled the whole Universe with a saturated solution of the right DNA bases and waited. The answer was that *no* triplets would be formed, even after a hundred times the present age of the universe. So it looked as though the formation of a nucleic acid triplet was extremely unlikely.

Creationists tend to seize upon such calculations as if they *prove* something – clearly only God could have produced nucleic acid triplets, since it's far too improbable for unaided chemistry to manage. That kind of argument falls flat, though. As theology, it fails for omphaloidean reasons: if anything on this planet is obviously impossible, then God's creation is flawed. As science, it fails for a more straightforward reason. Any theoretical calculation depends on assumptions, and the calculation may just be proving that the assumptions are wrong somewhere. Indeed, if correct calculations lead to totally unbelievable results, then the assumptions *must* be flawed.

They were. The calculation assumed that the molecules just bumped into each other when they were dissolved in water. However, the whole picture changes if we add catalysts: chemicals that promote certain kinds of reaction. What would happen, for instance, if the whole area of the bottom of earth's oceans (Isaac had that figure in his head, of course) was covered with a thin layer of clay, of a kind that would be a good catalyst for nucleic acids? Not only that: the surface of the seas (Isaac had that figure too) might also be covered with a thin layer of (the right) oil . . . There followed a lot of wild guesses and lots of scribbles on bits of paper – no point in using calculators, because we were making only ballpark estimates. After four hours we came up with a totally unreliable answer: the chances of getting a triplet are even after about 100 million years.

'About' here means six orders of magnitude up or down, we guessed. Jack used that episode for many years in his 'Possibility of Life on other Planets' lecture, the aforementioned POLOOP. A catalyst is a molecule that assists in the production of other molecules, without being used up in that reaction – so a catalyst can act over and over again, which enormously improves the odds in favour of the occurrence of those other molecules. POLOOP led to the conclusion that the creation of complex-but-unlikely chemistry was plausible, indeed likely, on a planet full of catalytic opportunities and with plenty of time available; moreover, there would be no trouble finding that one complex molecule. If it was life's precursor, then it would advertise its presence by multiplying until the seas were full of it and its descendants.

Those back-of-the-envelope calculations were fun, and indicative in a general sort of way – but we don't think like that now. Those ideas saw all of chemistry as 'school' chemistry, where the only kinds of reaction were 'terminal'. That is, something happened, then stopped, like a precipitate, or an explosion, or 'equilibrium': the mixture came to a particular colour, or level of acidity, or whatever, and *stopped reacting*. Those conditions were necessary, of course, for the school chemistry to happen within the time of a practical class, and for a theoretical explanation to be simple enough to be understood by a class of children. So the teachers set up experiments that could be explained by simple school-chemistry ideas, like mass-action equations and random collisions of molecules, and then they used those ideas to explain what had happened.

This kind of training had a disastrous side-effect: it led nearly everybody to think that real chemistry was always like that, so the chemistry of life had to be built by piling these simple reactions on top of each other until the mixture became self-propelled. The mental image here is like pushing an old car to start it: once enough momentum has built up, the engine can keep itself going. We should have seen that that the simple chemical model was inadequate, because it employed the words 'terminal' and 'equilibrium' – which mean 'dead' in biology. But chemistry was the authoritative science at that time, and its model was indeed molecules bumping randomly – and that model had a long run. Isaac, Jack and Lenny were using just that model to generate their magic triplet.

We now realise that real chemistry isn't like that at all. The equations we used to summarise chemical reactions were very inadequate summaries: even $2H_2 + O_2 \rightarrow 2H_2O$ is very misleading. (The chemistry teacher would quote that, then ignite a mixture of hydrogen and oxygen and it would go off with a pop, remember?) What he didn't say is that there are at least fifteen other steps to that reaction, involving unlikely radicals with formulas like $H_{13}O_5^{++}$. Moreover, the reaction does not proceed at all if the gases are very dry: it is *autocatalytic*, which means 'self-catalysing': some of the product must be there to begin with, or the reaction doesn't happen. It has turned out that many, perhaps most, chemical transformations are of this kind: they have many steps, and some of the later products are also involved in early stages of the process. Out-of-school chemistry, in the ozone layer or the chemical synthesis plant, in the explosion of diesel fuel or the setting of cement, always involves such *recursive* processes, where a late stage product is recycled through the system, making the process different the second time round.

There is a wonderful example of such reactions, whose recipe has been so refined that it can be demonstrated reliably and dramatically on an overhead projector. It is called the Belousov-Zhabotinskii (BZ) reaction, because it was developed by the Russian chemist B.P. Belousov and improved by another Russian chemist, A.M. Zhabotinskii. It interested these two scientists because it is an apparently cyclic oxidation/reduction reaction; that is, it seemed to go backwards and forwards, adding and removing oxygen. This was thought at the time to be contrary to the Second Law of Thermo-dynamics, which is normally interpreted as the statement that order cannot arise from disorder in an isolated system. (Traditionally, there are three laws of thermodynamics.) This made the BZ reaction seem as unlikely as a perpetual-motion machine, and Belousov had great difficulty publishing. Then Zhabotinskii refined the reaction to the point where, instead of oscillating back and forth, it progressed across a liquid film as a wave: each part of the liquid went blue as the wave passed, then returned to red, and then blue again, repeatedly, as the next wave passed. The resulting patterns were too vivid to be denied.

The main chemicals involved in the BZ reaction are sodium bromide and sodium bromate; a few others are needed to make the cycle work properly. Bromate has extra oxygen atoms compared to bromide, and the reaction shuffles this 'spare' oxygen to and fro. Each time the wave of chemical change passes, a small amount of the bromate loses some oxygen atoms and turns into bromide: the oxygen is taken up by another of the chemicals, malonic acid. The presence of bromide enhances the reaction, but carbon dioxide, from the malonic acid oxidation, inhibits it. So the wave is like a forest fire: it cannot go back on its tracks until the carbon dioxide diffuses away, when the next wave can pass.

However, the dynamic of the reaction is not just a repetitive oscillation back and forth, as the red-blue-red would indicate. It is more like a progression down stairs, with the risers showing blue and the treads red. As each wave passes, the reaction goes one step down, leaving less of the reactants in solution as carbon dioxide is given off, and the reaction 'dies' after about half an hour. (This, incidentally, means that the BZ reaction does not contradict the Second Law of Thermodynamics at all, but it is still very puzzling from the classical thermodynamic viewpoint, even so. We don't want to get involved in discussions about the Second Law of Thermodynamics, because this is very subtle and technical area which, we believe, is almost totally irrelevant to the question of life. See *The Collapse of Chaos* to find out more, including the chemical recipe and instructions.)

In general terms, three relatively simple chemicals are mixed together; a little later a fourth is added. They are shaken thoroughly and poured into a flat dish, where the mixture turns blue and then clears to a reddish brown. Blue dots appear and develop into expanding rings of alternating red and blue. If the dish is shaken to restore homogeneity, the process starts again. New dots appear, apparently at random times and places, and the process repeats. In addition, once the patterning has started, spirals can be formed instead of rings. This can be achieved by using a paper clip to break up the rings. Spirals soon dominate, taking over all of the existing patterns.

The BZ reaction exemplifies a change in chemical philosophy over the last twenty years, regarding complexity. It used to be believed that complicated and arcane processes are needed to make complex chemicals, for example the chemicals of life. Gradually the obvious came into the textbooks: add energy (for example heat, as in a frying-pan) to a mixture of sugar and oil and it caramelises and goes brown.

Many of the molecules in caramel are *far* more complicated than those that your eyes use to read this sentence, and very much more complicated than the sugars and oils that gave rise to *you*. In Miller's original experiments to see if the early-Earth atmosphere could make interesting chemicals for life, the glassware was hard to clean afterwards: it had nasty brown goo on it. This was what Miller called 'resins and tars' and, like caramel, it has very complicated molecules. Complicated molecules seem to be downhill to most organic chemistry – that is, they happen if you are careless, not only if you do everything exactly right.

Complexity alone is not enough for life: the complexity of lifeforms is also *organised*. So attention switched from raw complexity, which was a red herring, to organisation. Contaminating reaction mixtures with a bit of clay helps many reactions along, and Graham Cairns-Smith has developed an Origin-of-Life story involving clays. Clays are complex silicates – made from tiny crystals, with aluminium, magnesium, calcium, and traces of other metals like iron and copper. The clay crystals with different chemical compositions are all different shapes, and geologists have given different clays a variety of names (like 'montmorencite') because they have characteristically different solubilities, flow rates, and stickiness. These properties are important geologically, for house foundations, for silting up of estuaries – and for pottery. Clay crystals will grow, in a saturated solution of the major constituents, by making more of their own characteristic chemistry from the dilute chemicals in the solution, because that is what made them that shape to start with; so they replicate their shape and composition.

Cairns-Smith was impressed by how high-tech even the simplest living replication is. DNA replicates only with the assistance of a huge range of other molecules, whose complicated activity is coordinated by a cell, bacterium, or archaean. You'll be aware that DNA is a double-helix, formed by two strands of bases that wind round each other like a spiral staircase. Each 'tread' of the staircase is composed of two DNA bases joined together, and these joins are complementary: base A (adenine) always joins to T (thymine), while C (cytosine) always joins to G (guanine). If the two complementary strands can be separated, each can then be completed by adding the correct complementary bases, and now you've got two copies of the original pair of strands. However, achieving this separation is not easy – for a start, the two strands are wound round and round each other, and cannot be disentangled without a lot of molecular trickery, snipping them into short segments and taking care of the joins. And then they have to be entangled again in each copy. So DNA is no more 'self-replicating' than a car in a car factory.

For such reasons, Cairns-Smith believed that some kind of low-tech replication, like that of clays, must have preceded it. He compared living replication to a stone arch – *all* of it (nucleic acids, proteins, specific enzymes) must be there for the system to work. But perhaps, initially, there was clay replication, like the pile of stones on which the arch was supported as it was built, and which were cleared away. If clay-type replication, with catalytically attached organic molecules like amino acids and nitrogenous bases, came first, then the proteins and nucleic acids that they catalysed could make a complex interacting system (like the arch). Then the organics perhaps took off on their own. The clays would have catalysed polymerisation (the joining up of amino acids into proteins, and perhaps of nitrogenous bases into nucleic acids) and those polymers that catalysed each other's production would have formed replicating systems at least partially independent of the original clays. Cairns-Smith joked that our carbon-based life is but an interlude between silicon lifeforms: first came clay, then carbon-based life (including people at IBM), and then will come 'living' silicon computer chips.

The main factor that makes complex chemistry not just possible, but natural and easy, is the widespread existence of catalysts. Recall that a catalyst is a molecule that promotes the formation of other molecules without being used up in that process. The most obvious candidate for a self-replicating molecule would be one that is autocatalytic – catalyses its own formation. Such molecules are rare, if not non-existent: for all the hype about 'self-replication', DNA left in a beaker does not turn itself into more DNA. The replication of DNA involves a host of molecular helpers, with a neat twist: the helpers are themselves replicated with the aid of DNA. So the combined *system* of molecules, DNA plus helpers, is autocatalytic.

Lifeforms made from molecules ought, therefore, to be based on autocatalytic *systems* (and life made from other things, like radiation or magnetism, should be based on something analogous). How common, or rare, is an autocatalytic system of molecules, then? Kauffman has shown that whenever you have a sufficiently complicated network of chemical reactions, some sub-network is very likely to be autocatalytic. Roughly speaking, the reasoning goes like this. Imagine that the molecules are towns, and whenever one molecule catalyses another there is a road joining the corresponding towns. If there are lots of towns, and enough roads between them, then some collection of roads will form a closed loop – and that loop will correspond to an autocatalytic subsystem. The precise argument is not quite that simple (the roads are usually one-way streets) but it is similar in spirit, and is supported by solid mathematics. Since complex chemistry turns out to be easy, it follows that autocatalytic systems of molecules are easy too. (Easy for nature, with its vast laboratory and millions of years of time, that is. Not so easy for humans, in a tiny laboratory with very little time.) This means that one key ingredient for life, the ability to replicate, can arise in many ways – even if we restrict attention to lifeforms based on molecules. So replication as such is not a major obstacle.

Organising the autocatalytic system into a single entity, and controlling its behaviour so that that entity persists, is a lot harder, but even so, dozens of plausible suggestions can be found in the scientific literature. Many of today's alternative origins-of-life theories (note the plural) depend on surface catalysis. For instance, Günter Wächtershäuser has developed a very persuasive autocatalytic build-up of many of life's most basic reactions on iron sulphide or iron oxide particles: these arise as parts of submarine surfaces, with other chemicals – notably clays and their relatives – close by. So, rather than a 'primordial soup', today's models for biogenesis resemble primordial pizzas, in that they have active surfaces with very different properties in different places.

Proponents of the 'primordial soup' theory often argue that the sulphur chemistry around 'black smokers' or in the rocks of the oceanic crust merely adds further molecules to the soup. Not so. In 2000 a team led by George Cody showed that sulphur-iron chemistry naturally and easily leads to the production of pyruvic acid, a key constituent of the metabolic pathways of current lifeforms. The reactions involved are consistent with the known features of iron-sulphur metabolism in extremophiles, but the key step, the formation of pyruvic acid by a process called 'double carbonylation', occurs at high pressures and has not been observed in existing metabolisms. Yet. The reactions are highly plausible for primal lifeforms feeding on carbon monoxide, but they could also occur today under the right conditions of heat and pressure. Pyruvic acid is too unstable to accumulate slowly in a primal soup, so iron-sulphur chemistry and the primal soup are incompatible theories.

What does all this tell us about aliens? *Not* that they have to start from primordial soup, or clay, or iron and sulphur. There are at least thirty persuasive scenarios in the biogenesis literature that probably would, given time, develop lifeforms from their recursive chemistries. There are probably millions we've not yet thought of. Any of these could engender a kind of life on a suitable world, given half a chance. Some of them, like Kauffman's autocatalytic systems and Doron Lancet's 'replication domains', are mostly theoretical, but both authors have specified sets of 'likely' primeval molecules that would be able to do their tricks. Meanwhile, other more pragmatic workers have set up more-or-less 'primeval' experimental set-ups, fudging the chemistry a bit to avoid the necessity of waiting for millions of years and using all of Earth's seas.

It seems likely that life on this planet arose through some 'accidental' autocatalytic system. Once a successful replicator arises, there is soon rather a lot of it – because it replicates successfully. Nature had the whole planet to work with, and as much time as it needed: the geological evidence suggests it took a few hundred million years at the most. Humans can't wait that long, and can't try as many possibilities, so we don't yet have an 'artificial' – human-invented – autocatalytic system that works in a laboratory. However, we're getting close. Chemists have already synthesised catalytic proteins. A few years ago Günter Kiedrowski invented a self-replicating molecular system, a six-base length of DNA composed of a three-base sequence CCG followed by its complement GGC, but added in reverse order like this: CCG – CGG. When fed a 'diet' of the component sequences CCG and CGG, this strand lines up two of them next to it:

CCG – CGG ← original
GGC   GCC ← new

Then the two new strands spontaneously join, and separate from the original. The result is a copy of the original six-base strand; it looks as if it ought to be in reverse order, but once it has separated off, it is just the original molecule rotated through 180°.

In 1996 Reza Ghadiri and co-workers came up with a self-replicating peptide (a short protein). This takes the form of a sequence of 32 amino acids, composed of two fragments of lengths 15 and 17. When 'fed' such fragments, the peptide similarly lines them up, joins them, and sets them free.

Neither system described is truly autocatalytic – a supply of suitable fragments is needed as 'food': these are not part of the catalytic loop, and are too closely related to the replicating molecule to count as part of some generic 'environmental' background soup of molecules. But chemists have already achieved systems that come very close to being autocatalytic, especially with amino acids linked into peptides some of which catalyse the formation of other peptides and small proteins, and of nucleic acids and other clever polymers. Ribonucleic acid (RNA), which today's terrestrial life uses as genetic messenger molecule, has proved to be very versatile. Unexpectedly, it turns out that some quite short sequences of RNA have catalytic activity for synthesising and indeed replicating RNA – so-called ribozymes. Many people like the idea that there was once an 'RNA-world' which preceded the DNA/RNA one that we know today. As we write, our particular configuration of DNA and RNA has not yet been achieved from simpler precursors, so the RNA-world is unattainable from theory or practice; it is also not clear how to proceed towards our kind of life from the RNA-world. It has been suggested that the particular DNA/RNA/protein configuration used by most of today's lifeforms may have been a latecomer, improving on a seething mass of different autocatalytic systems. In that case we would not expect to find that configuration in the first-generation life chemistries, and perhaps experimentalists have done as well as we could expect in the circumstances.

We can't tell – perhaps we will never know – which of these was the historical basis for our kind of life on this planet. Or was it one of the thousand scenarios we haven't thought of yet? If we don't know what happened *here*, then it is foolish to follow the astrobiologists and insist that we know how alien life must get started many light years away. The lessons that we draw from Earth's history have to be general principles, not specific mechanisms. In any case, we can be pretty sure that if life hadn't arisen the way it did here, whatever that was, then it would have arisen by one of the other plausible routes. Life seems to be downhill to complicated chemistry, and complicated chemistry seems to be downhill to simple chemistry. So, if enough simple chemistry is going on, then eventually it will lead to life.

Many people have difficulty with this kind of thing, wondering where the complexity 'comes from'. The answer is that it comes from the unfathomable potentialities inherent in the laws (we prefer to call them 'rules' since nothing legal is going on) of nature. The BZ reaction demolishes the 'you can't get complications for nothing' argument, and 'life is too complex to arise spontaneously' suffers from the same misconceptions. Complexity, at least in the usual sense of 'this looks difficult to describe in detail', is not conserved between cause and effect. This isn't really a surprise: why should nature be governed by how simple or not something *seems* to be to a human observer? The origins of life, it is now generally believed among chemists who've thought about such issues, is a downhill bicycle race; the majority opinion is that we'll find some of these chemical scenarios, and perhaps more complicated lifeforms, in the seas of Europa, and perhaps even in the clouds of Jupiter.

Speaker for the Dead (Orson Scott Card 1986)

In *Ender's Game*, Andrew ('Ender') Wiggin fights off the Third Invasion of the aggressive, wasp-like Buggers, and destroys their homeworld, having been led to believe that he is merely playing a very elaborate computer game. He retires to a colony world, where he finds a tower, left for him by the Buggers. Inside is the pupa of a hive-queen. In an act of contrition, he becomes the itinerant Speaker for the Dead, moving from world to world, telling the stories of those who have died. And always he carries the pupa with him, searching for a planet where the hive-queen can awaken and re-start the Bugger race.

In *Speaker for the Dead*, he finds the ideal world. Humans of Portuguese descent have located a planet they name Lusitania, and find it to be inhabited by animals they call *porquinhos* – 'piggies'. Then they realise that the piggies are not animals at all: they are intelligent. Four piggy languages are identified: the Male's Language, the Wives' Language, Tree Language (used for prayer to the totem trees), and Father Tongue, which is expressed by beating various sticks together. The piggies use tools and houses; they are the first intelligent aliens found by humanity since the Buggers. Some of their ways are strange and terrifying.

The human colony on Lusitania is permitted to continue, but it is confined to a very limited area; the rest of the world is left for the piggies, who must not be disturbed. Ender, now on the ice planet known as Trondheim, is in telepathic communication with the hive-queen. She tells him that Lusitania is perfect for her kind. Ender replies that he will not risk destroying the piggies to atone for the xenocide of the Buggers; she assures him that piggies and Buggers can and will live in harmony. But he cannot forget that it was he who destroyed her race: perhaps she is lying as part of an elaborate act of revenge. She tells him he was an unwitting tool, and she has forgiven him: others destroyed her world.

Can he trust her? The fate of three intelligent races hangs in the balance . . .

Aqueous planets, in particular, make this kind of chemistry easy. Water dissolves many different molecules, and molecules in solution come into close contact and so have many opportunities to react with each other. So one important idea to take from this is that all aqueous planets will (eventually) have life. Perhaps not very interesting life, as Earth didn't for most of its history, but more interesting than school-fudged chemistry. Some of this early life will have been indistinguishable from the real complex recursive chemistry going on around it – only the persistence of its cycling processes would have marked it out as having a burgeoning heredity. Within a few decades we may well find ourselves having to distinguish these elements within the complex chemistry of Europa's seas, Jupiter's clouds, and Titan's methane-ammonia ice-fields.

There has been another related change of mind, but this time it concerns astrophysicists. The mind-set of the 1930s to 1960s saw the Second Law of Thermodynamics as *the* ruling principle of the universe, and wouldn't let Belousov publish anything that seemed to contradict that view (even though, we now see, it *didn't*). That same mind-set also saw the universe as 'running down', increasing in entropy – a thermodynamic quantity usually interpreted as 'disorder' – towards a lukewarm featureless 'soup' known as 'heat-death'. This model has its roots in the picture of a gas as large numbers of molecules that bang into each other, bouncing perfectly elastically, and thereby increasing the amount of entropy. Seeing the whole universe like that, starting with the Big Bang and irrevocably winding down to a much less interesting heat-death, was a major preoccupation of astrophysicists in the 1950s and 1960s.

Such entropy-dominated models had patterned the way physicists saw almost all natural processes, especially living ones, for fifty years before that. The original concept, however, had been much more limited: to heat engines, closed systems near overall equilibrium. The famous physicist Erwin Schrödinger, in his clever little 1935 book *What is Life?*, could explain life only in these physical terms by assuming that it 'fed upon negative entropy'. Food, in this view, was negentropy crystallised by other lifeforms, and organisms maintained their order only at the expense of creating more disorder in their environments: the net amount of entropy had to increase. This pessimistic attitude was rooted in the ideas of Ludwig Boltzmann, who invented that way of thinking about entropy. But Boltzmann – unlike the pessimists – understood the limitations of his reasoning: closed system, near equilibrium, all of the system in more or less the same state of order. These conditions simply do not hold for, say, a bacterium or a goldfish in a bowl, and the use of the simplified Boltzmann model of gas molecules bumping into each other for the whole universe was simply absurd.

The absurdity becomes rather obvious if we change the thermo-dynamic model of the universe to a 'gravitic' one, replacing elastic bouncing by the attractive force of Newtonian gravity. Now we find a clear tendency towards an *increase* of order, and a *decrease* of measurable entropy, as time passes and the particles attract each other and aggregate into stars and star systems. The reason is that gravity is a long-range attractive force, and so induces clumping on all scales, which thermodynamicists interpret as order. In contrast, molecular bouncing is short-range and repulsive, and causes matter to spread out uniformly, which thermodynamicists interpret as disorder. The different behaviour lies in the different nature of the forces, and conventional thermodynamics does not apply to a gravitating system of particles.

In particular, it is silly to expect the Second Law of Thermodynamics – disorder always increases – to hold in a gravitating system. Furthermore, a model of the universe that incorporates gravity has other properties 'up its sleeve'. As these gases, originally hydrogen and some helium, aggregate into stars, the conditions at the centres of these stars become far removed from the gentlemanly bouncing back and forth in free space that the entropy-mongers envisaged. Instead, the atoms are forced into encounters that produce new particles with new properties: new elements are formed. Starting with hydrogen and helium – and gravity – Dmitri Mendeleev's famous Periodic Table, the standard organisation of about a hundred elements, becomes inevitable. At least, this can be achieved as far as the element iron, using just the excruciating physics that inevitably appears in the middles of stars. To get from iron up to the transuranics (elements with bigger atoms than uranium) requires more extreme physics, in which those stars explode to form novae and supernovae. The universe complicates itself: it doesn't wind down, it winds *up*.

We're not suggesting that gravity is the magic ingredient for life. Gravity is just one of the forces that can make the universe clumpy, and the Second Law of Thermodynamics is irrelevant to that particular kind of 'order'. Chemistry has forces, too, and these are more subtle than gravity: attractive forces that join molecules together, repulsive forces that separate them. Life on Earth is complicated chemistry. So the thermodynamic model of bouncing billiard-balls is even less appropriate to living systems, and the Second Law is even more irrelevant.

In his unorthodox but brilliant book *Investigations*, Kauffman has conjectured that there might exist some kind of 'Fourth Law of Thermodynamics', complementing the usual three. This Fourth Law would give a precise mathematical form to the idea that living systems, and more generally, autonomous agents, complicate themselves by moving into the 'space of the adjacent possible' as rapidly as circumstances permit. Kauffman's central objective is to lay down a framework for a new, more 'general', biology, of which life on Earth would be just one instance. This new style of thinking is ideal for the question of alien evolution. It offers the prospect of discovering general rules – far more general than the universal/parochial distinction – for how alien evolutions should proceed. His thinking can be summarised in six steps.

First, he contrasts the traditional viewpoint in the physical sciences with that in biology and anthropology, specifically evolutionary biology and the evolution of technology. In the physical sciences, the possible results of processes and experiments can in principle be listed before they happen: the universe is the same at the beginning and the end of the experiment. Even if the experiment involves chaotic (divergent) or progressive (recursive) processes, like many investigations in subatomic physics or autocatalytic chemistry, it is possible to pre-state the system's phase space – the variety of possible outcomes – in a precise, simple manner. The phase space for a thermodynamic system of N particles, for instance, consists of all possible lists of $6N$ numbers – three position coordinates and three velocity coordinates for each particle. In contrast, in both evolutionary biology and technological advancement, the processes that occur change the rules governing subsequent events, so that the universe of possibilities at the end of the process is larger than it was at the beginning. An ecosystem of N organisms does not remain an ecosystem of N organisms, for instance: some organisms die, new ones are born, and on evolutionary timescales the whole system can change from a swamp to a forest or a desert. Gravitic systems are perched curiously in between: the phase space of a gravitating system of N point masses also consists of all possible lists of $6N$ numbers, and the difficult thing to pre-state is not the phase space, but the significant configurations within it.

Second, Kauffman argues that because the universe *can* perform such expansions of the phase space, because living things do so, then even physicists and chemists may be wrong to limit the phenomena that they are willing to consider to those that *don't* do so. And this means that the universe is not closed, with a fixed set of laws to which it conforms, but is progressive. The universe constructs more complex realities, and expands its own phase space, as it evolves. In a sense, the universe makes up its rules as it goes along. Chemical rules for making molecules 'appeared' only when stars had made different kinds of atoms; natural selection, as a principle, 'appeared' only when competing lifeforms (perhaps ones as simple as different kinds of nucleic acids competing for sites on clay crystals) produced different numbers of descendants. The contention that the universe makes up its own rules does *not* contradict the physicists' belief that the universe runs on a single, universal set of laws – a Theory of Everything. The 'laws' that Kauffman is thinking about would be *emergent* consequences of a Theory of Everything, should such a thing exist. 'Emergent' here means that there is no practical way to predict those laws from the laws of the Theory of Everything, except by running the universe and finding out what it does. Which is another way to say 'watch it make up its rules as it goes along'.

Third, he describes this evolutionary step as 'expansion into the adjacent possible.' This image is based on an early biological phase space, the 'fitness landscape' of an organism, which represents possible organisms (or their genetic makeup) by points in a horizontal plane, and the evolutionary fitness of a given phenotype by a height above the corresponding point. A fitness landscape is like hilly terrain, with the peaks of hills corresponding to organisms that are better at surviving than any small variation on them would be. In order to evolve their way from one hill to a higher one, the organisms would have to traverse a lower-fitness valley. One possible biological tactic for achieving this is to allow the organisms to vary by small amounts, each 'one mutation away' from the current peak. This limited kind of exploration of the adjacent possible led Kauffman to a much more general model that can encompass, for example, quantum uncertainty at the subatomic level.

Fourth, he generalises the concept of 'life' to any system with just two properties: replication, and the performance of a thermodynamic energy cycle. Recall that he calls any such entity an 'autonomous agent'. Flames are nearly alive, by this criterion, but they don't quite return to their starting point. Neither do sexual organisms, quite, but we can see the difference: the progeny are *equivalent* to the parents, even though they're not identical; on another occasion we might find that organisms like the parents are progeny, and vice versa. We can easily see that there could be many physical and chemical ways to achieve autonomous agents, and that terrestrial DNA/RNA/protein organisms are but one class of examples. Computer viruses are another, and Kauffman's General Biology would be as comfortable with these as it is with malaria parasites and HIV. Orthodox 'life = DNA' biologists were scuppered by BSE ('mad cow disease') because it seemed to be a living, transmissible parasite – but it didn't have any DNA or RNA. The agent of BSE is a *prion*, a protein that converts other protein molecules of the same general constitution into its own pathological shape. When we find alien life, genuinely alien from another origin of life and not simply splashed from one planet to another in the inner reaches of the solar system, it will be at least as different from Earth's DNA-life as prions are. Conservative biologists will be forced to deny that it *is* life, just as they do now for computer viruses and prions.

Fifth, life must replicate inexactly, so that its heredity forces it into the adjacent possible as rapidly as possible. Genuine life will very soon be selected to favour those mechanisms that do just that. Whatever the first, simple form of pre-life was, it was soon overtaken – probably 'eaten' – by forms of its own organisation that competed better, and were better at *changing* to forms that competed better. It is quite likely that the DNA/RNA/protein system that most life on Earth now uses was the minimum necessary to 'trump' those other early attempts.

Sixth, life self-complicates to levels of complexity that change the context for other life, so that most of the important context for life on a living planet is provided by other lifeforms. This contextualising process continues, especially via 'symbioses' in which two or more organisms join forces to their mutual benefit, and leads to intelligence and the use of technology. Then technology acquires its own kind of heredity through the life that wields it, and expands into its own adjacent possible, driven by intelligent 'invention' and by variation in its manufacturing methods.

The scientific standpoint of *Investigations* is as different from the astrobiology stance as it's possible to get: it looks outward to the general, rather than inward to the particular. Astrobiology looks at Earthly life, the best-documented exemplar that it knows, and asks whether the same can be produced elsewhere; Kauffman generalises from the parochial living processes on this planet to innumerable other systems that could exhibit the same crucial properties without being made from the same ingredients. Where the entropy physicists saw the Universe following the slippery slope downwards to oblivion, today's complexity thinkers see the universe as complicating itself and inventing new rules as it goes along.

Jack&Ian thinks that we get a more accurate picture of what the universe does if we think of it as making up the rules as it goes along. It does not believe that God put all the rules in a cupboard, to be taken out, dusted off, and used when required ('Whoops, someone's just lit a fire on that planet . . . Quick, where's the rule about hot air rising? Ah, yes, now we need the business about all the different chemical reactions in flames making light'). The excruciating physics determined by the interaction of masses and gravity in the middle of stars caused particular configurations of subatomic particles to make stable bigger atoms, in recursive processes, and these determined what chemistry would be possible when these atoms made molecules. Planets, their volcanoes and their clays, their atmospheres and their seas exploited only a very tiny fraction of possible chemistries – and chemistry teachers chose a very restricted fraction of that to explain the Laws of Chemistry to the rest of us.

Alien life, however, does not have to respect the educational difficulties posed by naked apes with oversized brains, and can exploit *any* autocatalytic system to generate complex 'lifeforms' if the environment permits. And the range of environments – and auto-catalytic systems – is huge.

5

POSSIBILITIES OF LIFE

*T*HERE ARE MANY *ecosystems within a hundred light-years that are more beautiful, more diverse, and more dramatic than any on Earth. But the one that the two alien prospectors have found is the most fascinating symbiosis this side of Betelgeuse. They slide down the beach into a shallow lagoon.*

A curious bright blue-and-yellow parrotfish comes up and takes a couple of experimental nips. 'Amazing,' says Abel, 'that all this coral-sand has passed through parrotfish like this one. Makes you realise what a long time it all took. When I was a bud there weren't coral reefs on Earth – enormous clams instead.'

'Only about six hundred species in the reef itself,' says Cain, 'not counting bacteria and viruses.'

'But ten thousand in the seawater,' says Abel. 'Many species send their babies up into the plankton.'

It is evening, and some of the coral polyps are spread out, catching plankton. The giant clams have stopped being plants, spreading out their mantles with algae in the sunlight, and are gently filtering too. 'It's a war-zone,' says Abel. 'The anemones are all trying to poison each other, the sea-fans, and the other soft corals, while the hard corals are fighting back. The tastes are wonderful. The corals are all leaning over to shade light from each other, like trees, and stinging each other where they touch. Here's a little congregation of sea-slugs, ready to eat the winner . . . no, they've been picked off by an anemone fish.'

Cain has found a long, thin moray eel, banded black on white; it takes three little damsel-fish as he watches. The events pile one upon another. So much is happening that it is almost boring, like a film that has been edited down just to the events; there is no relief from interesting items. Slowly Abel withdraws

from the water, up on to the fringing reef; Cain climbs up beside him.
　'What's that*?' asks a basking human female, but her companion doesn't know.*

The not-quite life of complex clays or RNA molecules provides an important stepping-stone from 'ordinary' chemistry to the complex recursive chemistry of life itself. We now look at how life might have evolved from not-quite life, and extract some general evolutionary principles that are broad enough to apply to alien lifeforms as well as terrestrial ones. In particular we emphasise a distinction between 'universal' features, which can be expected to occur many times and in many places, and 'parochial' ones that occur only once – though they may then be passed on to numerous descendants and thereby become very common on a given planet. The Darwinian concept of evolution through natural selection is a universal, but the neo-Darwinian obsession with DNA is a parochial.

What can we learn from present-day lifeforms about the very early history of life?

It is tempting to assume that very simple lifeforms, or almost-alive structures such as viruses, are somehow representative of the early stages of Earthly life. But it is probably a mistake to do this, for the same reasons that today's amoeba is not in any sense an ancestor of today's higher organisms.

A modern amoeba *can't* be an ancestor: it's alive *now*.

The issues are difficult, and there's always the chance that simple lifeforms now can give us clues about simple lifeforms billions of years ago, but most of the evidence is contradictory. And the simplicity of viruses depends on the existence of more complex bacteria, whose DNA replication system is subverted by the virus to turn out copies of the virus. Viruses on their own are not credible independent lifeforms. So it is probably best to assume that today's viruses are recent simplifications of other kinds of life, and that they do not give us a reliable picture of the earliest, simplest lifeforms.

However, we do have two good pieces of biological evidence that the formation of the early lifeforms was *easy*, in evolutionary terms, even if we don't know exactly what routes were taken, for example whether membranes appeared before nucleic acids.

The first of these concerns the oldest protein families, those shared among the most diverse organisms on present-day Earth, and which were therefore present in the most ancient of our common ancestors. There are excellent reasons to believe that when they originally evolved they were selected to function in hot water. For example, ribonuclease, the enzyme used to create and destroy RNA, remains effective today in heated cans of soup. Most recent proteins were invented and evolved while the seas were cool, and are destroyed by boiling. These ancient proteins are quite different; they probably originated either while the seas were still boiling hot, or at undersea volcanic vents. They are resistant to temperatures well above boiling point, but are found in creatures whose ancestors – as far back as we can go – have been in cool temperatures. The argument is that they must be survivals from the earliest lifeforms, which therefore lived at high temperatures, more than four billion years ago when the seas were still boiling, at least sporadically, as large asteroids still bombarded Earth. That means that these early lifeforms were there *right at the beginning*, as soon as there were seas to live in. So the Earth didn't have to wait very long for them to evolve. Therefore, the argument goes, they were *easy* to evolve, and they evolved promptly.

There is a danger of circular logic here. We are inferring that evolution must have been easy, because it didn't take very long. An alternative is that it was very hard for life to evolve of its own accord, if not impossible, so life could occur only because somebody or something (God, the Solid State Entity, the Galactic Mind, the Cosmic Computer) created it. However, we have already pointed out, in connection with the omphaloidean heresy, that a created world would have to look *as though* it wasn't created at all. We live in a universe that *looks* as if life can evolve easily, and we can't catch God out getting it wrong, so actually it has to *be* one in which life can evolve easily. Even if it was actually created ten minutes ago and was totally impossible before that. So the scientific clue is straightforward, unlike the quasi-theological reasoning: if you start out by assuming that life arises easily, you'll be on the right track. If you think the origin of life has to be hard, you are setting yourself up for all sorts of trouble.

The other evidence that life arises easily is absolutely concordant with this. If life had been 'difficult' to achieve, we would expect there to have been a long period with a barren Earth, before life finally managed to appear. But this was not so. The first traces of terrestrial lifeforms are more than 3.8 billion years old. These lifeforms have been detected through their products – there are no actual fossils, only some chemistry that is obviously of biological origin. There is genuine fossil evidence of actual organisms from more than 3.5 billion years ago: fossils of creatures like modern bacteria. So we would argue that on an Earth re-run, and on other aqueous planets, life will turn up quickly too.

The major criticism of this argument is anthropic. Perhaps life can be formed *only* at the initial steps of a planet's existence, in boiling seas. If that is so, then however difficult and rare it is, in the few cases where it happens we will find, as we have on Earth, that life arrives promptly on the planetary scene. So even if we were the only lifeform anywhere in the universe, we would still be using the argument we have just outlined, but it would be totally mistaken. The chemical evidence, however, strongly and independently argues that life is easy on planetary scales – by modern variations of the Asimov, Cohen and Dawidowicz calculations – so that, for the moment, we can avoid anthropic criticism. And the theological one, we've already dealt with.

Granted that originating life is easy, what happens afterwards?

In the last few years more of the picture has been obtained of what happened on Earth. Among the simplest free-living creatures we have on Earth now, there are two major kinds, archaeans and bacteria. The terminology keeps changing as new discoveries are made, reflecting the advance of the science: many archaeans used to be called bacteria, because that's what they look like. But now that we can analyse their hereditary machinery (proteins and RNA molecules involved in the basic processes of living and reproducing), it is clear that archaeans and bacteria are completely different evolutionary lineages – and probably have been for three billion years. This is the case despite these simple creatures' habit of swapping useful genes with one another, even across the great divide between their lineages. The genetic material is probably conveyed by the viruses that have always preyed on them both.

The bacteria include *Escherischia coli* (more familiarly *E. coli*), which lives in the human gut, *Salmonella* of food poisoning fame,

Staphylococcus and *Streptococcus* of skin and gut infections, and the bugs that make yoghurt out of milk and turn our excreta back into pure(ish) water. Greatly outnumbering the bacteria important to us on these local levels are the myriads in seawater, where marine photosynthetic bacteria and archaeans release more oxygen and absorb more carbon dioxide than all of Earth's forests. Similarly huge numbers of bacteria are to be found in soil and fresh water. Other bacteria form long, tangled filaments, like the so-called 'sewage fungi'. Some filaments are photosynthetic – the ones that we used to call 'blue-green algae', and are now known as cyanobacteria, familiar as the slimy bloom on freshwater ponds and lakes. The name changed because they're often not blue-green and they're not really algae. Sorry, science is like that.

The archaeans live in these 'ordinary' environments too, but some of them have amazing life stories, surviving in cracks miles deep in rocks by taking in a few molecules every century, or living it up in sulphurous hot springs. Many live around the volcanic vents ('black smokers') at the bottom of the oceans, some of them as symbiotic partners in the cells of other organisms. Some bacteria do that too, and some more complex organisms.

Archaean chemistry frequently involves sulphur compounds, and this ability recently solved a puzzle that had been around for over thirty years. The ocean floor contains vast reserves of methane – ten trillion tonnes (tons) of it, double the Earth's known reserves of coal and oil. It is probably the result of biology, like coal and oil, and not for example methane from the original atmosphere. The methane is stored in the form of crystalline methane hydrate, which is formed when molecules of methane gas become trapped in 'cages' of water, a reaction that readily occurs at the moderately high pressures found in the ocean depths off the continental shelves. Methane is a powerful greenhouse gas, twenty-five times more so than carbon dioxide – so if those methane reserves escaped into the atmosphere, they would wreak havoc. And in fact a lot of methane does bubble up from the ocean floor ... but it never gets to the surface. Where does it go? This is what William Reeburgh asked in the 1970s. It was clear that some group of micro-organisms was eating the methane and turning it into carbon dioxide – but what?

Methane-eating bacteria do exist, but they require plenty of oxygen and fresh water, very little of which is to be found on the ocean floor. And it turned out that these hypothetical organisms must also destroy sulphate, which was even more baffling. Then, in 2000, a team headed by Antje Boetius discovered some very strange organisms in the mud of the Eel River Basin, off the Californian coast. They were clusters of about a hundred archaeans, surrounded by a shell of bacteria, and Boetius suggested that these clusters were symbioses. However, for a time there was no direct evidence that the clusters could feed on methane. In 2001 Christopher House and colleagues showed that they did, and that the bacteria derived the carbon that they needed from the waste products of the archaeans. This strange symbiosis, until recently considered impossible, probably eats 300 million tonnes (tons) of methane every year, much the same amount that humans and their activities, especially agriculture, produce.

Today's archaeans come in many different shapes and types: some of them have no bounding cell wall, barely a membrane, while others have a very convoluted surface. Modern bacteria are mostly short rods or little spheres, but some take the form of spirals or long filaments. We have no idea what the primeval archaeans or bacteria looked like, because no record of their form has survived in the rocks. The earliest fossils that have been identified are aggregates of what look like today's filamentous cyanobacteria, mixed with archaeans and bacteria to form fibrous mats called 'stromatolites', much like those we find today in Australian tidal marshes and lakes where limestone makes the water hard.

Simple organisms are nature's starting point for more complicated ones. The theory of evolution argues that once life has got started, however simple it may be at first, there are routes by which it can change, and in particular, get more complex. The reproductive processes of life make occasional errors, or create new genetic combinations in other ways, and natural selection preserves those changes that improve the organism's prospects of surviving to reproduce its kind. Our evidence about paths of evolution comes from several sources, but it forms a very consistent picture. The main modern source is the DNA sequences of various existing organisms. In practice, the sequences used are usually those of some important RNA machinery, but the RNA is generated from the DNA and the code sequences are the same. Because these DNA sequences have diverged from the original sequence in the common ancestor of any pair of today's creatures, through regularly occurring mistakes in copying, the extent of the difference tells us how long it is since that common ancestor lived.

This 'genetic clock' is only approximate, however. Different DNA sequences change at different rates, different organisms have characteristically different rates of divergence even for the same DNA sequences; but now that so many of its branches of the Tree of Life have been documented by these divergences, they do make a fairly complete story. The first half of this story concerns the 'prokaryote' kingdoms of the primeval seas, which existed long before any life got on to the land, and before there were any complex cellular creatures. Prokaryotes are characterised by keeping their DNA as a loop attached to their outer membrane. They include bacteria, cyanobacteria, and archaeans, which derive from several very ancient lineages: see Colin Tudge's superb book *The Variety of Life*.

In contrast, 'eukaryotes' are true cells, with the DNA kept separate inside a nucleus; a eukaryote cell is far more complex than any prokaryote, more like a biochemical factory compared to a workshop. From about a billion and a half years ago, when nucleated eukaryote cells first arose from the evolutionary symbiosis of bacteria and archaeans, there have been many eukaryote evolutionary lines doing very different things. A surprising number of these have survived to the present day.

Many protists – unicellular organisms like the 'slipper animalcule' *Paramecium* and *Amoeba* which you saw in your school biology practical classes – have turned out not to be like *Paramecium* and *Amoeba* at all. The apparent physical resemblances between such organisms are belied by their DNA sequences. For example, the common human parasite *Trichomonas* looks a bit like *Paramecium*, but we now find that its lineage diverged from those of *Paramecium*, *Amoeba*, oak trees, and people some two and a half billion years ago. This makes it very clear that *Amoeba*, oak trees and people are much nearer to each other, in evolutionary/genetic terms, than they are to slime moulds or to *Trichomonas*. We have lots of lineages to learn from, but most of what we're interested in isn't back there in the remote recesses of time, when the *Trichomonas* ancestor was trying out its new way of life, whatever it was. There were no human guts to parasitise, then; indeed, no guts at all. Instead, most of what we're interested in concerns the complex animals with many nuclei, many cells, the Metazoa and, among plants, the big many-celled ones, known as Metaphyta.

If we're to discuss what might happen on other planets, we must have some familiarity with the events that occurred here. In many ways the best example starts with the 'explosion' of innumerable body-forms among the early Metazoan creatures of the early Cambrian period, some 543 million years ago. They have been made famous by Steven Jay Gould's *Wonderful Life*, which is mostly about the soft-bodied organisms fossilised in the Burgess Shale in Canada, which were re-interpreted by Harry Whittington and Simon Conway-Morris to reveal an astonishing degree of diversity. Even after the Cambrian explosion, from about 530 million years ago onwards, the fossil record is never complete, but is generally adequate for groups of organisms like molluscs, corals, and vertebrates, which have (some) hard parts. The fossil record is very poor for organisms that are all soft, like soft-bodied worms and flatworms; it is also poor for flying birds, which rarely fossilise. On the other hand, it is extremely good for organisms like beetles, sharks, and snails, where nearly every individual that wasn't eaten seems to have left a trace of its existence in the rocks.

Science's picture of the evolutionary diversification of life has changed as new techniques and concepts have become available. Nowadays we are beginning to document the divergence of DNA sequences, but not so long ago the main technique was to analyse the divergent anatomy of today's organisms. Anatomical differences between organisms give clues about the form of their common ancestors, and some indication of how long ago those ancestors lived. Since the time of Darwin, Richard Owen, and Ernst Haeckel, fossil evidence and the varied but related forms of modern organisms have helped us build a picture of life's evolution on this planet. This picture needed very few modifications when DNA evidence appeared. The main problems were with simple creatures like *Trichomonas*, whose divergence was a great surprise; this confusion had arisen because there had been no fossil evidence. The fossil evidence about Metazoa with some hard parts had given us a pretty good story before DNA sequences, and it wasn't necessary to modify it much when the new, independent, DNA evidence turned up.

The details of this picture can be found in numerous books, especially those of Richard Dawkins, Gould, and Tudge: we lack the space to say very much more about them, except to restate that they form an important database underpinning xenoscience. The main question is: how should we make use of that database? Astrobiologists tend to swallow it whole and assume that Earth's story is the only story, and that alien life *must* be very similar to ours. And so they search for aqueous planets, and assume that oxygen is a must and that the only way to get oxygen is via chlorophyll in plant-like organisms, and so on. SF authors realised long ago that far more fanciful scenarios can be conceived, even if we do not yet know whether they could really exist. A dramatic example is Robert L. Forward's 'cheela', who live on a neutron star in 67-billion-g gravity. In *Dragon's Egg* he provides a very detailed description of the physics, neutronic 'chemistry', evolution, and social structure of the cheela. It is a real *tour de force* in making the invisible book visible – and then he adds the real invisible book as an appendix, in case we still disbelieve the science.

Dragon's Egg (Robert L. Forward 1980)

Half a million years ago a red giant star two hundred times bigger than the Sun collapsed under its own gravity when its nuclear reactions ran down, and became a neutron star only 12 miles (20km) in diameter. It had a surface gravity of 67 billion g. The neutron-rich nuclei in its crystalline crust formed increasingly complex nuclear compounds. This process proceeded at blisteringly fast nuclear speeds instead of slow molecular ones. Every microsecond, millions of new nuclear combinations were tried, compared to only one or two chemical combinations per microsecond on Earth. Five thousand years ago, 'in one fateful trillionth of a second, a nuclear compound was formed that had two very important properties: it was stable, and it could make a copy of itself. Life had come to the crust of the neutron star.'

Neutronic life evolved with equal rapidity. By 100 BC a 'plant' structure had arisen that exploited the energy difference between the heat of the deep crust and the relative cold of its surface canopy.

'Animals' evolved, stealing seed-pods from their cousins; unlike the plants, they had to contend with the star's strong magnetic field when they moved.

In 2020, this neutron star is detected, approaching Earth. It is a pulsar, putting out a signal that repeats every 199 milliseconds. By 2049, Pierre Carnot Niven and the *Dragon's Egg* exploration crew have headed out to take a closer look. They start mapping the surface with a laser scanner, notice a curious flowerlike pattern on the surface, and realise that it must be a sign of intelligence, however unlikely that might seem. They send a simple numerical signal along the laser beam. It is detected by Commander Swift-Killer, who notices the numerical patterns. The cheela quickly figure out that they are receiving a message from some offworld intelligence, however unlikely *that* might seem.

Contact is made, and the Terrans inadvertently civilise the cheela during one twenty-four-hour period. The cheela continue to evolve, quickly outstripping the humans in technological achievement. They find five small Black Holes inside the Sun, and kindly remove them; the book ends with a helpful message to us slow creatures: 'Please don't despair . . . there are only a finite number of fundamental truths to learn about the universe, so eventually you will catch up to us.'

Of course, the story of Earthly life is the only story of life for which we have any observational evidence, so it would be easy to conclude that Earth's story is therefore the only 'scientific' one, and to dismiss alternatives like the cheela as mere speculation. In many areas of science, that approach pays dividends – though even then it always helps to have a few mavericks speculating about other possibilities. But when the topic is alien life, this apparently sensible approach is fatal. If we restrict our science to one very probably unrepresentative sample, and fail to understand how unrepresentative it is, we will make serious mistakes.

There are many ways to do science. The textbooks emphasise the interplay of theory and experiment, but in many sciences experiments are impossible. We can't experiment on dinosaurs, or on Black Holes. We can, however, observe and infer. And we can test theories in computer simulations, or by other indirect means. There is nothing unusual in this, but the issues are sharper in xenoscience. Imaginative proposals here can be perfectly 'scientific' even if you can't see what's involved through a microscope or a telescope. It is, for example, entirely feasible to do solid science on alternatives to DNA or on the complex systems that could occur in various 'exotic' habitats (see chapter 10 for both). How complicated can a network of magnetic vortices in a star become? Could such a network become self-reproducing? If so, how? All this is good xenoscience, and quite a lot of work of this general kind already exists in the scientific literature.

This is a book about science, not a book *of* science. Our aim is simpler: to map out some of the main areas that must go into xenoscience. Of course one major area has to be 'Earthlike environments and Earthlike life', and it is this area that astrobiology concentrates on. There's no harm in that as long as we appreciate that it's just one major area, and as long as we have a proper grasp of the true diversity of Earthly life. In this book we will also devote plenty of space to Earthlike scenarios, but we will do that because it is in this area that we can be most specific, not because we think that nothing else can happen. When it comes to evolution, for instance, we view Earth's story as just one sample from what is probably a far broader range of possibilities. And we can ask two things. One is: just how broad *is* that range? The other is: are there any general, unifying principles?

There surely are. For the moment, we will concentrate on one important example: the distinction between universal features and parochial ones in evolution. This distinction can be clearly seen in Earth's evolution, where there is a simple test to determine whether any given feature is universal or parochial. Later, we will extend the concept by providing it with a theoretical basis, which will allow us to apply the same distinction to hypothetical alien evolution.

Universals first. Some innovations in Earth's evolutionary story have happened several times in different branches of the evolutionary tree: flight is a good example. Flight has been invented independently by birds, bats, fishes, insects, and the extinct pterosaurs. Eyes are another such case: they have been invented separately by scallops, snails, octopuses, vertebrates, insects, and other arthropods. We know that these methods for exploiting the evolutionary advantage of being able to detect light have evolved independently because of the fossil record; moreover, eyes have a different structure, and work differently, in different types of organism. A third instance is photosynthesis – getting energy from sunlight to power your internal biochemistry – which has also been invented several times. Violet bacteria use a very different membrane system from the chlorophyll system used by plants, and this is different again from several archaean tricks, which use sulphur chemistries and pump up some of the chemical steps by using energy from light.

Anything that has evolved independently, several times, is likely to be a generic trick that offers a strong evolutionary advantage. We shall call such a trick a *universal*. We would expect alien evolution to exploit appropriate universals, precisely because the trick is generic. 'Universal' does not mean 'ubiquitous': not all terrestrial organisms fly, for instance. It means that in suitable circumstances, the generic trick is available for exploitation and is therefore likely to evolve.

In contrast, if something has happened only once, as far as we know, during the Earth's evolution, then all we can deduce is that it's *possible*. It may, however, be highly unlikely, the result of an accident that would not be repeated if evolution were run again on a thousand planets. The awkward way that the human airway crosses the foodway, so that every year a few of us choke to death, is an example. We acquired it 'accidentally' from an ancient ancestor: the lobe-finned fish that came out of the ocean and invaded the land. This unfortunate feature of our anatomy is totally unnecessary – so if it was created, then the creator made a silly mistake. Why, then, has it not been weeded out by natural selection? The answer is that it kills such a tiny proportion of us that selection against it is extremely weak; moreover, it is tied to various other developmental accidents that make it difficult to evolve towards some other solution to the air/food problem. Such a 'once-off' feature we call a *parochial*. By once-off, we don't mean that it occurs in just one species: we mean that it evolved just once, and was then inherited by descendant species.

On a re-run Earth we (that is, the new actors playing our universal role on the evolutionary stage) would probably not have inherited our awkward arrangement of airway and foodway in the first place. There were many other kinds of fish in the sea, and it is entirely reasonable to suppose that a differently-designed one might have come out on to the land instead. Many fish had dorsal lungs, that is, on the top, whereas our ancestor had ventral lungs, underneath. So the descendants of one of those other fish would not have coughed like we do.

By the way, there are some incorrect images associated with this 'fish that came out on the land'. For instance, it didn't flop across the mud by waving its fins ineffectually, until natural selection got in on the act and improved the fins to turn them into legs. Actually, the fins of that fish had already *begun* to be walking-limbs while the gills were still being used – so much for the limbs being developed 'for' life on land. This kind of change is called 'pre-adaptation', which isn't a very good name because at the time it occurs, it is unclear what it is a forerunner to. Indeed, that's the point: the new organ or ability evolves because it offers some advantage; perhaps, in this case, scuttling about in the shallows. Then, *because it now exists*, some totally unexpected evolutionary pathway opens up, and further changes occur. Most features of organisms originally evolved to carry out different functions from those that their later modifications now perform.

The skeletal pattern of those early limbs, initially with eight (or more) digits, evolved into the familiar 'pentadactyl limb' pattern in the earliest amphibians: five digits, knees and elbows initially held out from the body, then resolving in opposite ways. But the skeletons could have done plenty of other things. Jointed limbs are clearly a universal, but the pentadactyl limb is a parochial derived from that particular fish. The same goes for the 'face' with nose above mouth. Yet our drawings of aliens are hung up on knees-and-elbows and recognisable faces. These may be justified, by the artist but not by the scientist, as fitting the circuits we have in our brains for recognising land vertebrates, so that the artist communicates like that. Ditto the marketing consultant. However, we would be very surprised indeed to find any aliens with our suite of such immediately understandable terrestrial land vertebrate characters. So surprised, indeed, that we would be sure that they must originally have *been* terrestrial vertebrates, sharing our ancestry, and that they had somehow journeyed out to wherever we found them.

The difference between parochials and universals is that we cannot confidently expect parochials to occur again at some other place or time. However, we *can* expect universals to recur: the fact that they have evolved independently on several occasions implies that they are *likely*. If we re-ran Earth, the default would be that they would happen again;

we would have an oxygen-containing atmosphere, and in time there would be creatures flying in it, some of them eating others. They would not be insects this time, or birds, or bats, because the contingent, accidental features determined by which gene changes and combinations were available, and which resulted in those groups first time around, won't happen again. All of our *exact* kinds of organisms must be parochial – found only here on Earth, and only this time around. But the multiple inventions, like flight or fur or photosynthesis, are universals: we should expect them each time we re-ran Earth, and, by the same token, on all of the other aqueous planets out there around other suns. Given enough time, of course.

The distinction that we wish to make, between universal and parochial features, is a cornerstone of xenoscience. However, that distinction will be too limited for such a purpose if its definition depends solely on what has already happened on Earth. The difficulty becomes immediately apparent when we begin to run through key stages in the evolution of life on Earth, looking for the universals. Our planet had only prokaryote organisms for more than half of its history. Bacteria and archaeans were the major lifeforms for two and a half billion years. Perhaps there were other creatures of which we have no record, but these would almost certainly have been of the same grade of construction. These creatures do not develop as they grow: they only grow in size, then divide when they get big enough.

During that long reign of the prokaryotes, though, some were photosynthetic, and their excretion of oxygen caused problems for their peers in the ecology after about a billion years. Oxygen in the atmosphere, and dissolved as a corrosive gas in the oceans, began to be problematic only after all the hydrogen, methane and ammonia in the original atmosphere had been oxidised, and indeed after many minerals had been oxidised too. So the planet could 'take' that pollution for a long time . . . before oxygen finally overwhelmed its 'defences' and an oxygen-containing atmosphere forced the biggest change on Earth's ecology that there has ever been.

Eukaryote precursors probably developed independently, many times, so 'eukaryote' may also be a universal. But what is the status of 'oxygen atmosphere'? Since an oxygen atmosphere developed only once on planet Earth, our current test makes it a parochial. However, there are strong theoretical reasons for believing that other aqueous planets are also likely to develop oxygen atmospheres: we'll review these in a moment. If we are right, then 'oxygen atmosphere' ought to be a universal. But is it?

There are three different questions here. The first is 'Did it happen more than once here?' The second is 'Would it happen again on Earth, if it was re-run from the beginning?' And the third is 'Is it likely to happen on some other world altogether?' Our current test is based on the first question, and if the answer is 'yes', then we infer that the answer to the other two questions is also likely to be 'yes'. However, if the answer to the first question is 'no', it's still possible that the other two questions have a different answer. Anything that would happen again on a re-run Earth would probably happen on any aqueous planet under reasonably common conditions, so in the context of aqueous planets the second and third questions should have the same answer. In other contexts, though, they could have different answers. In order to resolve this problem, we must generalise the universal/parochial distinction to cover evolutionary features that might arise on another planet, or indeed any other alien environment.

The most useful definition of the distinction must therefore be theoretical, not observational. We will call a feature of an evolving system universal if it corresponds to a generic trick – one that exploits a context that is likely to be common – whose adoption offers clear evolutionary advantages over organisms that lack it. The trick could involve exploiting a new resource, or 'solving' a widespread problem, for instance. In contrast, any feature that seems to be special, relies on an accidental context, or fails to offer any evident advantage, is a parochial. Our previous examples fit this theoretical distinction, and the theory tells us *why* the universals are universal. Eyes are universal because light exists throughout the universe and can be exploited to detect predators or prey: eyes exploit light as a resource and solve the problem of observing surroundings more distant than those you can touch. Flight is a generic trick because any large enough planet will retain gases and thus have an atmosphere, as do most of the planets in the solar system. The advantage of being able to fly, compared to others that can't, is obvious: it gives airborne prey a new way to avoid ground-based predators, and airborne predators a new way to catch ground-based prey.

What about airborne predators and airborne prey? Now the outcome is less clear. This is a subtle feature of the distinction, which is about *innovations*. Our theoretical test compares 'new' organisms that possess the innovative feature with 'old' ones that do not. It does not, and should not, compare new organisms with other new ones, because that is a question about how evolution goes after the initial innovation has appeared. The universal/parochial distinction is about why an innovation appears to begin with, and only about that.

Ring (Stephen Baxter 1994)

This is one of several of Baxter's books centred around the Xeelee, aliens so advanced that humanity can neither understand them nor communicate with them. The Xeelee's concerns are *big*; for example, they are trying to fend off the end of the universe.

The central problem in *Ring* is more immediate. Lieserl, who has been modified by nanobots to render her immortal, is equipped with a refrigerator-wormhole and sent inside the Sun to investigate its physics. The Sun is dying, from an excess of photinos – fundamental particles of 'dark matter', supersymmetric partners of ordinary photons of light, particles of the kind invoked by cosmologists to explain why galaxies don't rotate the way their visible matter would dictate. But as she approaches the Sun's core, nothing looks right. 'According to the Standard Model, the temperature should have fallen rapidly away from the fusion region ... But in fact, the temperature drop was far more shallow.' Later, 'There were still divergences from the Model, she saw. There were divergences *everywhere*. And they were even wider than before.'

Deeper and deeper into the Sun she penetrates, far beyond her original mission's objectives, further and further into the Sun's future. In the core, she observes moving photino-objects, which must be responsible for the dark-matter cancer that is smothering the Sun's nuclear fire. The objects form flocks, they move with a sense of purpose. They are *alive*. She names them photino-birds. They are aware – but probably not intelligent. They are feeding on the Sun's nuclear reactions. They breed by a kind of cloning. All the stars in the galaxy, maybe the universe, are infected by flocks of photino-birds ... The universe is being *eaten* to death by mindless creatures made of dark matter ...

It will be useful later if we now characterise the distinction between universals and parochials in terms of evolutionary phase spaces – the metaphorical spaces of all potential organisms or ecosystems, in which the actual organism or ecosystem sits. An innovation is universal if it lies in a region of phase space that is adjacent to the region currently being occupied, but 'pointing in a new direction'. That is, its adoption would expand the phase space. A parochial, in contrast, just explores another little bit of the existing phase space. Recall that Kauffman, in *Investigations*, refers to this situation as the 'space of the adjacent possible', and suggests that organisms and ecosystems expand into that space as rapidly as physical and chemical constraints will permit.

With the theoretical basis for our key distinction now firmly established, we can apply the associated test for universality to the question of an oxygen atmosphere. We will argue that this feature is a universal for aqueous planets, but probably not elsewhere. First, we must show that oxygen could reasonably be a generic resource (or problem) on an aqueous world. The argument here is that photo-synthesis is surely a universal. Indeed, we've already established this by the observational test, but let's go over the same ground again from the theoretical point of view as a check. There are plenty of stars, and plenty of planets circling them, so light is a universal energy resource. (Indeed, if Stephen Baxter's photino-birds in *Ring* are taken into account, it may be even more universal than we think, and exploited in radically different ways.) Innumerable chemical reactions are driven by light, so life can hardly fail to make use of this generic resource. The reactions that occur in Earthly photosynthesis involve elements and compounds that should be common on aqueous planets. In particular, there must by definition be water on an aqueous planet, and water molecules contain oxygen, which the action of light can liberate. The 'useful' aspect here is not the oxygen, but the energy produced by the reaction. Photosynthesis lets life steal energy from the stars. The oxygen is a by-product, and at least to begin with it is a pollutant, not a useful resource. Problems, however, are also opportunities: oxygen changes the game, potentially, and it does so in a big way. Anyway, given all this, it will surely be common for aqueous planets to produce a variety of simple organisms, some of which will excrete oxygen.

Next, let's recall how Earth's lifeforms dealt with that poisonous, corrosive pollutant. This development constituted the next big step for life on this planet, which was the production of real cells, cellular eukaryotes with the DNA in a nucleus. This development was forced by the presence of oxygen. Cells have several kinds of 'organelles' – specific parts that have specific functions. We now know that the ancestors of these organelles were once free-living prokaryotes. Indeed, today's eukaryote cells are the descendants of associations between several kinds of archaeans and bacteria; this evolutionary symbiosis was achieved by several different kinds of organism, and we have different kinds of eukaryotic organisation today. Several different protists have 'alien' mitochondria, and the *Actinomycetes* and red algae (*Rhodophyta*), and our friend *Trichomonas* are all descended from different aggregates of prokaryotes.

The Earthly organisms that later became mitochondria were probably good at 'using up' oxygen, so their immediate environment was depleted of that element, and thereby made *more* comfortable for their oxygen-hating colleagues. Which was to everyone's advantage. That makes evolutionary symbiosis a universal innovation – we would expect to find the same process occurring again on a re-run Earth, and on other planets. We would expect to find more sophisticated 'cells', sharing the technologies of several previous prokaryotes, in the oceans of these planets – *if* we wait until there is an oxygen-containing atmosphere. But we would have to wait a long time.

Photosynthesis is such an effective trick, on an aqueous planet, that it has virtually taken over the running of our own world. And not just through plants. When it comes to photosynthesis, the divide between plants and animals has been crossed from both sides very often on Earth. Plants make their own food photosynthetically, although some bacteria and archaeans do it 'chemosynthetically', that is, by direct chemistry without using light energy, especially at undersea vents and hot volcanic springs where the heat is an alternative energy source. Then animals eat plants.

There is an interesting – universal – alternative to eating the plants. A diverse list of animals – among them one species of *Paramecium*, most coelenterates (*Hydra*, jellyfishes, corals), a few flatworms, bivalve molluscs like the giant clams – have photosynthetic algae in their cells that give them sugars when they're illuminated, and receive amino acids, minerals, and protection. Pogonophoran worms, marine creatures that look more like fragile flowers when they poke out of their tubes, have chemosynthetic bacteria and archaeans in their cells, and no gut, and the biggest of these worms live around undersea volcanic vents. So on an aqueous planet that pursues the photosynthetic universal, we expect to get independent synthesisers (plants) and grazers on them; then filter-feeders on the grazers, and predators on them ... and creatures that take in productive lodgers after the major symbioses have resulted in effective cells, defended against that nasty corrosive oxygen. The sequence here, we're sure, is not a universal. Worms with symbiotic prokaryotes might pre-date real plants, for example.

Even today, there is a great variety of all kinds of photosynthetic organisms on Earth, so we'll doubtless find micro-plankton in alien seas, doing a lot of the continuing photosynthesis even when there is a respectable variety of higher 'grades' of lifeforms.

Another universal is some kind of cell, with the possibility of becoming organised into colonies and genuine many-celled creatures. None of the parochial details of our cells, mind you, but the same general kind of set-up. On the whole it seems that evolution on an aqueous planet will start off in the oceans – that's why being aqueous is important. But life adapts, so once it has got going in the oceans, it doesn't need to stay there. Invasion of the land by marine and freshwater organisms is surely another universal: in particular, it has happened here many times, independently, which is strong evidence of universality. As we've said, flight is another classic universal, of course.

To sum up: the most likely scenario that we expect to find on another planet like Earth, if we arrive randomly in time, is that it will have only simple life in the seas. Its oxygen production by life will not yet have oxidised the atmosphere and minerals, so there'll be no free oxygen – yet. As soon as the free oxygen appears, we expect that successive grades of more complicated lifeforms will appear. However, that could take a long time; it took nearly three billion years here. They will not be just like those we have here, because they are likely to have a different genetic system – and even if they had the 'same' genetics, using DNA, RNA, and proteins, there would not have been the same mutations to exploit the same environments. There will be no vertebrates, no molluscs, no insects. Nonetheless, many generalities will be recognisable, because they embody universals. Different kinds of creatures will do, we expect, much the same kinds of things. There will be plants competing for light, grazers competing for plants, predators competing for grazers, just like here.

If evolution is drama, then the actors will be different, but the play will be the same.

What about the stage on which the play is enacted? What other environments might be suitable for life? The question of life on other planets has changed as chemistry has changed, and as astrophysics has changed. It has changed because our conception of 'life', 'other', and 'planet' has changed. And planets may not be the only suitable places for life to form. In the last decade new scientific paradigms have come into being, based on the viewpoint of complex systems. Countless simulations and other 'experiments' show, repeatedly, that systems composed of many simple units interacting in quite simple ways have the ability to generate large-scale patterns and structures, to self-complicate, and to evolve. And even to behave like rudimentary lifeforms, with reproduction, parasites, social behaviour, and sex. This kind of thinking has supplanted the old heat-death models. We no longer expect the universe to run itself down: instead, we think that it is most likely to crank itself up. Given this new viewpoint, do *all* planets, not just aqueous ones, build up their systems towards complex life?

Surely they do, but not all will succeed. On those that do, the paths that life follows could be much more various than those of the one evolutionary story that we have seen (so far) on Earth. It is as if we have watched one chess game, and we have to work out the rules, as distinct from one possible series of moves within those rules. Are there guidelines, deducible from what we have discovered about evolution of life on Earth, that will permit us to infer what other chess games might be like – for example, as we've said, what might happen if we re-ran the Earth-system? Perhaps we could extend the metaphor: 'Earth running again' might be compared to another game between the same opponents, whereas 'What do alien evolutionary scenarios look like?' could be compared to other games, by different players. Then the question becomes 'Do we know that they must be playing by the same rules?'

To think about this question more deeply we must consider what the 'rules' do, what laws of nature are. Here again, the physicists have set misleading standards, and there is another physics-given practice that we have to criticise. It has been called the Newtonian paradigm, to be contrasted with, say, the Darwinian. Newton's discovery of the law of universal gravitation led to a 'clockwork' view of the solar system; not only did it explain apples falling and water running downhill, it showed us a planetary orbit as a continually-falling trajectory around the parent body, but modified by the attraction of many other bodies. This allowed people to argue from the laws to the facts; so Uranus not doing what it ought to, according to the laws, was evidence for the existence of Neptune. Physical laws apparently worked both ways: from the facts to explanation by the law; from the law to what the facts should be.

Contrast this with Darwin's discovery of natural selection. Take some standard example of evolution in action: for instance, Darwin's finches. About five million years ago a tiny group of finches arrived on the Galápagos Islands, all of the same species. They thrived, in the absence of competitors, and have since diversified into thirteen species, plus one more on the Cocos Islands. We can explain Darwin's finches by natural selection. But we can't tell, from natural selection, whether there should be another species of finch, in the way that we predicted, and then found, Neptune. Natural selection explains specific examples and makes general predictions, but it cannot lead to predictions with the precision we expect in astronomy . . . and that's one of the *least* exact of the 'exact sciences'.

Unfortunately, the successes of physics have persuaded us to believe that *real* laws should work both ways. Chemistry (old-style) was fully convinced that all the best chemical equations went both ways. Relativity was all right, the facts could come from the equations as easily as *vice versa*, in principle, anyway; for anything more complicated than a point mass orbiting a sphere, monstrous calculating facilities would be required. Quantum theory was different, and it gave physicists problems: it specifically denied the possibility of tying the facts to the rules in both directions . . . yet it worked to twelve decimal places: it was beautifully accurate when you asked suitable questions, but it was causally ambiguous.

Most physical discussions in the last fifty years have been attempts to justify quantum thinking as real, law-like argument. These attempts range from David Bohm's concept of the 'implicate universe' with many hidden patterns, to nutty ideas of SF fans who can't find the roots of a quadratic equation but think they can solve quantum gravity problems. Jack was editor of the journal *Speculations in Science and Technology* for five years, and he received about twenty crackpot submissions a month in this area. Similarly, recursive chemistry, like that of the Belousov–Zhabotinskii reaction or clay crystallisation or the production and destruction of the ozone layer, has built in one-way-reading: we can't argue what really happens from our understanding of the laws.

The requirement that the laws work both ways must be relaxed for biological laws, because biology is *like that*. As Kauffman says, we can't pre-state the phase space of the biosphere. We have to be able to look at history, at what the real world has done, for guidance. We know about only one actual history of life on a planet, but we know about that example in an awful lot of detail, and we have some pretty good guidelines (such as natural selection) about how the organisms live, and what they did and when. So, can we answer the scientific question that an enlightened physicist might ask: 'What would happen if you ran the Earth's evolutionary system again?' This is a very interesting question, but not because there is a clear answer. It's really several questions rolled up into one, and we have to disentangle the different interpretations.

We might begin by explaining to the physicist that such questions aren't very happily answered even in astrophysics: 'What would happen if we ran evolution again? Would we get people?' is just like 'What would happen if we ran the solar system again? Would we get Earth?' And of course it all depends on how closely Run Two resembles Run One, and what counts as 'the same' outcome.

If in fact we live in a clockwork universe, and if we reproduce the initial conditions exactly in both runs, then of course we must get the same results both times. Even if a cog wears out and falls out of the clock, it will wear at the same rate on Run Two, and fall out at the same moment. Quantum indeterminacy probably doesn't make a big difference here, because 'in the wild' the indeterminacy averages out in a statistical way and leads to the same macroscopic physics whatever the fine quantum details may be. All this suggests that our question is more interesting if Run Two of the universe is similar to Run One, but not identical. Which phenomena are *robust*, in the sense that they will probably show up both times?

The physicist could reasonably reply that the answer to the astrophysical question is much more robust than the likely answer to the evolutionary one. We have plenty of confidence in the astrophysical answer, because we have pretty good models of what happened to the solar nebula, and there are plenty of solutions to the equations that result in big planets Jupiter-wards and out, and small planets like the Earth, cuddled in closer to the Sun, just like the solar system. So the physicist would be able to answer the astrophysical question with 'Yes', and feel fairly happy that she was right. If biological laws are like physical ones, she might expect the evolutionary question to have the answer 'Sure, we know how evolution works, so we'd get people again. Humans are a robust solution to Lummox's Laws of Life.' But biologists can't say anything like that, and not just because Sir Lennox Lummox has not yet been born.

However, the biologists wouldn't admit defeat that easily. They would point out that astrophysics is much less sure than it used to be about what the laws that govern solar systems are. We have now found lots of alien solar systems: fifty-eight at the last count, with new ones being discovered almost weekly. There are two main ways to locate them. One is to observe tiny wobbles in the movement of stars, as they are whirled round by their planetary dancing partners. The other is to spot the slight dimming of the star that occurs when a planet passes across its face, which has a characteristic 'signature'. Both methods require very delicate and very accurate observations, and they are currently limited to finding big planets – Jupiter-sized or larger.

A slightly worrying feature of some of the alien solar systems found so far is that a few of their planets are much too close to their suns, or much too massive, to be obeying what even very recently we thought were general laws. In 2001 Geoffrey Marcy's group detected a giant planet orbiting the star HD 168443, which lies 123 light-years away in the constellation Serpens. The planet has at least seventeen times the mass of Jupiter, and circles its parent star in just under five years. But planets that form from swirling discs of gas and dust, the standard theory, should not exceed eight to ten Jovian masses: above that, they sweep their system clean of dust. But the same star also has a planet of at least 7.7 Jovian masses, orbiting once every fifty-eight days. That's also a teaser, because planets that close to a star should never have formed to begin with: any matter that started to aggregate ought to have been pulled apart again by the star's gravity. Moreover, above thirteen Jovian masses the core of a 'planet' should become so hot that deuterium starts to burn up in a nuclear reaction, and the object becomes a brown dwarf. But the giant planet is not one of those either.

To some extent this problem may arise because large planets are the only ones that we can detect at this stage of the art: limits on our observations may be highlighting the rare exceptions. But it is a little worrying, particularly for our Newtonian view of astrophysical laws, that right now these planets look spectacularly wrong. The same team found another strange system in the same year: Gliese 876, fifteen light-years away in Aquarius. This star has two planets, both roughly the mass of Jupiter, locked in a 2:1 resonance. That is, one takes exactly twice as long as the other to orbit the star. Resonances are common, but this one is on the very edge of instability, so it's a surprise to observe it.

If something as law-like as astrophysics can be seen as deriving laws from reality, but is presently much less sure about predicting reality from the laws, then our physicist would understand when we claimed that understanding evolution *doesn't* allow us to predict what will happen.

Doesn't that stop evolution being science? Isn't science about predictions?

Here we run into one of the most pernicious misunderstandings of the scientific method that there is. The word 'predict' has two meanings. One, appropriate to gypsy fortune-tellers with crystal balls, is to foretell the future. The other, appropriate to science, is to state that *if certain conditions apply*, then certain consequences will follow. For example, a valid 'prediction' about the solar system is that if the Earth and an asteroid occupy the same position at the same time, there will be a rather large bang. It is not necessary to predict *when* such an event will occur: it is still a prediction in the scientific sense – and one that now worries most of Earth's governments. As it happens, we can probably predict such an event a few years ahead, maybe a century or so, provided we know enough about all likely asteroids. But we can't predict the date *now*.

Evolutionary biology makes plenty of predictions in this sense. For example, it predicts that species will change in response to changes in environment. It can add a certain amount of detail, and we know that sometimes these predictions work – they do, for instance, for Darwin's finches over the last twenty years or so. What we can't do is predict how the environment will change.

For this reason, it makes perfectly good sense to sift the history of Earth for the law-like rules of biological evolution, as distinct from the purely contingent events. That is, we want to find the rules of the game, not the actual moves made in one historical instance. And by doing so, we will find that it is possible to give a robust answer to the question about humans as a constant of evolution. Though we will be led to a more interesting answer, to a slightly different question – and a more sensible one.

The way the question is posed makes it reasonable to confine our guesses to the re-run of an aqueous planet like Earth or early Mars. A major political problem on today's Earth is 'global warming' caused by excess (man-made) carbon dioxide in the atmosphere; this leads to the 'greenhouse effect' in which solar energy becomes trapped, warming the planet. On an aqueous planet, the greenhouse effect cuts both ways: if the temperature is too cold, the ocean freezes; if too hot, it boils. Greenhouse warming, whether natural or not, changes the 'default' temperature – the temperature that would occur at that distance from the parent star purely as a result of its radiant energy. If a planet has an orbit just outside the distance at which the default temperature would cause water to be liquid, then the greenhouse effect can warm it up to the point at which oceans can exist. But if the planet is too close to the distance at which the default temperature would turn water to steam, then the greenhouse effect can kick it over the edge. So enough greenhouse warming can help Earth-like life get going, but too much may create unsuitable conditions.

The greenhouse effect is one of the things that made life on Earth, at least life of the present kind, possible: without carbon dioxide, Earth would be a frozen lump of ice. The story was very different for Venus, which probably hit the problem of runaway 'greenhouse gases' very early, and got so hot that it never had aqueous seas, but we're not talking about Venus. One of the conditions that must be satisfied for our 'prediction' to hold is that an aqueous planet appears, and that's the planet we're talking about: for purposes of discussion, its existence is a 'given'. It is clear, by the way, that aqueous planets should occur in many solar systems, so our prediction is not vacuous. Hydrogen is the commonest element in the universe, and there's plenty of oxygen to react with it to form good old H_2O. Matter aggregates into planets without great difficulty, probably round most stars; if the matter is too far from the star then the H_2O will be ice, and if it's too close it will be steam, but in between are cases when it can be 'just right', and give water.

What counts as 'just right' depends on all sorts of factors: the shape of the planet's orbit, the type of star (is it variable or not?), the presence or absence of a cloud layer . . . as we've said, the naive story of a 'habitable zone' around each star, a hollow sphere inside which H_2O forms liquid water, is nonsense, but the messy reality that it approximates is qualitatively similar. For probabilistic reasons, there will be (will have been) zillions of aqueous planets around, many of them *remaining* aqueous for geologically long periods.

Given that we're looking at an aqueous planet, we have already explained why we are sure that life-like chemistry will appear in drying-up tidal ponds, by volcanic flows into the seas, at rifts in the sea-floor where hot sulphurous gases are forming black smokers at oceanic vents. So this will happen again on a repeat Earth-run, or elsewhere on any other aqueous planet.

Yes, but will life there evolve into humans? That's trickier. Actually, if taken literally, the answer is almost certainly 'no'. Sorry, but that *Star Trek* universe just isn't on the cards. However, that's a silly question: it's like disallowing an otherwise Earthlike planet because the continent that ought to be Africa is the wrong shape. Will *human-like* lifeforms evolve? To answer that, we have to decide what the important, general features of human beings are. Five fingers, two legs, and sexual organs mixed up with excretory ones will not be among them. Intelligence might well be. If you will accept 'intelligent aliens' as a substitute for 'humans', then the answer to the re-run question is 'yes'. But those aliens won't look like us, won't act like us, and won't *be* like us in most specifics.

Same play, different actors.

6

THE DRAKE EQUATION

*F*ROM THE OUTSIDE *it is an ordinary three-bedroom house in St Albans, in a road of three-bedroom houses like a thousand others. Inside, there are subtle differences from the neighbouring dwellings. Cain is pleased to see a new garage for a thousand flying saucers, and even more pleased that the suite of twenty new offices to handle the life-support requirements of the chlorine-breathing Illensans is up and running.*

The tour buses, like the house, are unremarkable from the outside. At the moment they are disguised as lawnmowers, but their devices for cruising among the grazers on a grain of sand, or taking passengers down to the Maracot Deep, use roughly the total annual energy consumption of Iceland. It has taken much persuasion to get the non-corporeal aliens to share with the heavy-gravity ones. Two of the secretaries have tentacled in their notice, claiming that they are being required to work non-social years. Worst of all, someone has filled in the Terran Council Tax form with '1,023' in the box for 'How many normally resident in the dwelling?' Nevertheless, Cain is managing, and the Earth Prospectus *is bringing in the customers like hot flies.*

There have been 'incidents'. One group of Denebians accidentally bred while on the coral reef – but just as the progeny were planning to take over the Earth's oceans, they were drawn in and digested by a giant clam, so that was all right. Several of the staff have brought their progeny along for the 'educational experience', and the youngsters that have been eaten have provided ample education for the others – a definite plus.

Abel has taken over the management of the next lot of tourists, a special group getting bargain rates because they are locals. The Titanians have had space travel for only a hundred thousand years, so are terribly unsophisticated. The technology that speeds up the Neptunians to reasonable metabolic rates has the disadvantage that three thousand generations pass for the tourists for each lunchtime at home. The friendly Jovian blimps are still astonished by the idea of life appearing in the oxygen pollution and liquid ice of Poisonblue. The Mercurian metallomorphs are a trial, though: four very different species that have to be kept apart because the existence of each is a heresy in the religion of the others.

All of our rhetoric so far leads to the conclusion that there are vast opportunities for alien life. Can we quantify that statement, and add some numbers to the rhetorical mix? How many intelligent races exist in our galaxy, right now?

This is an ambitious question, given that we don't have a particularly accurate idea of how many *stars* there are in the galaxy, but it's basic to the whole enterprise of xenoscience. If, for example, the answer is two or higher, then there must be somebody out there. If the answer is one, then we are alone. If the answer is zero, then something has gone wrong with our theory. So if we can answer this question, we'll learn something important.

As yet we can't just let observations determine the answer, by going out and seeing how many intelligent alien races we can find. However, there are theoretical ways to estimate the numbers. The most important feature of these methods is not the answers they yield, but the tacit assumptions that are revealed if you stop trying to do the sums and think about what the sums *mean*.

The archetype of all such methods is a mathematical formula called the Drake equation, introduced by the astronomer Frank Drake around 1960. This was, at the time, pioneering work – a real scientist was taking the possibility of aliens *seriously*, for heaven's sake. Like most pioneering work, it was oversimplified and suffered from major flaws. It was also devised for a specific reason: to give some idea of the chances of success in Project Ozma, a 1960 forerunner to the SETI project, the Search for Extraterrestrial Intelligence. The idea of this celebrated project is to look for radio signals from distant aliens. It's been going for a while now, and it hasn't found anything yet – but at least someone is looking.

The Drake equation did a useful job for Ozma and SETI, by

providing ballpark estimates of the chances of detecting the kind of alien civilisation that thinks it's a good idea to broadcast powerful radio signals intended to alert other intelligent beings to your presence. It's worth realising that even if they do that, we may have trouble picking up their transmissions: according to Michael Lachmann, M.E.J. Newman, and Cris Moore of the Santa Fe Institute, efficiently coded communications look exactly the same as black-body radiation. Their paper on the topic originally bore the title 'Any sufficiently advanced communication is indistinguishable from noise', but eventually became the more prosaic 'The physical limits of communication'. We'll refer to their work as 'Lachmann's theorem'.

Unfortunately, the Drake equation has become an astrobiological icon, and is now being used for very different purposes: estimating the likelihood of alien life of (allegedly) any kind. As we mentioned in chapter 1, Ward and Brownlee's *Rare Earth* is a typical example, and we'll discuss it later in this chapter. We will see that the Drake equation does a better job of solving the problem confronting Ozma and SETI than it does of answering anything sensible about aliens. One very positive feature of the Drake equation is that it pioneered quantitative xenoscience. But all pioneering work has flaws, which become apparent because the pioneer did enough to get everyone else thinking critically; so another positive feature is that we can get an improved understanding of how to think about the prospects for alien life, by understanding what's wrong with the original Drake equation and working out how to improve it.

Contact (Carl Sagan 1985)

For countless aeons a gigantic polyhedral structure, orbiting a blue-white star, has been broadcasting messages into space. Dr Eleanor Arroway, Terran scientist, picks up a signal from the direction of Vega and recognises it as the series of prime numbers. This is followed by images of Adolf Hitler opening the 1936 Olympic Games in Berlin. At first the signal is thought to be a hoax, but then it becomes clear that the aliens must have picked up an early terrestrial television broadcast, and are sending it back again. The long message from Vega continues with instructions for building a complex machine, dodecahedral in form.

Several machines are built. Five humans, Ellie among them, enter the Japanese Machine, which is activated. It takes them by way of a Black Hole to Vega, but this turns out only to be a cosmic shunting-yard, and they continue, jump after jump, out into the galaxy . . . They dock at the celestial equivalent of Grand Central Station, leave the Machine, and find themselves on . . . a beach. Ellie is visited by an alien disguised as a young version of her long-deceased father. The Hitler broadcast has alarmed the aliens: Earth civilisation is in dire need of help, and they will do what they can. The alien shows her their transportation system, which was built long before any of the existing races came into being. It is the same in numerous other galaxies. Billions of years ago, the Builders packed their bags and left – no one knows where.

The five Terrans return home, to arrive the instant they left. Their story is not believed; their taped recordings are blank. But the aliens have told Ellie that there is a hidden message in the digits of pi . . .

The idea behind the Drake equation is to try to put some figures to the chances that aliens exist by dissecting out the various component factors and estimating *their* chances. In effect, it tries to aggregate all the various uncertainties into one pot. It is expressed as a complicated-looking mathematical formula, but that's little more than window-dressing: what matters is what the equation means. It is a recipe for estimating the number of intelligent alien races that exist right now in our galaxy. By 'intelligent', here, it is traditional to mean 'capable of interstellar communication'. The formula tells you that this number can be calculated by multiplying together seven other numbers:

- The mean rate of star formation in the galaxy, in stars per year.
- The proportion of stars with planetary systems.
- The number of planets per system capable of producing and supporting life.
- The proportion of those planets on which life appears.
- The proportion of life-bearing planets on which intelligence evolves.
- The proportion of intelligent-life-bearing planets on which those lifeforms attain interstellar communication skills.
- The lifetime of that technology, in years.

Let's carry out the arithmetic and see what we get. The figures we'll use are reasonable, but have been chosen to illustrate some pertinent issues in a simple setting: later we'll consider whether other figures would make more sense. For the moment, then, we'll assume that the mean rate of star formation is 10 stars per year; the proportion of stars with planetary systems is 0.1; the number of planets per system capable of supporting life is 3; the proportion on which life appears is 0.1; the proportion on which intelligence evolves is 0.1; the proportion with interstellar communication skills is 0.1; and the lifetime of that technology is 1 million years. Multiply all these numbers together, and we get 3,000 intelligent alien races in the galaxy. There are about a hundred billion stars in our galaxy; so if the equation and the numbers are right, then at this moment (or any other) about one star in thirty million is the abode of intelligent aliens. Plenty of aliens, then – but spread rather thinly.

That's the kind of calculation you have to do if you take the Drake equation seriously. However, when we take a critical look at the equation, and the assumptions behind it, we discover some major flaws. A crucial feature is that it tells us to find seven different numbers and multiply them together, and that gives us an immediate handle on how accurate the answer is likely to be. For example, we've estimated the likely lifetime of interstellar communication technology as 1,000,000 years. The reasoning, which we'll present later, is rather loose: the figure might as well be 100,000 or 10,000,000. If the first, then the answer reduces to 300 alien races; if the latter, it shoots up to 30,000.

There are similar uncertainties for every number that we want to plug into the formula. Suppose that every one of those seven figures could be somewhere between one tenth its size and ten times its size. In astronomy, anything 'within one order of magnitude' is generally considered spot on. In particular, we do not yet have a definitive estimate of something as straightforward as the number of stars in the galaxy: it might well be a trillion, ten times as big as the current orthodoxy. If you multiply 10 by itself seven times you get ten million. If you multiply 1/10 by itself seven times you get one ten-millionth. So the range of values you might have got for the number of alien races, by making the seven numbers on the right-hand side ten times as big or ten times as small, lies between one ten-millionth of the answer you got and ten million times as big. That's a range of *fourteen* orders of magnitude. Even astronomy might baulk at that kind of tolerance.

Specifically, that nice comfortable 3,000 could reasonably be anywhere between 0.0003 and 30 billion. The first figure says that even *we* don't exist (though with the assumed definition of 'intelligent' we don't, so no problem there). The second says that about one star in thirty (not one in three, because the calculation leading to 30 billion would have to use the 'trillion' estimate for the number of stars in the galaxy) is home to intelligent aliens, which seems rather unlikely.

This 'numerical instability' of the Drake equation is a real pain. It means that even if the equation is right, it's no use whatsoever until we refine our understanding of intelligent life to the point where we can get those numbers pretty much spot on. It's hard to imagine a universe in which it's possible to do that without already having a very clear idea of the distribution of intelligent alien lifeforms. In which case, we don't need the Drake equation. However, that's by no means the worst flaw in the Drake equation. The most serious flaw is that it's wrong.

The reasoning behind the Drake equation is fun and instructive. The idea is that there is a constant turnover of stars and alien races, and that this is in a steady state – the rates at which stars appear and disappear, and ditto aliens, are constant. We can think about what comes in and goes out over any convenient period of time, then – it doesn't matter if it's a year or a million years, because the reasoning leads to the same answer. So let's take one year. During that time, we get a certain number of new stars to think about – the first number in the list. Of those, only the proportion given by the second number in the list will have planets, which cuts the number of star systems down to what you get by multiplying those two numbers . . . and so on. Finally, such technology, once it has evolved, will be around for a certain number of years, so you have to multiply by that.

Buried in this argument is a pernicious assumption: that there are no alternative ways for alien life to operate. For example – we invent – maybe intelligent life that evolves on an Earthlike world, with a fair amount of land, is very likely (let's say 'certain' to keep things simple) to invent communication technology. Land provides the metals that you need to build radio antennas, it provides the kind of environment in which you can discover radio . . . It provides the kind of environment in which you can write with chalk on blackboards to discover the mathematics needed to discover radio. And so on, the details don't matter and don't have to be accurate. If, however, it is an ocean world (assuming for the sake of argument that this is possible) then the chance that intelligent life can make the step to interstellar communication is much reduced – say, zero. Because these two routes to interstellar communication are different, and the proportions are different, it is not sensible to amalgamate them mathematically into a single route with fixed proportions. Whether you make it to interstellar communication doesn't just depend on whether you make it to intelligence. It depends on *how* you make it to intelligence, too.

In other words, each possible route to interstellar communication has different values for those seven numbers. A land-based alien race has different chances of becoming intelligent and inventing high technology than an ocean-based one. The chance that life evolves on a gas giant planet like Jupiter might be very different from the chance of it evolving on one like Earth. Therefore the proportion of gas giants with life will be different from the proportion of aqueous planets with life. What this means, mathematically, is that the Drake equation needs to become even more impressive. Instead of one set of seven numbers, there ought to be many sets: one for *each conceivable route* to interstellar communication. You should multiply out each set of seven, and add all the results together. Just *one* of those sets represents the answer given by the original Drake equation. That equation assumes there is only one route to interstellar communication: *ours*. (Leave aside that we've not actually got there yet.) The assumptions built into the form of the equation are that in order to evolve an alien civilisation with interstellar communication capability, you must carry out the following steps: get yourself a star, find one with a planetary system, choose a planet that is capable of producing and supporting life, wait for life to appear, wait until it evolves intelligence, and wait until that intelligence learns how to build an interstellar communicator. Moreover, because the numbers to be multiplied at each stage are fixed, once and for all, there is no significant variation in how that stage happens. Which means, since the above route was our route, that ours is the only route – because any different route would inevitably involve different numbers.

The whole philosophy of the Drake equation is to start with the wide reaches of the universe, and then narrow everything down until all that is left is aliens just like us living in environments just like ours with technology just like ours . . . only much more advanced, of course . . . though not *too* much more advanced. It is a philosophy of successive *restriction* of the options for life and intelligence. An Easter Islander could invent a 'Drake Equation' for intelligent life on other islands (you do see that it *would* be 'islands'?). The factors involved would be the mean rate of volcano-formation in the ocean, the proportion of volcanoes that form islands, the proportion of islands capable of supporting life (which needs fresh water, coconuts, fish from the sea, trees and rocks for building . . .) and so on. The Easter Island 'alien communication' project would be SEEIC – the search for extra-Easter-island canoes. And then Europeans arrive from a continent, on a big ship with not a coconut in sight . . .

There are many other problems with the Drake equation. For example, it assumes that nothing significant changes with time. If the lifetime of the communication technology is not too big, this doesn't matter much; but if (as many people hope) a technologically advanced civilisation is likely to hang around for a long time, then it does. If we go back a mere 3.5 billion years, our lovely blue-white oxygen planet was incapable of supporting the lifeforms that it does today. Go back another half a billion years and it wasn't capable of supporting any kind of life at all. 'Planet capable of producing and supporting life' has many different meanings, depending on the stage that the planet has reached in its own history. And it's not simply a matter of time. The early Sun was much cooler, so the Earth needed more greenhouse effect – and provided it; now it needs less, and is providing less.

Another, glaring assumption is that technologically advanced aliens stay at home. However, interstellar travel is not that much harder than interstellar communication. Indeed, it may be easier: humanity is already capable of building a workable starship; what's lacking is the political will to spend that huge amount of money it would entail. It wouldn't be a very *good* starship, but it might get us to the nearest star in a century or two; it would have to be what SF calls a 'generation starship', in which people have children and pass on knowledge of the ship's operating systems to them. In contrast, we're nowhere near working interstellar communication . . . except to pop a letter in an envelope and send it along on board the generation starship.

Now, aliens that have interstellar travel don't have to sit at home and wait for their planet to turn into the right kind of planet for something or other to happen. They can go looking for suitable planets. Once *one* planet 'accidentally' evolves aliens with interstellar travel, they will shortly occupy *all* remotely suitable planets in their neighbourhood. This upsets the assumptions of the Drake equation beyond salvation. Not only will the aliens occupy all the suitable planets: they will occupy all the *un*suitable planets too, because they will have 'xenoforming' technology, just as we are close to having terraforming technology that would let us live on Mars. (See Kim Stanley Robinson's series *Red Mars*, *Green Mars*, *Blue Mars*, for a moderately realistic scenario.) The effect of this alien diaspora, taken to extremes, renders the Drake approach totally irrelevant. In essence, and with a few caveats, which you can think of for yourselves: once a star-voyaging race appears within a galaxy, very rapidly it occupies the *entire* galaxy. This raises another traditional question, asked by the physicist Enrico Fermi about 1950. Advanced aliens ought to have the technology for interstellar travel, so – if they exist – they should have visited us by now. The Fermi Paradox poses the question: 'Why aren't they here?' We'll come to that later.

The idea behind the original Drake equation is actually a little subtler than we've indicated. Drake knew that different routes involved different proportions, different probabilities. But to keep the formula simple – and, after all, its main function was as a back-of-the-envelope calculation to give some rough idea about the chances of success in Ozma – he assumed that at each stage the different routes can be 'lumped together' by using some kind of average, ballpark frequency. There is a term for this kind of theory, and for the thinking that goes with it. It is called a *mean field theory*. The idea behind a mean field theory is to replace a lot of complicated interacting things by a single smeared-out average. One of the chief exponents of mean field theories was the geneticist Ronald Aylmer Fisher. He built an entire theory of population genetics on such a foundation – how genes flow through an ecology. It is widely used to this day, partly because most of its practitioners do not appreciate its severe limitations. Instead of thousands or millions of individual organisms, each with its own characteristic genetics, Fisher saw the ecology as a homogenised 'gene pool' in which all that matters is the frequency with which each gene – strictly, each gene alternative, or *allele* – occurs. Interbreeding mixes up the gene pool by creating new combinations of existing genes, and all that matters, said Fisher, is how the proportions change from one generation to the next.

Mathematically, this is like throwing all the organisms into a genetic blender, switching it on, and studying the gooey mess that results. For some problems, the method works like a charm. But it's terrible when what counts is *combinations* of *alleles*, and when the way they interact is not independent. That is to say, for most problems about real ecologies. The more complex the interactions, the more complicated the players, the more likely it is that a mean field theory will produce nonsense. The mean field human being, for example, has one testicle, one ovary, and one breast – along with half of various other vital organs.

When Drake lumped all sorts of variability together into one number – a single frequency for 'able to develop intelligence', for instance – he was setting up a mean field theory. He was assuming that all routes to alien intelligence and communication are effectively equivalent. His formula is reasonable if that's actually the case – but for the general question of alien intelligence, it's not, as already explained. It does, however, make reasonable sense in a more limited arena, the one for which the equation was originally conceived: the Ozma/SETI experiment, which searches the stars for signs of extraterrestrial intelligences *that choose to communicate by radio waves*. The most likely route to the evolution of such intelligences is similar to ours; similar enough, perhaps, that a mean field model isn't too far wrong.

In that spirit, let us return to the original Drake equation, pretend that it works, and argue the numbers more carefully. We're not going to get a good estimate for the number of alien communicators out there, but we probably are going to get a good estimate of the number of alien communicators out there who are more or less like us. Which are the ones SETI or its ilk might be able to contact, and the ones we might be able to understand. A bit. The figures we cited earlier were chosen to make a debating point without getting tied up in extraneous complications, such as 'a frequency can't be bigger than 1'. We'll now argue for some figures that give the same answer – 3,000 communicating alien races in the galaxy – but are closer to what we think is sensible.

There is general agreement among astronomers that the rate of star formation in the galaxy is about ten stars per year. If in fact the galaxy contains ten times as many stars as we currently estimate, which is entirely feasible, then this figure might have to be changed to 100.

There has been a big change recently in the consensus view of the proportion of stars with planetary systems. Until the late 1990s, scientists who thought alien life was likely preferred a fairly high figure, while those who didn't want aliens to exist were convinced that the number was ever so tiny, maybe one in a thousand or one in a million. On current evidence, a value of one in ten is pessimistic. As yet we can detect the existence only of *big* planets, and already they look pretty common. That was always the smart bet, because it fits the prevailing theories of star formation: small planets, for statistical reasons, should be more common than big ones – much as small incomes are more common than big ones in human finance. Now there is relatively direct observational evidence that other stars have planets. We can't detect small planets yet, because our instruments aren't sensitive enough – but the smart money says that they're out there, in greater numbers than those planets that we *can* detect. For all these reasons, we'll settle for a figure of 0.1.

What about the number of planets per system capable of producing and supporting life? The answer for the solar system is clear: one. Yes? No. Actually, that's wrong. We should also include Mars, which at some time in the past may well have had extensive oceans, or at least big lakes, and so was probably *capable* of supporting life. (Not *now*, of course, but that's not part of the Drake conditions; issues of timing are dealt with by the final factor, the lifetime of the relevant alien technology. And please note the word 'capable'. Whether life *arises* is dealt with by the next factor on the list, the proportion of such planets on which life appears.) The number goes up to two, then. But – and realising this, with serious evidence, is a major new development – planets aren't the only bodies where life might exist. Satellites of planets might also harbour lifeforms. It's not relevant that the Drake equation talks only about 'planets': once we know that satellites might also provide alien habitats, we must broaden the definitions to include those too. As already mentioned, there are reasons to suppose that Jupiter's satellites Europa (definitely) and Ganymede and Callisto (possibly) are, or were, or will be, capable of supporting life. Maybe the same goes for Saturn's moon Titan and Neptune's moon Triton. So we're being conservative if we put the number of planets per system capable of producing and supporting life at three; it could easily be seven.

Jack&Ian is a maverick when it comes to the next two factors: the proportion of suitable planets on which life appears, and the proportion of life-bearing planets on which intelligence evolves. We believe (it believes?) that as long as you are prepared to wait – and this is permitted, indeed required, by the derivation of the Drake equation – then both of these frequencies are 100 per cent. That is, the corresponding numbers in the formula are one, and not, as we assumed earlier, 0.1. We'll justify this view later.

What about the proportion of intelligent-life-bearing planets on which those lifeforms attain interstellar communication skills? We'll go along with 1 per cent, a figure of 0.01. If anything, it's on the low side. In chapter 14 we'll argue that *most* intelligences eventually evolve technology. Now, technology is an autocatalytic process: it feeds on its own successes and grows explosively. Interstellar communication is then inevitable as long as the lifetime of the technology isn't ridiculously short.

But surely many technological civilisations are likely to blow themselves to bits before they get that far? That may be true, but the factor that builds in the likely lifetime of the civilisation – and the only factor where this is a consideration – is the final one: the lifetime of the technology. So here we'll be pessimistic, and go for 100,000 years rather than a cool million – Hold it. Human civilisations typically last a few hundred years; maybe (China, say) a few thousand. Why do we select such a big number? Because what counts is not the lifetime of any particular civilisation. Civilisations rise and fall, but their technology goes on. There is an unbroken thread from a caveman squatting beside a fire to us watching TV with the central heating on. The technology of fire has been around for well over 100,000 years, even though no single culture has lasted that long. To err on the safe side, the conservative side, we've settled for a lowish figure, because we do think that it's entirely likely that technologically advanced civilisations will develop more ways to destroy themselves.

Anyway, putting all those numbers together and multiplying out, again the answer we get is 3,000.

And that is just those aliens who followed the same kind of evolutionary track as us, who live on planets like ours, who have technology like ours. The modified Drake equation, with lots of similar terms added together, must surely account for a good many more aliens who went some other way. We're not saying they are definitely out there. We're saying that on any reasonable back-of-the-envelope estimate, they surely ought to be.

The Drake equation has become something of an astrobiological icon. *Rare Earth* by Peter Ward and Donald Brownlee, extends the Drake equation by refining the list of conditions needed to produce intelligent aliens, and concludes that previous estimates have been over-generous. This is the exact opposite of our criticism of the Drake equation: we think it is too narrow, while Ward and Brownlee think it is not narrow enough. *Rare Earth* is a self-avowed book about astrobiology, and it advances the view that complex alien life (that is, above the level of bacteria) is extremely unlikely. We are obliged to counter its arguments, which flatly contradict almost everything we say. So we'll take a look at Ward and Brownlee's position now. They call it the Rare Earth Hypothesis: the name is self-explanatory, perhaps more so than they intended. The whole game is given away in two tabulations, which occur right at the start. The first is headed 'Dead Zones of the Universe'; the second is a list of 'Rare Earth Factors'. These two sections list, respectively, reasons why entire regions of the universe cannot possibly evolve life, and special features of the Earth that have played significant roles in the evolution of complex life here, and which Ward and Brownlee therefore consider to be essential for complex life anywhere.

In order to keep the discussion within bounds, we'll focus on their second table, which is entirely typical of their reasoning. This table lists eighteen features of planet Earth that we now realise have had major influences on the evolution of life, and we start by summarising these features. Our planet had to be the right distance from its star, otherwise there would be no liquid water. The star had to be of the right mass, to make it develop into the right type – long-lasting and not putting out too much ultra-violet. Our planet's orbit had to be stable, staying much the same for billions of years. Our planet had to have the right mass – enough to retain an atmosphere. There had to be a Jupiter-like neighbour to clean out stray comets, and a nearby Mars-like world 'to seed Earthlike planet, if needed'. Our world had to have a structure that permitted continental drift, to assist the growth of biological diversity. It had to have an ocean, 'not too much, not too little'. It needed a Moon to stabilise its axial tilt. The tilt had to be the right amount to avoid severe seasons. Impacts from large asteroids should be very unusual. There should be enough carbon for life to form; but not so much that there is a runaway greenhouse effect (which is what probably happened on Venus). The atmosphere had to have the right mix of gases for plants and animals. There had to be a 'successful evolutionary pathway for complex plants and animals'. Photosynthesis had to evolve, to put oxygen into the atmosphere, and it had to do so at the right time. Finally, we had to be in the right kind of galaxy, with enough heavy elements; we had to be in the right part of that galaxy, and then – on top of all this – we had to get very lucky, and be subjected to some crucial wild cards – the Cambrian explosion when life diversified into complex forms, and other equally unlikely and unusual events that we know occurred and had big effects on earthly life.

Impressive. It makes you think that we only made it by the skin of our teeth. All those obstacles to overcome. All those ways for incipient life to fail. How likely is it that life could gain a toehold anywhere else? Earth is perfect, ideal for life. It may well be the only Earthlike planet in the universe.

But are you *really* impressed by the list? We're not. We had to work hard to resist adding a nineteenth requirement: 'animals need legs exactly the right length to make their feet reach the ground'. Ward and Brownlee clearly don't realise the intellectual trick they are playing. But we can use the same kind of thinking to convince you that you can't possibly live in your own house. You see, in order for you to live in your house, you have to be in the right galaxy (not just the right *kind*), in the right place in the galaxy, in the right solar system and on the right planet, on the right continent, in the right country, the right county, the right city, the right street, the right number. Oh, and be there at the right time, and be the sort of person who has a house . . . and be in the sort of city where people build houses, sell houses, and buy houses. In a country where there is clay for bricks, wood for roof timbers, and brass for door-knockers. And then there's the catflap issue . . . This may sound flippant. Actually, it's deadly serious. *Rare Earth*'s basic thought processes are precisely of this nature; they just operate in a less transparently silly way.

It is sadly predictable that the very first Rare Earth Factor should be simplistic, outdated nonsense: 'the right distance from the local star'. This is the hoary old chestnut about 'habitable zones', which was demolished long ago, as we've already seen. Yet Ward and Brownlee use exactly that phrase, and give a long, straight-faced explanation of the apparently obvious but actually false point involved. Because it's their first point, we'll explain what's wrong in some detail, including repeating some of our earlier objections. The basic idea of a 'habitable zone', not *totally* false, is that a planet too close to a star gets too hot, and one too far away gets too cold. Life cannot tolerate these extremes, so it has to exist in the acceptable zone in between. We know the mathematics of stellar radiation – it is the same in all directions and falls off with the square of the distance – so we can deduce that each star is surrounded by a region between two concentric spheres, like a hollow ball. The inner surface is where things become too hot, the outer surface is where they get too cold.

It sounds sensible, but the real world isn't like that, and the reasons why are good xenoscience. The temperature on a planet's surface does not depend solely on how far it is from its star. What matters is how the heat transfers to or from the planet, and that depends on the planet's own make-up. Heat can be reflected back into space by light-coloured materials: the ice at the Earth's poles helps keep us cool by this method. During ice ages it overdoes the cooling in a positive feedback loop, but this is eventually disrupted by other factors. Recent research on global warming has revealed that whitecaps on oceans also reflect heat. An ozone layer can keep out ultraviolet energy. Heat can also be retained by a cloudy atmosphere and the greenhouse effect . . .

If Earth's own make-up was different, it would be *outside* the habitable zone right now. Venus could have been inside the zone, but a runaway greenhouse effect has heated its surface to 800 degrees centigrade. Mars could also have been inside the habitable zone, and probably once was, but it lost its atmosphere long ago. We now think it was blown away by the solar wind.

Ward and Brownlee see the current states of Venus and Mars as evidence that the Sun's habitable zone is much smaller than everyone used to think. We see them as evidence that 'habitable zone' is at best a metaphor, and not a very good one. However, it's probably worse than that.

Take Mercury. We used to think that it rotated once on its axis for every revolution round the Sun, so that it always presented the same face to the Sun – just as the Moon currently does to the Earth. In fact that's now known to be wrong (actually it rotates three times on its axis for every two revolutions round the Sun). But for present purposes, just suppose that Mercury actually was like we used to believe. Then one face – the one pointing towards the Sun – would be incredibly hot; the other incredibly cold, shielded by the bulk of the planet. And somewhere in between would be just right. A habitable zone. Closer to the Sun than the orbit of Venus, which is *not* in the habitable zone. So, merely by thinking for a few minutes about the four innermost planets of the solar system, we see how misleading the naive notion of a habitable zone becomes. And the contention that there is a 'right distance' to be from a local star is in ruins.

The ruin of 'habitable zone' as a useful concept becomes even more apparent if we consider the next planet out: Jupiter. Jupiter is a gas giant, with an atmosphere of hydrogen, helium, and various other gases in lesser quantities. No oxygen. The planet produces a lot of radiation, has either a small rocky core or no core at all, and is almost entirely atmosphere. It is much further out than Mars, so is too far from the Sun to be in the habitable zone if you believe in such things. We'll speculate about life on Jupiter-like planets in a moment, but for the time being we'll focus solely on the prospects for Earthlike life. The 'habitable zone' concept is completely demolished by Jupiter's satellite Europa, with its underground ocean. The Europan ocean could easily contain as much water as all of Earth's oceans put together, and conditions at the bottom are similar to those in the terrestrial ocean depths: high pressures, no light, but plenty of heat. And those conditions, we now know, are entirely suitable for life to evolve. Indeed, one currently respectable theory is that Earth's lifeforms started around mini-volcanoes, the black smokers, on the ocean floors. These early lifeforms were bacteria or archaeans, and their metabolism worked on iron and sulphur, not oxygen and hydrogen. At any rate, the 'habitable zone' appears to include a hollow ball of water under the surface of distant Europa. Ganymede and Callisto also look as though they have under-ground oceans. Saturn's moon Titan has a thick atmosphere and might also be suitable for some kind of life. Neptune's moon Triton is unexpectedly active volcanically, so probably has underground heat and therefore might also have an ocean. Almost certainly it has liquid nitrogen about 100 feet (30 metres) below the surface.

The Sun's habitable zone is looking much less like a hollow ball centred upon it. And we haven't even begun to ask 'habitable *by what*?' In *Wheelers*, we indulged in some fictional exploration of various other potential habitats for alien life in our solar system. We'll describe three of them now, just to open up a few of the less orthodox possibilities. The most plausible of these habitats, we suspect, is Jupiter's atmosphere. Many SF authors agree that the most likely ecosystem on a planet like Jupiter is a 'balloonist' one, with aliens based on hydrogen or 'hot air' balloons. In effect, Jupiter's atmosphere plays the role of an ocean, and balloon-organs correspond to swim bladders. The second exotic habitat is the photosphere of the Sun, its bright outer layer. Ordinary matter comes to bits at the temperatures that prevail there, but our 'plasmoid' aliens are not made of ordinary matter. They are woven from magnetic vortices; their 'genetics' encoded in the topology of the weave. Magnetism is a major feature of the Sun, and it 'survives' the high temperatures and intense radiation without difficulty. Indeed, the solar environment is 'just right' for magnetic aliens. The third habitat is empty space, where herds of magnetotori – rather dumb beasts that propel themselves on the ramjet principle by scooping up traces of hydrogen gas and generating energy from them by nuclear fusion – roam. They probably evolved in stars, but left their ancestral home long ago.

Could such aliens really exist? An SF story, however plausibly written, is not evidence. But dismissing such possibilities without taking them seriously is not science. Astrobiology pays no attention to even moderately imaginative proposals like these. Xenoscience will be worthy of the name when it develops enough understanding to take them seriously and figure out whether they could really work. For example, the Jovian system is a high-radiation environment. Could balloonist lifeforms really cope with such levels of radiation? (We can imagine them having evolved something suitable, and a bunch of intelligent Jovians asking, 'Could hypothetical terrestrial lifeforms really cope with such a serious radiation deficit? And what about all that corrosive oxygen?')

You've got the idea by now, and if you now go back to the list you'll probably spot several other items that are open to similar challenges. Anything else with 'right' in it: right mass of star, right mass of planet, right kind of galaxy, right position in galaxy. Even if there really was a single 'right' set of conditions, we must recognise that real solar systems and galaxies are hotbeds of diversity, and local conditions can always be very different from the average ones in similarly situated places.

Let's look at a representative sample of the other seventeen Rare Earth factors. We have similar criticisms of them all, and of Ward and Brownlee's table of 'dead zones', which is simply a galactic-level repeat of all the mistakes they make on the level of solar systems.

'Right mass of star'. The mass matters because it is the main thing that determines what sort of star evolves, what sort of radiation environment arises around it, and how long the star survives. Ward and Brownlee are on fairly safe ground when they say that the star ought to have a reasonably long lifetime. But what do we make of 'not too much ultraviolet'? First, there is the assumption that ultraviolet light is bad for life. It's certainly bad for *our kind* of life, but who knows what might have happened elsewhere? Life evolves to suit the local conditions, and in a high-ultraviolet environment it would find ways to cope with – indeed, to make good use of – all that nice energy. Archaeans survive temperatures above 160°C in the seabed beneath Earth's oceans; they have been found several miles (kilometres) underground, where temperatures are similarly high. Recently, live bacteria have been captured 25 miles (41 km) up in Earth's atmosphere, a height at which there is so much ultraviolet light that you could use it to sterilise surgical instruments. (Chandra Wickramasinghe has suggested that this discovery proves that bacteria come from space, verifying his and Fred Hoyle's theory of 'panspermia', that life began in space, not on Earth. The jury is still out; nonetheless, it's a very striking finding.) Ultraviolet-resistant aliens on a planet of a high-ultraviolet star, studying our own Sun through their telescopes, could well be saying to each other, 'Well, we can rule *that* one out. Not enough ultraviolet.' Leaving that little difficulty aside, we ought to bear in mind that planets possess many tricks for keeping out radiation. Earth's ozone layer keeps out a lot of ultraviolet. Its magnetic field keeps out all sorts of high-energy particles, too (and we wonder why Ward and Brownlee failed to include 'having the right magnetic field' in their list). But what of putative Europan creatures, or their more distant counterparts? Underneath a thick layer of ice, what do they care about how much ultraviolet their star puts out? And what of creatures on a world like we thought Mercury was, with one side eternally shielded from the star's terrible ultraviolet by several thousand kilometres of planet?

Helliconia Spring (Brian Aldiss 1982)

This is the first book in a trilogy, the others being *Helliconia Summer* and *Helliconia Winter*. The evolutionary history of Helliconia has been shaped by its unusual orbit. In a system that contains three other planets, it revolves around a star, Batalix. This whole solar system revolves in turn in an elliptical path around a more distant star, Freyr. Freyr is fifteen times as big as the sun and 60,000 times as bright, and it takes 3,000 years for Batalix to complete one revolution of the giant star. The climate of Helliconia ranges between extremes of heat and cold. Seasons last for centuries, and entire civilisations can rise and fall within one Great Year – a single circuit of Freyr.

There are aliens on Helliconia that resemble humans in virtually all respects. Tribes of them hunt the shaggy, horned yelk, which travels in great herds. In among them are often found the larger biyelk, also horned and shaggy, and the excitable gunnadu with small heads on long necks.

All three species are necrogenes, which must die in order to reproduce. They are hermaphrodites. After they have mated, their sperms develop into maggotlike forms, which begin to devour their maternal parent from inside. They start with the stomach and spread like wildfire once they reach a main artery. When the host dies, they continue to eat the corpse, and each other, until eventually just two or three survivors emerge from one or other bodily orifice.

The hunters, though, are far more interested in more challenging prey, the indigenous, intelligent phagors. Their horns have two sharp edges, like a curved sword, and form coveted trophies.

The Earthmen in the orbital Earth Observation Station watch them all. They know many things. They know about the helico virus, which strikes twice every 1825 years: once during the period of twenty eclipses that marks the start of spring, and later in the Great Year during a period of six eclipses. Climate change triggers viral hyperactivity, and two diseases result: bone fever in the spring, and the fat death later in the Great Year.

The virus is carried by the phagor tick. Eventually, humans and phagor will have to come to an accommodation. But not yet.

A more subtle item on the list is 'stable planetary orbits'. We need to explain the background here. To a first approximation, planets revolve round their stars in elliptical orbits. If all that existed was one planet and one star, such orbits would be stable, in the sense that the shape and size of the ellipse wouldn't change (much). But planets typically occur as just one body in an entire solar system, and the gravitational pulls of the other planets can disturb those neat ellipses. The most interesting potential consequence of such perturbations is 'chaos', in which the orbits become irregular and acquire elements of randomness and unpredictability. The amount of chaos in a solar system depends on the exact details of its planets – masses, distances from the star, inclination to the plane of the system. The wrong kind of giant planet – 'bad Jupiters', in Ward and Brownlee's revealing terminology – can create a lot of chaos. But if a planet's orbit is too chaotic, then it makes climatic conditions so variable that life has a much harder task to get going.

This is one of the better items on the list. However, there are several things we must take into account when assessing its relevance. The first is that solar systems do not get put together piecemeal, one world at a time. The whole thing condenses out as a system. If it's *too* chaotic, it hasn't yet finished putting itself together. Anything that survives for a long enough time will have ironed out most of the chaos. The second is that chaotic orbits have their own kind of stability: they are emphatically *not* just random. The shape and size of the orbit can change, and it can change unpredictably, but it will stay within certain well-defined bounds. Those bounds *may* be tight enough to keep the planet's climate inside a reasonable range. For example, the entire solar system – ours – is actually chaotic (on a timescale of tens of millions of years). That didn't stop us evolving. The third, as always, is that lifeforms may be protected against the effect of an unstable orbit – or may even depend upon such instability for survival, as in Aldiss's *Helliconia* trilogy.

'Jupiter-like neighbour'. Only recently did we realise that Jupiter plays a very effective role in our own solar system: it sweeps up comets. The spectacular consequences of this made themselves vividly visible in July 1994, when comet Shoemaker-Levy 9 broke into twenty pieces and they all smashed into Jupiter. If the whole comet, or any of those pieces, had hit Earth head-on, there would now be no life on this planet above the level of bacteria, if at all. So a life-bearing planet needs a Jupiter? Could be. Subterranean life, as is envisaged for Europa, would be better able to survive a major impact than life on the external surface, but even Europa would have problems with a comet. The big flaw in this criterion is that Jupiter's gravitational pull also disturbs asteroids in the asteroid belt, and can send them towards Earth (with a bit of help from Mars). The K/T meteorite that hit the Earth and destroyed the dinosaurs, 65 million years ago, may well have been sent our way by Jupiter. (The letters K and T come from the German words for 'Cretaceous' and 'Tertiary', two of the standard periods of geological time that delineate the major events in Earth's history. The demise of the dinosaurs happened on the boundary between the two.) Anyway, it looks as if Jupiter helps with comets, but causes problems with asteroids, and so cancels itself out.

'Plate tectonics.' This item is very speculative. Plate tectonics is the process that supports the theory of continental drift, that the Earth's great landmasses have moved over the aeons. The east coast of America is suspiciously similar in shape to the west coast of Africa, and strong evidence now exists that they were once joined, and then split apart. They are still drifting, slowly, as liquid wells up along the mid-Atlantic ridge on the ocean floor. The movement of Earth's continents may have assisted in the development of biodiversity, which is the reason for this item being listed as a Rare Earth Factor. Equally, it may not have done; and even if it did, lots of other things might do the same job. For instance, imagine a world that is mostly ocean, with innumerable volcanic islands. Lots of Hawaiis, but no Americas. The volcanic islands pop up here and there, and erode away again. This makes a pretty good substitute for shifting continents, and if anything it is likely to promote even more biodiversity. No clear reason to need plate tectonics, then.

In fact, primeval Earth probably was like this; then a few of the islands became nuclei for the accretion of 'plates' around them. We are now in a Glacial Phase, with water locked up in ice caps and continental shelves exposed. In the more prevalent Greenhouse Phase, Earth's surface is warmer and the continents are smaller; there are plenty of shallow seas, but the deep seas lack oxygen, making black and green slate as sedimentary rock, not sandstone or limestone.

'Large Moon.' There has been some controversial speculation about the role of the Moon in making life possible, but even astrobiologists are unsure whether to believe it. Here's a quick outline. Planets rotate, and if our solar system is anything to go by, the axis of rotation is typically tilted relative to the plane of the ecliptic – the average plane of the planetary system. In 1993 the French astronomer Jacques Laskar and his collaborators F. Joutel and P. Robutel did some wonderful computations on the possibility of chaos in the axial tilt of various planets. It turns out that Earth, virtually alone in the solar system, has a very stable axis whose tilt changes only slowly, and within close bounds. It is kept stable by its companion, the Moon. Clearly a stable axial tilt helps to prevent rapid climatic variations ('rapid' here means 'taking about ten million years'), and that makes it easier for life to evolve. However, the case here is very weak. First: how come 'moderate' asteroid impacts are good for life, because they encourage diversity, but changes in axial tilt (which would also offer novel challenges to evolving lifeforms) are bad? Even land creatures can easily adapt to changing climatic conditions by migrating to wherever the climate is suitable. An axis that changes its tilt over ten million years poses no problem: the creatures wouldn't even notice. On Earth they'd have to change position by about a mile (1.6 km) every 10,000 years. Beetles have been moving around like this on Earth for many tens of millions of years. If you look at the beetle fossils in any fixed location, you find species suddenly vanishing, as if they went extinct. Sometimes they suddenly reappear again, too, which is puzzling. But if you look at the fossil beetles over the entire world, you realise that the beetles were just moving around. If North Africa got too cold, they nipped down towards the equator. If the poles moved so that the equator wasn't the equator any more, the beetles had ample time to move as well. Ward and Brownlee try to strengthen this already weak case by pointing out that the Moon's origins are highly unusual. The Earth-Moon system is near enough a double planet: it is unusual to find a planet with such a large companion. But this similarity of size is what lets the Moon stabilise the Earth's axial tilt. Moreover, the Moon seems to have been created as the accidental result of a massive impact, with a body about the size of Mars. This is surely a rare event. *Rare Moon*, then? To cut a long story short, we'll merely observe that there is a second double-planet in our own solar system: Pluto and Charon. Enough said.

'The right amount of carbon.' What about silicon-based life, very plausible scientifically, or other alternatives to carbon?

'Evolution of oxygen.' The scientific consensus is that the early bacteria on earth evolved in the oceans when the atmosphere was mostly hydrogen, water vapour, methane, ammonia, and sulphur compounds, and they produced oxygen as a side effect of photosynthesis (once they'd evolved *that*). Some geologists think that the oxygen arose from ordinary chemistry in rocks, but they're a small minority. Oxygen is a poisonous, corrosive gas – this is part of the reason why iron rusts and lots of things catch fire. Eventually – it seems to have taken about two billion years – earthly life came up with chemical tricks to protect itself from the bad effects of oxygen, and exploited the stuff for its energy-giving features. A new network of metabolic pathways came into existence, beautifully adapted to the poisonous oxygen atmosphere. Ward and Brownlee's point is that this new metabolism was one of the key steps towards complex life on Earth. However, they deduce that oxygen is a *necessary* step everywhere. This is poor logic, and equally poor science. To choose an alternative at random, complex life on Earth might well have arisen *faster* if there had been some other way to get rid of the oxygen – say by sequestering it into rocks, or combining it with sulphur, or whatever. Entirely different kinds of complex lifeforms might have evolved instead of the ones we now see; their metabolic pathways would also have been entirely different, and they wouldn't have needed – or tolerated – oxygen. After all, that's exactly what life was like (albeit not very complex) when the bacteria were producing the oxygen as a waste product. And it's what life is like in Earth's biggest ecosystem, the deep ocean, when the Earth's surface flips over into its (more common) warmer Greenhouse Phase and the ocean depths become anoxic.

'Wild cards.' On this planet, various potentially nasty events occurred, which could easily have killed off life altogether. But it survived, and as a result it diversified – it seems to be good evolutionary biology to think that mass extinctions generate new bursts of diversity. *Rare Earth* is ambivalent about natural disasters. Some, 'wild cards', are asserted to be essential for complex life to develop. Others are asserted to be inimical to life, and their *avoidance* is essential for complex life to develop. You will not be surprised to find that the wild cards are the things that have happened here, and the rest are those that haven't. Since the particular parochial sequence of events that occurred here is highly unlikely to repeat *exactly* anywhere else, Ward and Brownlee conclude that we've been very lucky, and that even on Earth, life's existence has been tenuous and fragile. Life is clinging desperately to the edge of existence.

Our own inclination is to read the runes very differently. Our life-bearing world has sailed its way, entirely successfully, through between four and seven mass extinctions, brought about by a variety of gigantic mishaps. The diversity of species if anything *increases* after each mass extinction. This is a mildly controversial view, but the fossil record supports it, and the reasoning is compelling: species that survive the disaster initially experience a reduction in competition, because most competitors are dead, so their evolution gets a kick-start. We think that terrestrial life would survive the loss of 99 per cent of its species: the 1 per cent that wasn't killed off would quickly restart the process. Those bacteria 2 miles (3km) or more down in the rocks would be enough – though it would take another couple of billion years or so to get back to the kind of complexity we see now.

Of course, some events *can* kill a planet. Its star going nova would be one possibility. But that sort of disaster corresponds to other factors in the *Rare Earth* version of the Drake equation. Our world has thrived for nearly four billion years, despite everything that the universe has yet thrown at it. It has just been realised that the earliest period of life's evolution on this planet overlapped the 'early bombardment' phase of the solar system, perhaps for a hundred million years. About 4.5 billion years ago the Moon was splashed off the Earth in a collision with a body the size of Mars. Within a further 100 million years, the inner solar system quietened down. However, about 3.9 billion years ago the Earth and Moon were hammered by the 'late heavy bombardment', a hundred times as intense as anything before or since. The lunar maria, which create the characteristic 'Man in the Moon' markings, probably date from this event.

The cause of the late heavy bombardment is of course unknown, but there are several proposals. The original evidence for this event was controversial – analysis of Moon rock from the Apollo expeditions. In 2000 strong confirmation came from lunar meteorites. Barbara Cohen, Timothy Swindle, and David King studied these meteorites – material ejected from the Moon's surface and taking about a million years to reach the Earth. Within these meteorites are formations created when an impacting body melted the Moon rock, and these can be dated by measuring the ratio of two isotopes of argon (argon–39 and argon–40). The ages of these formations cluster around 3.92 billion years ago. It is estimated that the Moon suffered at least 1,700 impacts big enough to make craters 12–750 miles (20–1,200km) across. The Earth suffered at least 17,000 such impacts, and large quantities of surface water were boiled away. Coincidentally (?) the earliest isotopic evidence for life dates to the same period. So despite (because of?) the late heavy bombardment, life *still* managed to gain a hold.

Ward and Brownlee see the occurrence of life on Earth, despite all these apparent disasters, as an indication of how lucky Earth's living creatures have been, and how narrowly they have escaped total extinction. We think that it demonstrates the exact opposite: that life is extremely resilient, that it can survive almost any cataclysm, that the harder it is hit, the more effectively it bounces back. Species come and go, but life – even complex life – remains.

The climax of *Rare Earth* is a modified Drake equation with a lot of extra factors, eleven instead of seven. The authors make a lot of fuss about this equation, but they never suggest numerical values for anything in it. Instead, they try to impress us with the enormous list:

> With our added elements, the number of planets with animal life gets even smaller. We have left out other aspects that may also be implicated: Snowball Earth and the inertial interchange event. Yet perhaps these too are necessary.
>
> Again, *as any term in such an equation approaches zero, so too does the final product.*

Mathematically speaking, that's true, but in the present context it is also completely irrelevant. New factors can't change the overall proportions. A new factor can only approach zero if the old factor, of which it is a part, already approached zero beforehand. For instance, the Drake equation uses one number to represent the proportion of planets per system on which life appears: suppose the figure is 0.1. If we replace that number by a product of ten more refined factors, this product still has to equal 0.1. Admittedly, we might estimate the number more accurately by getting really good estimates for those ten new factors – but those estimates would have to be *very* good, because of the way errors mount up when you have a lot of factors.

What Ward and Brownlee are telling us here, probably without realising it, is that the more stages you introduce into your mental model of a process, the more chance there is for some stage to be highly improbable, and if so, the whole process is highly improbable. But that's complete nonsense: *any* process can be refined in this way, to as many stages as you want – think of manufacturing a car, for instance. The level to which you refine your mental description has no effect at all on how many cars actually get made. *Rare Earth* goes on to say:

> How much stock can we put in such a calculation? Clearly, many of these terms are known in only the sketchiest detail. Years from now, after the astrobiology revolution has matured, our understanding of the various factors that have allowed animal life to develop on this planet will be much greater than it is now. Many new factors will be known, and the list of variables involved

will undoubtedly be amended. But it is our contention that any strong signal can be perceived even when only sparse data are available. To us, the signal is so strong that even at this time, it appears that Earth indeed may be extraordinarily rare.

'Signal' is a curiously confused metaphor. What stops a signal being perceived, or allows it to be perceived only with difficulty, is noise – not sparseness of data. You can extract signals from noisy data by various clever methods, but when you have sparse data, the only recourse is to acquire more. That quibble aside, we dispute that the 'signal' is as strong as they claim. In fact, we dispute that it exists. It's all smoke and mirrors.

Of course, the existence of alien life does not depend on whether our criticisms of *Rare Earth* are correct. It depends on the aliens, if there are any. We'll know who is right if and when that unprecedented moment of First Contact happens. If *Rare Earth* is right, it won't happen for a very long time, until humanity has developed effective interstellar travel, gone out and explored a 99.999999 per cent barren universe, and totted up a few million bacteria-infested rocks. When we finally encounter intelligent aliens, they will live on a planet like ours in a region of the galaxy like ours; they will look much like us, behave much like us, run on DNA like we do, metabolise oxygen like we do, have technology like ours, and communicate like we do. We will have little difficulty understanding them (though it is questionable why we should bother when they are so like us).

If *we're* right, the scenario is entirely different. First Contact could happen any time, and is likely within, say, another 100,000 years. Don't hold your breath, though. The aliens will have evolved somewhere incredibly unpleasant to us, will resemble nothing we have yet conceived, will be sexually bizarre or have unrecognisable reproductive tricks, will behave in incomprehensible ways, will use technology we don't understand, and will communicate in a manner that we've never thought of, and which we can't even recognise as communication. They will be here for reasons that make no sense to us, and they probably won't find humans the least bit interesting. We will find it almost impossible to understand them.

However, until then (if ever) we have to do what science always does when it can't make direct observations: argue by analogy, inference, simulation, and theory. What we're saying is that unless those analogies, inferences, simulations, and theories make a real, imaginative attempt to open up the possibilities of the universe, instead of narrowing them down to a carbon copy of life on earth, then they will do nothing beyond equipping us with blinkers. 'Astrobiology', as we've said, is a self-betraying name. It stands for astronomy, as seen from Earth, *plus* biology, as experienced on Earth. It therefore has two flaws. The obvious one is that it is hopelessly Earth-oriented. This is fatal when the topic under discussion is alien life. It is like trying to understand the terrestrial ecosystem from the point of view of a limpet on a rock on Easter Island.

The second flaw is that astrobiology goes no further than astronomy *plus* biology. It uses biology to work out what the requirements for life are – water, oxygen – and astronomy to work out where we might find such things. There is no significant interaction between the two disciplines. As we said earlier in connection with anthropic reasoning, this is like trying to explore London by walking along Oxford Street and Charing Cross Road, or New York by staying on Wall Street and Broadway. In the space of ideas, astronomy and biology are like the two axes of a Cartesian coordinate plane. Every point in the plane is a *combination* of those two axes, but if all you explore is the axes themselves, you don't cover the whole plane.

In *The Collapse of Chaos* we introduced the word 'complicity' to refer to any process by which two different factors recursively complicate each other by repeated interaction. There is complicity between an actor and the audience as the performance proceeds, for instance. Evolution is complicity between genotypes and their phenotypes in a progressive ecosystem. An understanding of the possibilities for alien life requires *complicity* between astronomy and biology, not just lumping the two together. Such a complicity has to form the core of xenoscience. Astrobiology gives us only the two axes, but not the boundless plane that those axes encompass. No spark of interaction. In place of the endless potential of a rich and unfathomable universe, we look only for places just like the ones we live in, like a British tourist scouring foreign beachfronts in search of fish and chips and a pint of lager.

7

THE EVOLUTION OF ALIEN LIFE

*C*AIN AND ABEL *have received a new model VRM from the suppliers in Cassiopeia, and it has just finished constructing itself on the lunar surface. It has facilities for six thousand more species, and they need it to be functional for their next tourists; the new gimmick in this model is an 'If' button, very popular with visiting historians. A large meteor is noted by Earth astronomers, falling to the north of London. It has hardly cooled when Cain is inside, with part of Abel, running the 'If' facility with 'K/T meteorite missed' to see what other extelligence might have emerged if humanity had not.*

There is a glitch (there always is . . .) when the machine needs to have 'Which past?' specified. Tentaclebook p 341, column 28, under 'Futures Available', informs them that in the set-up process the 'Past-to-be-Accessed' has to be loaded through the Infinite-Improbability-Access-Port, and that this has not been done. Abel hauls out the unglitcher, which argues with the VRM for several minutes before kicking it very precisely under the IIAP. 'Your humble servant is at your disposal; there will be no further insubordination' appears on the side of the machine, and Cain climbs in again.

. . . A Yilané warrior lounges against a hot-tree, against a background of a herd of docile Triceratops grazing a field of maize. Her city, almost at the horizon, has a steamy haze rising from it as its multiply engineered plants carry out all the infrastructure required by an advanced civilisation. Flying pteros are being piloted by other Yilané, and the VRM's viewpoint locks on to one of these as the creature turbos into supersonic mode. Arriving at the base of a space elevator, it deposits its passengers before greeting a friend and joining a crowd of others in the ptero bar to refuel. Its passengers include a

male Yilané, and he has been fussed over by several officials before all of them gather to meet a deputation of Jovians and Mercurian-2's from the very impressive portal at the base of the elevator.

'Well,' says Abel, 'we might have done a lot better with that lot; they would have been far easier to deal with than humans.'

'Yes,' says Cain, 'but we might have been forbidden the use of the planet altogether. Remember what happened on Europa? Humans are pre-technical, but that has its advantages from the tourist-agency viewpoint. We can keep the visits covert.'

Recognising the flaws in astrobiology is one thing. Getting rid of them will be much harder. However, we can make a start. In order to predict (or postdict) how life might have evolved on another planet, we must apply our understanding of *general* evolutionary mechanisms from the Earth's particular story. This is where the distinction between universals and parochials comes into its own. We must always assume that contingent – accidental – events will have occurred in other ways. That is, universals will probably have happened elsewhere, but with many differences in detail, whereas alien lifeforms will have their own parochials, about which we can only guess. We can conjecture *instances* of universals, though: these will be 'possible' parochial realisations – but because they are still parochials, they should not be treated as detailed predictions. 'Here's the kind of thing that could occur . . . but of course it won't be exactly like that.'

We have seen that the answer to the question 'What would happen if we re-ran the Earth-system?' is equivalent to 'What could happen on another aqueous planet over much the same time-span?' As we've seen our answer can be summarised as: 'Same play, different actors'. This is a good, useful, orthodox bit of theoretical biology, which follows from the universal/parochial argument. But it's not much help to the SF author in search of a backcloth for a plot, or the special-effects team employed to create credible aliens for *Lost in Space 89*. Neither does it tell would-be xenoscientists what they might expect to see if – when – humans reach other planets with liquid water on the surface. It tells us what we *won't* find – vertebrates, humanoids, insects, any of our specific organisms – but not what we *might* find. It argues that the same contingencies won't produce the same organisms, but it doesn't provide any rules for how the different patterns of evolution might proceed.

There is a lesson from Earth that is not encouraging, if we're to think about SF scenarios in which the humans meet creatures of much our size, with technology, a language in sound or light, and a reasonable conversational style. We like the old cliché of meeting the alien lifeforms, discussing 'fundamentals' like the hydrogen atom and Pythagoras's Theorem, and then programming our Universal Translator to discuss how their faster-than-light-drive works or the motives of the characters in *Hamlet*. However, it's highly implausible. The big lesson from Earth's history is how long it all took, and the clear message is that when an 'Earthlike planet' is found, it will most likely have little more than microbes in the seas, not much oxygen in the atmosphere, no life on land . . . altogether not much excitement there. This is a legitimate *Rare Earth* point. But there should be *lots* of aqueous planets, and at an appropriate period, life should appear on many. For more than half of its history, Earth has been like that. So the chances are that if we bump into a planet like Earth, it will still be in that phase of its history: not much of interest happening yet . . . keep an eye on us about every billion years to see how we're doing. To continue the metaphor of the play, the theatre will still be empty.

We can avoid this impasse in the discussion of alien lifeforms by assuming that we won't just charge out to *any* aqueous planet. If the SETI philosophy has any reflection in reality, that search will only turn up those planets with our kind of intelligence, contacting us by radio. But even if the alien world is not selected like that, we are already looking for planets with an oxygen atmosphere, and these will probably have prokaryote assemblies, with the old-style anaerobic creatures cuddling up to oxygen-users like the proto-mitochondria on early Earth, which depleted local oxygen levels and made life more comfortable, less effortful for the soon-to-be-symbiotic anaerobes. Theatre again: the curtains are just trembling before they open to reveal the first scene. But this scenario is not yet redolent with SF plot themes.

If we can do this much even before we visit a target planet, then it is probably not too much to ask that we could especially consider those planets with high levels of carbon-compound combustion products, anticipating that the inhabitants might have developed a rail or road network. Later we will discuss the concept of 'extelligence', the cultural counterpart of intelligence: on Earth, books and the internet are part of our burgeoning extelligence, and it is extelligence that has made us Lords of the Planet. We are only marginally more intelligent than a chimpanzee; extelligence is the way we amplify our limited intelligence, for it not only lets us pool our intelligences – it allows us to transcend them. For now, we need only the word. The point is that much depends on how many other planets there are, how many we find, how far away they are, and how many have extelligence.

So . . . we go to one of these planets, we meet the extelligent inhabitants and we exchange translators . . . What would the aliens look like?

We won't know for sure until we meet them, but there are three ways to think about the question before that unprecedented day arrives. The first, popular in the 1930s and in cryptic form up to the present day, asks, 'What would *we* look like if we had developed on Mars?' Because of the period in which it originated, this question assumes the 1930s view of Mars, when it was thought to be covered in a network of 'canals'. These surface markings started out with the name 'canali', Italian for 'channels', but transmuted into 'canals' through mistranslation. Astronomers soon became convinced that the canals were of artificial construction, evidence for life on Mars. When the Pioneer space-probe got to Mars, its images showed only craters – the canals were optical illusions at best, self-delusion at worst.

The most famous Martian SF of the pre-Pioneer period is Edgar Rice Burroughs's Barsoom, elaborated in a whole series of rather trashy novels, with princesses that lay eggs, fierce tribesmen, even fiercer monsters, and an Earthman hero. These books incorporated (maybe led to) the accepted answer to 'What would *we* look like if we had developed on Mars?' Namely: we would be tall because of the low gravity, we would have enormous chests because of the thin atmosphere, and so on. Similarly, if we had developed on Jupiter (again 1930s version, with a surface) we would be enormously squat and strong, with special respiratory mechanisms to deal with the poisonous atmosphere.

Both question and answer are bad astrobiology, let alone bad xenoscience. In both cases the assumption – not just in the answer, but in the argument that supports it – is that we would be humanoids with added adaptations. Too many TV SF plays, especially 'juveniles', are constructed around this skeleton. It would be as sensible to ask, 'What would a fish be like if it evolved on land?' and to deduce that it would have huge fins for walking and carry a water-filled bulb on its head for its gills to work. Fish evolve in seas: they don't evolve on land, so the answer to the 'what if?' is that it would never happen. The human form evolved *because it works on Earth*, and it makes no sense to imagine it somehow transported to another world, along with all of the parochials that make sense only on its home planet, and then modify those parochials to work on the new world.

There *is* a sensible question here, and it is fuzzily confused with the evolutionary one. Suppose that you had to re-engineer humans to live on another world, while keeping them as close to their current form as possible: what would they look like? But that's not what evolution does. Every stage works in the environment that it inhabits. This is why, in *Wheelers*, we made our fictional plasmoids from magnetic vortices: those are exactly what the solar environment provides.

The second route to guessing what aliens would look like follows the great biologist Conrad Waddington in supposing that the highest form of life on any planet would resemble Waddington. This argument goes: well, obviously an intelligent organism has to be bigger than a molecule and smaller than a planet. So we humans are obviously just the right size, and any respectably evolved alien will clearly follow this argument. Then, we've got to have manipulatory appendages at the end of each limb, and two is too few and a hundred is too many, so five is clearly the optimum. This argument can go so far as to justify Christianity on other planets, too. However, it lacks persuasiveness, even when linked, much more sensibly, to J.B.S. Haldane's essay 'On being the right size', which explained that volumes and areas scale differently with size, that both predators and prey are under evolutionary pressure to get bigger, that eyes big enough to detect moonlight would be very useful, and so on. This essay, too, contains a lot of special pleading. At any rate, route two doesn't work either.

West of Eden (Harry Harrison 1984)

The time is 65 million years ago, and a large meteorite does *not* hit the Earth. The dinosaurs survive, and the tiny insectivorous mammals that dared to come out only at night are not released from dinosaurian thrall.

At least, not *then*.

Mosasaurs, a less familiar contemporary of the dinosaurs, develop intelligence; these are the Yilané, the most advanced race on the planet. They are lizard-like, with scales, a stumpy tail, and a crest; they are bipedal, and have clawed hands not greatly different from those of humans. They live in cities, which they grow from plants. The plants are like trees with multiple trunks, and their structure is controlled by the selective application of hormones.

The Yilané are genetic engineers and masters of the ecosystem. They open up new regions of jungle by spreading larvae, which turn into caterpillars, as long as an arm and covered in bristles, carefully crafted to be voracious feeders on jungle vegetation. Only skeletons of trees remain after the caterpillars have passed, and the ground is foul with layer upon layer of their droppings. But lurking in their genes is a death-switch, and when this activates, they die and rot into the bed of their own faeces.

Nematode worms now turn the repulsive mass into good, fertile soil, helped by bacteria. Beetles eat the dead trees. The Yilané sow grass seed and plant thornbushes as barriers. What once was jungle is now an arable field. Next, the fields are populated with herds of grazing animals, of many kinds, such as the armoured omnivores they call urukub, twice the size of an elephant.

This Eden cannot last. Each year, the climate becomes colder. An Ice Age is on its way, and the Yilané are forced to cross an ocean, where they discover a strange continent and begin to colonise it. But there they encounter a new kind of mammal, one that did not remain small and mindless, one that walks erect, hunts, speaks, and possesses stone tools.

And weapons.

The third route to guessing about real aliens, and to designing credible fictional aliens, is to take the universal/parochial distinction seriously. If we 'ran Earth again' on an alien planet, we would expect to see most of the universals, but they would appear as specific *instances*, implemented as parochials. These would mostly be like those we've seen on the standard run, but would happen to different organisms. For example, Jack thought it was reasonable to give the Yilané of Harry Harrison's *West of Eden* a penis-pouch, which incubated eggs laid directly into it by the socially dominant females. The Yilané were extelligent reptiles descended from mosasaurs – 'fish-lizards' contemporary with the dinosaurs – on the different Earth that resulted when the K/T meteorite missed and the dinosaurs, along with much else, survived.

The reasoning behind the choice of penis-pouches was based on today's biology: female seahorses lay eggs into the male's pouch, and modern lizards, descendants of near-mosasaurs, have a penis-pouch with prominent veins. In the novel, these parochials are combined into a persuasive new reproductive strategy, not seen on our version of Earth. The second volume in Harrison's trilogy, *Winter in Eden*, includes a new trick carried out by brown algae – they can melt their holdfasts (the plates that form their base) a little way into the edge of the Arctic and Antarctic ice-sheets, like snowdrops melting their way through spring snow, and so exploit a whole new habitat. Enormous browsing mammals were then invented to exploit these algae, just as sea-cows exploit rampant fresh-water vegetation. These creatures were not pretend-alien; instead, they were exactly and credibly what *could* happen if we ran the last bit of Earth's history again.

The big practical problem with using modern terrestrial parochials to justify – make credible – alien or alternative parochials, is not only that the author should have a quiver-full to employ as needed, but that the reader/viewer must be also be informed about the terrestrial examples. And you can't easily put such information into the story: 'Gbzzt,' said Swampmother Superior, 'methinks yon penis-pouch bears a strong evolutionary resemblance to one that I have heard tell in the space-bars can be observed among the oceanic-equines of mythic Terra . . .' Sorry, no.

The big theoretical problem, of course, is the question of the universality of our biology and arguments from it: will there be penises on other planets? Earth's history shows them being invented tens of times in different branches of the evolutionary tree; we have no doubt that we'll find sex, copulation and penises on most planets with cellular (multicellular) life. They are universal. The *shape*, there again . . .

That brings us to another question. Will life 'progress' on other planets – or on the re-run Earth – as it has done here? Will it necessarily go from the simple to the complex, possibly with occasional re-simplifications? That, broadly speaking, is what has happened here. When we think about this question, it is unfortunate that the suite of parochial features carried by the fish that came out of the water, the species that founded the land vertebrates, seems so 'natural' to us. Unavoidable, of course: we've got used to it, over the millions of years, and our minds contain a 'body-image' map against which we check to make sure we're OK. ('I've suddenly grown an extra nose, I don't think I ought to have done that.') All the land vertebrates, amphibians, reptiles, birds and mammals have, for example, our 'airway crossing foodway' trait. We hardly ever think about how silly this is, except when we choke on a peanut, but as we've already told you, we inherited it from a parochial feature of an ancient lobe-finned fish: our ever-so-distant ancestor that adventured up on to the land.

When we say that complex life evolved from simpler life, we don't mean that the complex forms *replaced* the simpler ones. A common image of evolution is that of 'succession', in which early, primitive creatures are replaced by more sophisticated ones. Culminating, as every informed Victorian knew, in *us*, but let's not go into that again. However, as many biologists have pointed out, perhaps most persuasively Gould in *Wonderful Life*, there has been *no* succession of lifeforms on Earth. Bacteria are still with us. Indeed, they are with us very intimately: each human being has many more bacteria in their gut than the total of their own cells. And bacteria are still much more common and prolific out in the ecosystem than any 'higher' organisms. Several popular writers, including Dawkins and Gould, have tried to make the point that this means that we are still in the 'Age of Bacteria' and that the existence of higher, more complex organisms is not ecologically important.

One answer to that argument is that the mineral world has barely noticed the thin skin of life on this planet: the core and the mantle have not been affected by the thin skin of oxidised minerals at the surface. However, it does seem that limestones, produced by tiny marine organisms, lubricate the processes of continental drift and make it go a little quicker. Does the Earth 'notice' the extra velocity, though? Should we then say that Earth isn't out of its Stone Age? Surely we should classify successive times with their *highest* accomplishments? We do that with human accomplishment, having Iron Age succeed Stone. Can we really claim to be in the Electric Age, or the Information Age, when we *still use* iron – and stone, for heaven's sake – in our most obvious artefacts, buildings?

Although the *variety* of species is much greater now than it was early on, the vast majority are still the simplest forms. It has been calculated that the viruses in the seas outweigh the whales at least twenty times. Whales have been on the planet for only thirty million years or so – though there were reptiles nearly as big, ichthyosaurs and plesiosaurs and mosasaurs, from 140 million years until 65 million years ago. The fossil evidence indicates that before 300 million years ago there was no creature anything like as big and complicated as a 20-metre (65-ft) whale or a 12-metre (35-ft) ichthyosaur. There were some pretty big amphibians and early reptiles, then, but no birds or mammals. Five hundred million years ago there was little animal life on land, some plants but no trees.

We humans think that it's important for an ecosystem to have angiosperms like oak trees, and mammals like elephants, monkeys, and whales, as well as marine crustaceans of greater numbers and biomass, and a vast plethora of insects of greater diversity. We think we're at the end of the Age of Mammals, and well into the Age of Managing the Ecology. We are amused at Haldane's response to a woman who asked him what he had learned of God's beneficence from his study of Natural History: 'Madam, that He has a special fondness for beetles.' Some people deny that he ever said this, but the Oxford Dictionary of Modern Quotations has this from a report of one of his lectures: 'From the fact that there are 400,000 species of beetle on this planet but only 8,000 species of mammals, he [Haldane] concluded that the Creator, if He exists, has a special preference for beetles, and so we might be more likely to meet them than any other type of animal on a planet which would support life'. At any rate, there are more kinds of beetle than all other kinds of insect together, and they contribute between one seventh and one fifth of all named organisms – but this doesn't mean that we have to force beetles on you. When it comes to alien life, beetles are pretty much irrelevant.

When Jack invented the ecology of the successor to *Legacy of Heorot*, with Larry Niven, Jerry Pournelle, and Steven Barnes – it was called *Dragons of Heorot* in the UK and *Beowulf's Children* in the US – he invented exoskeletal flying forms because the authors had put flowers on to the island where Heorot was set. Pollination *could* be a universal – we don't know; it might just be an unusually common parochial. But the propulsive system of these exoskeletal fliers was quite unlike that of terrestrial insects. The appendages in one pair were aerofoils, and those in the rear pair were propulsive. The alien creatures looked like beetles, but they worked in a rather different way. They were different parochial instances of a biological universal.

On Earth, as we've seen, the development of an oxygen atmosphere led, eventually, to the evolution of the eukaryote cell. Unlike bacteria, these cells have a complex, very convoluted membrane surrounding them, and a nucleus into which they segregate their genetic material. They also partition themselves to form genuine many-celled organisms, whereas bacteria can manage only loosely knit colonies. So is oxygen *necessary* for multicellular organisms, hence 'higher' lifeforms? *Rare Earth* assumes it is, though it never makes out a case for this assumption because it never thinks to question it. The evidence, of course, is that on our planet oxygen was *sufficient* for multicellular organisms to evolve. Necessity and sufficiency are logically very different, so Earth's history cannot answer the question of the necessity of oxygen.

We've suggested that oxygen atmospheres are parochials in general, but they are universals when the planet is an aqueous one. Necessary, sufficient, or neither, they will nevertheless be *there*. Oxygen will accumulate in the atmosphere of most aqueous planets as they age. A variety of 'higher' organisms, combining the histories and specialities of their prokaryote symbiotic ancestors, will graze the prokaryote ecology – and then, in turn, be grazed by larger cellular or multicellular forms. In fact, though, oxygen probably is *not* necessary to the evolution of complex organisms: an obvious alternative would be to use sulphur chemistry instead.

How would the structure of complex aliens evolve? On Earth, some of the less familiar organisms, such as rotifers and nematodes, have a very well-defined structure on the cellular level: they have a constant number of body cells in each organ, and mostly cannot regenerate or even repair themselves when wounded. Perhaps there was a separate origin of this different, rather extreme kind of cellularity, suggesting that cellularity is a universal. Most cellular creatures, even small ones, don't have such specific cellularity, and can heal their own wounds. Whether we will find other evolutions to produce larger cellular or multicellular creatures, both of these routes have been taken several times on this planet, by animals and plants and fungi, and so both can probably be counted as universals. So we'll find larger, cellular creatures with differentiated regions on a re-run Earth, and on other planets – as well as an extended prokaryote biota carrying out the decay side of the ecology as well as much of the photosynthesis. But we will need to be more careful of words like 'cell', 'multicellular', and 'prokaryote' if we are to argue about the universality of these processes. And we must be prepared to analyse what might have happened in the evolution of other lifeforms before we visit Europa, let alone more distant worlds.

On this planet there was a wave of mostly radially-symmetric larger creatures, the Ediacaran fauna, before the massive diversification into many different body forms that occurred at the start of the Cambrian Period about 570 million years ago. The Ediacarans were probably Earth's first cellular experiments. It has been seriously suggested that the great advance upon these Ediacarans was the invention of the anus – a pass-through gut instead of a simple sac – which came into being with the evolution of 'triploblasts', creatures with a three-layered body-plan. This invention is thought to have had great ecological consequences, because the accumulation of faeces on shallow sea-bottoms (instead, presumably, of occasional corpses) opened up a new region, the sea floor, to specialised bacteria. These could then be grazed by specialised protozoa (ciliates like *Paramecium*, on this planet today, but similarly universal and differently parochial protozoa elsewhere and elsewhen), which were then predated by larger organisms. The whole sea-floor ecology was then ripe for filter-feeders like clams, and some surprising worms with crowns of tentacles to filter-feed with; then *their* specialised predators, leading to the enormous productivity of many shallow sea-bottoms on Earth. The phase space of possible phenotypes had suddenly expanded into a new dimension.

This process has been documented by the amazing variety of fossils in the Burgess Shale, and more recent discoveries of other ancient well-preserved biota in China and Australia. These organisms certainly diversified very quickly, in geological terms: about twenty million years saw the origins of *all* the major kinds of cellular animals we have now, embedded in an ecology with many other body plans that have not survived. There are, however, other possible interpretations: perhaps they were evolving in a little land-locked sea whose remains we haven't found, for 100 million years; perhaps they were initially very, very, variable, producing many 'monsters' before the awesome ability to develop reliably was evolved.

Taking that route in the sea may have paved the way, via worm-types and little arthropod-types (each of several different body-plans, so possibly universals), via the degradation of seaweeds and other marine debris on beaches, to soils on land. Soils are extremely specialised ecosystems for degrading difficult woody plant material – they are emphatically not 'just dirt' – and they made the diversification of land plants, especially tall land plants, possible. Soil fungi do not only break down wood. They also form many different symbiotic relationships with plants, especially the mycorrhizae on tree roots that help them find and take up minerals. Recycling plant material on land, by the invention and complication of soil ecology, was one of the most under-advertised, but one of the most important, steps in Earth's evolution of life on land. It moderated erosion, by permitting forest-clad hills, and this modulated water-use in valleys and their rivers and lakes.

This brings us to a very difficult question. So far, we have taken the easier route of seeing Earth's evolution from the viewpoint of the different anatomies of the organisms concerned, making different functions possible. We have stressed the universal/parochial distinction without placing these functions in the context of a full-blown ecosystem. In our metaphor of evolution as a theatre, with a play to be performed, but employing different actors on successive nights (re-runs of Earth) or in different towns (other planets), we have made tacit assumptions about the scenery (ecological context).

However, it must surely be the case that ecosystems evolve, not individual organisms. This has not been generally recognised in the teaching of biology on Earth; in fact, in many circles it is considered fallacious, if not seditious. Nonetheless, there is no question that it must be true, because each organism lives or dies in the context of all the others. The difference of opinion here revolves around a confusion of two different questions: 'How does each tiny step in evolution work?' and 'How does the overall dynamic of evolution "flow?"' The steps take place on the level of organisms, which die (or not) depending on how effective they are at surviving in whatever environment they find themselves in. But the overall pattern of evolution happens on the ecosystem level, because every organism's environment is mostly *other organisms*, either of its own species or of others. The background is made up of individuals, but those individuals are evolving because of the game they play against that background, so the *scenery* evolves along with the players.

You may be surprised that we haven't mentioned DNA here, especially given the 'selfish gene' view that an organism is merely a vehicle for the replication of DNA, with the gene being the true 'unit of natural selection'. In *The Collapse of Chaos* we argued against the selfish gene viewpoint, which we saw as a needlessly narrow interpretation of a much broader and more interesting complicity between DNA and organisms, between genotype and phenotype. If it were true that the genome of an organism determines every feature of its body-plan and behaviour, then it would be reasonable to adopt a gene-centred viewpoint. However, many other things go into making an organism, and most of them depend on context.

The genome plays a very subtle role, but its most important effect is to lubricate evolution, not to control it. DNA makes it possible for organisms to change from one generation to the next: it makes phenotypes 'fluid', able to respond to a changing environment. What determines the form and behaviour of those organisms that survive to reproduce is natural selection, and this acts on organisms; its action on DNA is very indirect and very cryptic. The dynamical flow in phenotypic space – how species change in their evolving ecosystems – is an emergent feature of a complicity between organisms, their genetics, and their environmental context. The 'unit' of evolution must be ecosystems, because those *are* the main context for everything else, themselves included.

It is therefore much more difficult to understand whether the ecological evolutionary route followed by Earth's early triploblasts – feeding and producing faeces – would be available on another planet, than it is to compare anatomical/functional adaptations and classify them as parochial or universal. Evolution is about populations of organisms specialising into ecosystems and making new niches – ways to make a living – for themselves. There is rarely an 'empty' niche for organisms to exploit, except by the accident of transport to a new land, like rats, goats, rabbits, deer, gorse, Japanese knotweed, Canadian pondweed, and a thousand other 'alien' species introductions. Organisms complicate their ecosystem, and the ecosystem constrains and evolves the organisms, much as a river winds across the landscape, changing the local geography but having its own path determined by the substrates it finds.

So, on other aqueous planets, the familiar things will happen on the level of ecosystems. We will find ecological system-architectures that translate directly into architectures that we know from our own world, with predators and prey, herbivores, 'grasses' with tough leaves fed upon by specialised grazers, 'trees' with the ability to catch any light there is and to shade other plants. But we cannot know whether those 'trees', like so many of ours, will encourage 'birds' to build their 'nests', and fertilise the tree roots with minerals from afar; or whether the 'grasses' will, as they have done here, prevent lakes and ponds lasting very long, by growing in from the edges. Presumably they will sometimes do those things, in some places and on some planets. From our experience of Earth, we won't expect to find 'desert planets' like Herbert's Dune (Arrakis) or 'jungle planets' like Harrison's Deathworld; in most eras, most planets will have diverse ecologies in different places, because *that* is a universal.

Over the last twenty years a whole new view of evolution has arisen – controversial, but fascinating, and with a solid core of good sense: it looks like a universal, and it should therefore apply to alien ecosystems, if suitably generalised. This is James Lovelock's concept of 'Gaia', of the Earth (and, by extension, any alien ecology) as a self-regulating complex system. To call Earth an 'organism' in its own right, as some do, is to take the idea too far – or possibly not far enough, for Earth's organisation is different from that of any single living creature. The Gaia view of evolution is now widely accepted, as regards its basis that 'the environment' is controlled by feedback from the organisms evolving in it. Some New Age romantics claim that Gaia is *conscious* of this process, but this extra dimension has few adherents among scientists.

Gaian evolution requires the fostering of heterogeneity, for example the diversification of life on to the land and the invention of whole new ways of life, such as dinosaurs, trees, and insects. Whether it is possible for one species to take over an ecology and turn it into a monoculture, as has been suggested by several authors, like Gerrold in his *Chtorr* series, is not yet clear. The reverse seems to be the major theme in evolution here, but monocultures do seem to work in limited regions of Earth, so perhaps they never got a real chance to strut their stuff on this world. (Maybe Gaia stopped them.) Our ecology diversifies and the most complex organisms grow ever larger and ever more complex. This must be *a* universal, but surely not *the* universal.

We are in good company in not considering the larger contextual evolution – the evolution of ecosystems on Earth as progressively more complex creatures come into existence. Most authors of alien scenarios give the ecology one or two quick brush-strokes, and don't call the reader's attention to it again. Even where the ecology is the point of the whole thing, as Herbert attempted in the *Dune* series, the author's thinking about it runs out very soon. However, the series of prequels currently being written by Brian Herbert and Kevin J. Anderson look as though they will fill in many of the gaps here. Those gaps include some straightforward questions, like: What do the enormous sandworms eat? Apart from their mystical attachment to their riders, what do they get out of the symbiosis with humans? Come to that, where does the water come from on Arrakis? What's the meteorology? Herbert is by no means alone, nor is he a worse than average sinner, in this regard.

Dune (Frank Herbert 1965)

Eight thousand years in the future, humans have colonised the galaxy and created a vast empire with numerous Dukedoms, each in charge of one or more planetary systems. Baron Vladimir Harkonnen has manoeuvred his enemy Duke Leto Atreides into becoming administrator for the desert planet Arrakis, colloquially known as Dune. Leto does not realise that he and the Baron are part of a breeding programme controlled by the reverend mothers of the Bene Gesserit, who plan to create the superhuman Kwisatz Haderach, able to foresee the future, by uniting the Atreides lineage with its deadly enemy. But Leto's wife Jessica, herself a Bene Gesserit and the natural daughter of Baron Vladimir, disobeys orders and conceives a son, Paul; the Kwisatz Haderach arrives a generation ahead of schedule.

Dune is the sole source in the galaxy of *spice*, a drug widely used by the aristocracy, and crucial to the hyperspatial transport system run by the Spacing Guild. There is a strong but unidentified link between spice and Dune's giant sandworms, which can be up to 400 yards (metres) long. The sandworms normally inhabit the deep desert, living far beneath the sand, but they sense vibrations and will rise to the surface to engulf unwary humans, or even entire spice harvester factories. Creatures known as little makers, half plant and half animal, are intimately associated with the sandworms. Apparently, no other non-human creatures occupy the desert.

The indigenes, the Bedouin-like Fremen, make use of spice in their rituals, and have devised a way to attract a sandworm and ride it. Special hooks are used to climb the sandworm's sides and prise open a gap between its segments; the worm then turns the sensitive exposed region to the top, and remains on the desert's surface.

Harkonnen forces invade to take over the valuable spice production. Paul Atreides flees into the desert, where he is found by the Fremen. Normally any outsider would be killed, but the Fremen sense that Paul is a long-prophesied messiah figure. For the remainder of the story, Paul struggles with an internal conflict: as a messiah, he can save the Fremen culture, but if he succeeds, every future that he can foresee leads to a galactic jihad, a terrible Holy War . . .

In both xenoscience and SF, it is not just individual alien organisms that must 'work' biologically. The ecosystem in which they live must also make sense. Herbert's *Dune* is an excellent example of an impossible fictional ecology that nevertheless provided the framework for an enormously popular series of stories. In both the books and the film of *Dune* the ecology is unbelievably impoverished: a desert planet. Little attempt is made to do more than portray the human inhabitants as very lightly-disguised Bedouin, since they are intended to be human-descended, not alien, and these 'Fremen' have come to some kind of symbiotic relationship with the great sandworms that live under the desert.

The whole scenario is terribly mystical, so lots of relationships are not made clear, and all kinds of resonant supernatural events occur. The sandworms are gigantic; they apparently eat something that becomes the mystical drug known as 'spice', found in their excreta. But, quite specifically, there is not enough vegetation on the planet to keep *anything* alive, never mind a big sandworm (very impressive in the film version – hungry, too, and we're not surprised). However, there is an oxygen atmosphere in which humans can survive without breathing apparatus: where did it come from? It is also never made clear what the sandworms get from the symbiosis with humans – though what Herbert gets is some great scenes.

However, some authors do have a natural feel for ecology. Harry Harrison in his first *Deathworld* book gets a jungle planet as right as one can (but this is a spoof, or at least an artistically exaggerated version, of a heroic adventure story: see the amateur film *Breathworld* if you can). His second book in the series, *Deathworld II*, takes us to an exaggerated Arrakis planet (or Southern California, in some respects, including the cowboys), and again the ecology works pretty effectively.

Clement, a high-school physics teacher, sets himself the task of getting the astrophysics right, and then lets the ecology ride on its coat-tails. So, in his most famous book, *Mission of Gravity*, the 700g poles of the planet Mesklin have totally different vegetation from the 3g equator. In Clement's *Close to Critical*, the physical assumption is that the atmosphere consists of sulphuric acid, sulphur dioxide, sulphur trioxide, and water, and conditions are very close to the critical point where all these 'phases' merge – so that enormous raindrops, metres across, settle gently to the ground. The metabolism of his animals and people is reasonably invented, with plants that segregate sulphur and animals that burn it, so the chemistry and energetics work reasonably well. But he leaves us to imagine the ecology, the evolution, and the other-than-intelligent beasts, with little more help than a couple of throw-away lines.

There is a useful analogy between the evolution of technology and evolutionary innovation and development in biology, which helps us avoid making unwarranted assumptions – even when they seem unavoidable. The analogy will help us to explain how living systems have left chemical and physical constraints behind, and have now moved into a new sphere of mechanism. Our example leaves rocket-fuel tecnology behind, and evolves stepwise into systems with new possibilities, where the old constraints no longer apply. The example's relationship to space travel is, for this purpose, irrelevant. But when we invented the analogy we had in mind that people with an interest in aliens might find analogies using space travel congenial. We will refer to this idea as the 'bolas' or 'space elevator' analogy, and we sneak up on it with a bit of the history of space travel.

In the 1950s and 1960s people who were interested in space travel thought in terms of rockets, orbits, 'escape velocity', liquid oxygen and hydrogen, hydrazine and nitric acid, solid fuels, specific impulse, two-or-three-stage rockets. It was *all* rockets, for those who wished to be realistic, and the chief constraints were Newton's Laws of motion and gravitation – especially their consequences: conservation of energy, of momentum, and of angular momentum. Most rocket enthusiasts could demonstrate how much fuel you have to burn to achieve a particular orbit, or to leave the planet, with a variety of options (different specific impulses from different fuels) including how much fuel you have to carry to do things once you've got wherever you want, or to come back down from whatever orbit. We could tell you, with complete certainty, the minimum energy you had to expend to get a 200-pound (100kg) man into Low Earth Orbit, or into the synchronous Clarke Orbit 22,000 miles (35,000km) above the equator, where the Earth is rotating exactly as fast as you are so your position appears stationary when seen from the ground. TV satellite transmitters are there now. The absolute minimum was, of course, the difference in potential energy of the man on the ground and the man 22,000 miles up, falling laterally around the world but staying over one place. But in practice you couldn't even get close to that, because of the aforesaid metal, and the fuel to lift that, and the extra fuel to lift that . . . till you had something the size of an Atlas rocket, which in the 1950s most people thought was simply too big to work.

Then came Yuri Gagarin, and everyone started to realise that the whole space travel game could be 'on'. The rest, as they say, is history. And with a twenty-year gap, for all kinds of political reasons, we now, for no very clear reason, have a manned Space Station. Perhaps people will go to the Moon again soon, perhaps we will achieve a continuous presence there. Perhaps we will follow our probes to Mars and to Europa. However, there were two notional inventions on the way, which have not yet been built and aren't even in prospect, but which provide wonderful 'evolutionary' technical steps on the way to real space travel. The first could, technically, be built tomorrow if the capital was available, but the second needs slightly stronger tensile materials – about three times as strong as we have at present. These are not suggestions, they are thought-experiments, but they are well grounded in workable physics. Our analogy does not require that they work, but that we can think about their working.

The first invention is the so-called space-bolas. A bolas is a weapon used by South American natives to catch rheas, running birds about half the size of an ostrich. Three 6-foot (2m) ropes are bound together at one end, and heavy stones are tied to the other ends. These are twirled like a lasso, and thrown at the bird's legs: the weights make the ropes wind around the legs and bring the bird down. The space-bolas resembles this only in that it has three cables attached at a centre point; the stones are replaced by passenger cabins with airlocks. The cables are, say, a hundred miles long. Bundle this up, and take it out to 120-mile (190-km) orbit. While there, unreel the cables and hold the centre point (its centre of gravity) in the orbit. Spin the whole thing vertically, like a Ferris Wheel, at about one revolution every thirty hours to achieve two-thirds Earth gravity in the cabins. You could power this gadget by dropping asteroid material or moon rock, or by returning astronauts to the cabins when furthest from the Earth's surface and releasing them at the lowest point. An astronaut wishing to get to a fairly high orbit would get into a cabin when it was 20 miles (30km) up (perhaps from a regular jet aeroplane with some rocket-assist), ride around the half-circle to the bolas's highest point, and get out. She would be travelling a little too fast for the 220-mile (350 km) orbit that she is now in; depending how fast the bolas was spinning. So she should leave it earlier to get a boost outwards . . . and then catch the next bolas in a vertical stack of the things . . . and then to the third bolas in the stack, connecting to synchronous orbit.

The bolases need not be powered by direct Newtonian action, mass coming down. A clever solar-powered gadget holding the three cables in the middle could 'pump' each bolas like a child's swing, reeling in or lengthening the cables at different stages of the cycle. Once the capital investment has been made, the cost of getting a 200-pound (100 kg) astronaut into orbits above 500 miles (800km) – from 20 miles (30km) up, and moving at about 15 miles per hour (25kph) relative to the Earth's surface – would be less than a twentieth of that for the best rocket. The Space Shuttle in our day is fairly good, but this method would be a hundred times cheaper. The main catch is that today's space traffic is too small to warrant the huge capital outlay required. However, we're not proposing the space bolas as a practical, economic solution to spaceflight in today's politico-economic climate. Our point is that Newtonian limits, which we thought were an absolute constraint, can be considerably bettered by rethinking the technology. Newton's laws are not *violated* here: they are applied in a different context.

The second invention is cleverer and better. In *The Fountains of Paradise* Arthur C. Clarke called it a space elevator; Charles Sheffield called it *The Web Between The Worlds*. We already have the beginnings of an effective system for delivering goods to synchronous orbit: at the moment this exists to service the communications and entertainment industries. Once you have such a system, the next step is relatively easy. All you need is material with a very good tensile strength: it doesn't yet exist, but it almost certainly can be invented, so we'll assume it can exist and call it 'asilon'. At some point in synchronous orbit, accumulate a lot of heavy junk that you can lash together, and a 22,000 mile (35,000km) long coil of asilon. Tie a rock or a small rocket to the end of the asilon cable and launch it backwards and towards Earth, so that it lands in Colombo, Sri Lanka, or somewhere else friendly on the Equator. While the asilon reel is unwinding, throw your heavy junk outwards – all tied together and to you – to keep your centre of gravity in Clarke orbit. You're going to need a lot more asilon for the next step, too; perhaps you should get some Moon rock, or catch a few meteorites. Tie a pulley with a pair of cabins to the bottom of your cable, and haul it up. Or build a cabin upstairs, and one downstairs, and have them climb up and down releasing a much stronger cable behind them, one that can take the weight of about 10,000 miles (16,000km) of itself plus the (comparatively trivial) forces and masses of the cabins. You will also need a massive counterweight, swinging outside synchronous orbit, to keep the top end of the cable synchronous. What you have now is a space elevator, 22,000 miles high. An astronaut can probably get up to synchronous orbit in a week. If other astronauts are also returning, the only cost is friction.

The point is not that we could or should build one or more of these, but that in principle, taking into account Newtonian laws, it is possible to have a system delivering astronauts to synchronous orbit and returning them, for peanuts compared with rocket costs. Evolving technology, whether in engineering or biology, can move itself out of the old set of rules and constraints into a new realm of opportunities, by changing the context in which the rules apply. This is a universal trick for expanding into the space of the adjacent possible – by making new possibilities become adjacent.

There is a metaphorical lesson here for xenoscience, and it's important. When chemists try to apply the old rules, like those of the Second Law of Thermodynamics, to the workings of a living mitochondrion, this is like asking about the rocket-fuel used in the space elevator. The question is nonsense, because the space elevator doesn't use rocket-fuel. In different contexts, you often have to ask different questions, or ask the same question in a different way.

When physicists try to model the efficiency with which a jumping kangaroo uses oxygen by assuming that the animal is a Newtonian sack of coal being thrown at each step, then lifted and thrown again, they come out with the answer that it's impossible to get enough fuel and oxygen to run a kangaroo. In point of fact, the kangaroo *bounces*, storing its energy of momentum in its elastic leg ligaments (and in clever body positioning) and uses less energy than we use running.

By analogy, we can refer to the sack of coal as a rocket, and the kangaroo as a bolas or space-elevator. In the same analogy, biology has had billions of years to invent and improve bolases, and about half a billion to invent, improve and diversify a great variety of space elevators, avoiding all the obvious constraints of simple-minded physics ('rockets') by investing material-and-energy capital into building each body. Yolk, milk, all the other investments which complex cellular eukaryotes make in their offspring, to enable those offspring to start with equipment a little more sophisticated in each generation, are called 'privilege' by reproductive biologists. The proton pump in the mitochondrion, which is the power-engine for most of our cells, is much more efficient than even our electric motors; muscle is much better than our technology, too. The poor job our surgeons make of replacing, say, a severed human arm is clear evidence that our technological knowledge is inadequate to explain all the space elevators in cell biology.

We are used to SF stories that have aliens 'of lizard stock' or 'very like enormous domestic cats' (these two mainly from authors with a particular brand of insecure femininity, in our experience – but McCaffrey was spoofing). No real alien will have a recognisable Earthlike lineage: no reptiles, no vertebrates, nothing terrestrial-parochial in its ancestry. So we will not be able to understand its biology. Indeed, we can't really understand the tricks, like proton-pumps, which our own mitochondria use and which are crucial for every aspect of sub-cellular metabolism. Not for bacteria, however, because they don't have mitochondria: instead, they have a suite of space elevators of their own.

We list a few of the bolas/space elevator tricks used by terrestrial life, to show you the scope and scale of what we might expect in other evolutionary histories:

- Cell membranes, which have proteins laced through them, so that pushing a receptor button on the outside rings a chemical bell inside;
- Tubulin, a fibre-assembled and disassembled protein, is used to move other molecules, and chromosomes, around;
- Antifreeze proteins are used by Arctic and Antarctic fishes to stop their body fluids freezing;
- Mammals use the dangerous technique of hanging their testes outside the body, to keep them cool enough to produce functional sperm.

Look at that familiar last example another way: to remain warmer and quicker than your prey/predator, you need your muscles to be warmer than your testes can bear – but the risk is clearly worth it. But what do hot females do, then? There is accumulating evidence that mammalian ovaries are at body temperature, but their Graafian follicles, with eggs inside them, are chemically cooled by about 2.6°C.

Any alien lifeform will have lots of cute tricks like those, not at all obvious and not at all transparent, because they will be exploiting tricks that aren't in our well-investigated lifeforms. Those tricks expand the phase space of potential alien lifeforms in ways that are totally unimaginable. The well-investigated lifeforms are actually only a tiny proportion of terrestrial organisms: think, for a start, of all those extinct species. But our lifeforms must have thousands of tricks, which may be expressed only in a rare deep-sea prawn that hasn't had buckets of cancer-cure money spent on it to understand the clever and unique way that it repairs its DNA faults – or whatever. We know only a tiny fraction of our terrestrial biology. A different, alien, evolutionary tree may be only marginally more mysterious, in that we will understand virtually none of it without a lot of work.

Some people who have thought about these matters are very unsure how many alien lifeforms will fit our criteria for life, even though these have broadened considerably in the last twenty years! Prions, crystallised viruses, a bacterial economy in the deep rocks, which might be comparable in mass to all Earth's marine life; these have stretched our understanding of our own global ecosystem in innumerable unexpected directions. In 2000, Russell Vreeland and colleagues reported that they had isolated bacterial spores, preserved in salt crystals for 250 million years, and that these came back to life when placed on a standard agar culture medium. The researchers took stringent precautions to avoid contamination. The bacteria were of an unknown species, but their DNA was similar to that of the modern species *Bacillus marismortui* and *Virgibacillus pantothenticus*.

Simple lifeforms seem to be virtually immortal. And the same goes for some more complex creatures, for instance tardigrades, often known as 'water-bears'. These are about a millimetre long, and are found on sphagnum moss in swamps on mountainsides, and in the muck in rain-gutters. They can withstand drying-out, boiling, and a variety of other usually lethal insults. Some species can survive at temperatures close to absolute zero and well above the boiling point of water. They remain alive (or potentially so) for 1000 years, if dried out: just add water.

More generally, biological development is itself a space elevator: instead of growing then dividing, growing then dividing, as most prokaryotes do, most larger organisms have a succession of stages that they pass through before they produce a creature that can then produce the spores, eggs, or sperms for the next generation. You may think that our reproductive sequence egg →blastocyst → implant → embryo →fetus →baby → infant → toddler →adolescent → adult is fairly complicated. Not so: some marine creatures have twenty or more different shapes and ways of life to follow their developmental trajectories. And they also have branches, with several different possibilities – often to become male or female, for example, according to circumstances, like our fictional Europan caretaker.

These tricks do not alter the universals of xenobiology: they are either parochials, or they exploit universals by novel routes. This means that on a planet with a similar history to ours, we would still be able to recognise which creatures are the higher lifeforms, despite parochial differences and local biological space elevators. The 'creepy-crawlies' would probably look similar to ours, to the untutored eye, too.

Indeed, the variety seen in such creatures here is immense, and many terrestrial organisms look distinctly alien. Pycnogonids – so-called sea-spiders – look like incredibly emaciated spiders: they tuck their guts into their limbs because there isn't room in the tiny body. Some moray eels, which live in coral, are a hundred times as long as they are thick. Leafy sea-dragons, found in a few parts of the Australian coastline and a major attraction in those few big aquaria that know how to keep them alive, are simply unbelievable, a cross between a seahorse and rampant seaweed. On land, wind-scorpions are the most leggy, hairy, fangy, scrabbly creatures you could ever hope not to find in your desert tent: the one that ran up Jack's shirt seemed to him at the time to be about the size of his head. It wasn't, it was only the size of his fist, with a body like a walnut with jaws like a pair of pliers. If we weren't so used to . . . elephants? Kangaroos? Cobras? Sharks? If these could be seen with alien eyes, as the unlikely denizens of our world, we could be more prepared for the strange creatures we will find elsewhere. 'What, it pushes over trees with its nose? What do you think I am, to believe that?'

As we've said, real aliens will never look like terrestrial vertebrates, so the 'highest' form will not look like us. Its anatomy will be – must be – different in every detail, because all our details are derived from parochials of that lobe-finned fish: two arms, two legs, pentadactyl (five-fingered), knees and elbows, airway crossing foodway, bone, teeth (which were once the scales of fishy ancestors), feathers . . . even mammalian hair is a very special development. A real alien won't have any of those, but it might well have jointed limbs, fur (of a different-from-mammal kind), airways and foodways – it might use its airways to modulate sound, or it might draw its comb across the edges of its wings, or have an electrically-driven diaphragm.

Most fictional aliens have not been presented as the results of credible evolutionary histories. The *Alien* series of films serves as a warning example here. The dragonesque Mother lays her eggs, which are apparently about the size of a football, in the open, near her gothic living caves, where they apparently wait for thousands of years for a spaceship to land near them. When it does, any that have survived hungry egg-eaters for all that time hatch out. Presumably they are sensitive to just that smell of rocket exhaust – advanced spaceships won't have to worry. They have the immediate ability to invade terrestrial mammalian hosts and live inside them, where the nutrients are just right for them. How did they become able to avoid our tissue-recognition immune system? Or how to design just the right local anaesthetic so that the host doesn't *know* he's got an object the size of his heart – extra – in his chest? Are they tuned to people, in fact, or are they general-purpose parasites – a concept that would make any parasitologist scream?

One general issue here, which we mentioned earlier in connection with Verne's *Twenty Thousand Leagues Under the Sea*, is how much effort the author makes to answer the questions that are likely to occur to readers. But there is a more subtle point behind that issue, and it's this: any scientists who talk to the public, and most science teachers, know that it is never enough to give the answers. The audience has to be shown what the *questions* are.

Scientific questions are usually not obvious ones. Our favourite example here is the sunbeams that spread out into a fan at sundown and sunrise. In lectures, Jack often puts a picture up on a screen and says:

'There are two theories of where the Sun is. Either it's ninety million miles away and the rays are damn-near parallel because the whole Earth only intercepts a tiny thin pencil of all the Sun's light . . . or we're really on Pratchett's Discworld, and the Sun is a small bright body going round the planet not very far above the clouds. All of you have seen sunbeams like that, and not questioned it. Look again, now. That appearance is good evidence for the *second* theory.'

And everybody's jaw drops.

They hadn't thought of that before. They hadn't realised that every sunbeam poses a *question*. Then, and only then, are they interested in the answer, and able to understand what it means. If the lecture had started by presenting the answer ('Let me tell you about the geometry of sunbeams, and how parallel lines seem to meet when they are projected . . .') the audience would all have gone to sleep or left. Instead, they're spellbound.

It is necessary, then, to explain what the question is. To dodge the issue, as in the *Alien* films, by *not* saying, 'Hey, how does this parasite manage to do these biologically impossible things?' is as dishonest as having Robin Williams as an alien on daytime TV who can do all kinds of magic. But *Alien* pretends to rationality, indeed to science. So it *should* expose problems and solve them, not paper over the cracks, even though only a couple of people in each audience will be upset; it is discourteous to the audience to pretend that such a fantastic scenario is possible.

We want to be helpful as well as critical. We want to encourage scriptwriters, producers, and special-effects teams to give us more films as enjoyable as the *Star Wars* and *Star Trek* series. But please understand that there will be no vertebrates, reptiles, mammals, or giant ants on other planets. However, there may well be high-level similarities in the siting of alien brains, eyes, mouths, and limbs, a nerve cord if there is one; there may even be similarities in the use of colour. And jointed limbs are probably a universal, hence ubiquitous, invention. So think about what might be necessary, desirable, or evolutionarily plausible in your alien designs – show us the courtesy of giving their design as much thought as you give to the main (human) characters' futuristic clothes or transport . . .

Um . . . Perhaps we mean *more* thought.

8

Dragons, Teddy Bears, and Toddlers

*N*ON-MANIFESTATION WARNING
Valued guest, for your own safety and to fulfil legal obligations, we do warn you that any tourist who does manifest his, her, its, their, or any other form of self to the pre-technical humanoid inhabitants of planet Earth, will be held liable for prosecution by the effectuators.

We do assure you that there is a reason for this warning, a peculiarity of the way these humanoids build their minds. The process is confined to the early stages of their lifespans, and it involves the use of animals as iconic images. Any tourist who manifests his, her, its, etc. self is liable to enter into the culture as a new icon, with unpredictable results.

Do not fail to make the scheduled voluntary inspection of the Toy Room at the base in St Albans. There you will observe primitive constructs known as 'teddy bear' and 'womble'. On no account be tempted to purloin one of these exhibits and convey it to a local nursery or open-air crèche with the intention of providing slightly more response than the infant humanoid is expecting. The joy of infants at a toy becoming as interactive as a pet may be delightful, and it may indeed reinforce the cultural message the child would normally get. However, a favourable outcome cannot be guaranteed, and the penalties for infringement are therefore severe.

The penalties are even more severe (if that is possible) for inflicting mischief on humanoid children by programming them inappropriately for their culture. Manifesting one's self in the form of a teddy bear is expressly forbidden, even in the unlikely event that its behaviour remains within cultural norms. Manifesting one's self in one's habitual form is even more strongly forbidden, because it is impossible to predict what seed it might plant in the infant brain. (This directive is phrased metaphorically.

Parasitic herbiforms should note that its literal interpretation is also forbidden.)

(Earth Prospectus, *page 88734.)*

We saw in the last chapter that aliens will not, almost cannot, look like Earthly lifeforms such as vertebrates or insects or octopuses – in detail, that is. Some of the features of these animals, though, such as their size, the appearance of their sense organs, their disposition to have the sense organs near the brain, and their locomotion, may be like that of some Earthly creatures. But the *combinations* will be different: extelligent aliens might have exoskeletons where we have endoskeletons, they might have wheels or fur or electric organs or ⚏ ⛰ ⚘ (which we ignorant Earth people don't know about – this category no doubt includes most of the universe's tricks).

However, when we invent aliens for films, even for books, the temptation is to use familiar characters in familiar combinations. Earlier we gave some consideration to the biology of real aliens, and had a quick look at some invented alien ecologies. Clearly there were reasons for picking out the mobile trees of Epona; as Jack recalls, they were the result of some argument among the 'Contact' group that was designing that world. The near-adult-stage giant-caterpillar predators of the Chtorr invaders of Earth were like that so that when the adults appeared, the caterpillar image could justify our being 'fair' with the readers.

In *Legacy of Heorot,* Niven and Pournelle invented the grendel, a monster distinguished by its astonishing speed. The harmless 'samlon', which swim in the rivers and lakes and are eaten by Terran colonists for food, turn out to be grendel children. Grendel biology (but not speed) is based on an Earthly organism, the frog with nasty habits. This frog, *Xenopus,* lives in Africa. It is to be found in dish-shaped ponds containing only *Xenopus* adults, which are carnivores, *Xenopus* tadpoles, which are algivores, and algae (plus a bacterial decay recycling circuit, of course). The adults eat the tadpoles, having provided themselves with the next step down in the food chain. The grendel variants of the sequel were determined by the earlier invention of the simpler island grendels in the first book.

Ray Bradbury's Martians in *The Silver Locusts* were fanciful and philosophical because his elegant prose was well adapted to that kind of faerie. This is in total contrast to the Red and Green Martians of Edgar Rice Burroughs, who were noble-but-warlike savages, appropriate to his culture; they were alien transplants of sword-and-sorcery hero Conan the Barbarian. Burroughs's Red Princess of Mars was near enough human, so that John Carter could first fall for her, then shack up with her. Adolescent Jack was worried about sexual intercourse between a human and an egg-laying female alien – the shell-making equipment might have given the man quite a shock.

Spock, in the first series of *Star Trek*, was portrayed as a human-without-emotion, a completely rational person; he was supposed to be a half-breed between a human and a Vulcan, humanoids from totally different evolutionary trees. The *Star Trek* office, when informed that this was unlikely in the extreme, invented an Ancient Alien Race that had spotted the galaxy with DNA of such kind that it would inevitably progress to organisms indistinguishable from Hollywood actors with a bit of facial wax and funny ears.

In this chapter we look at some of the fictional aliens that humans have invented. We argue that many are rooted in the nursery tales examined by Shepard in *Thinking Animals*, or the supernatural myth-creatures of other cultures, from which they have often metamorphosed by adoption into western folk tales or even the serious theatre. Some of the aliens that our race has invented relate to more biological phobias, like our irrational fears of spiders, mice or non-poisonous snakes. Many of the odd-but-iconic 'aliens' that we known and love (or hate, or fear) and their odd-but-iconic behaviour come straight out of our own psyches. Understanding these origins will enable us to distinguish icon from reality, and be more objective about real alien biology.

To see how the choice of alien morphology has discernible cultural roots, consider the dragon that became Mother Alien, teddy bears that turned into Ewoks in the film *Return of the Jedi*, and the extent to which ET is the exaggerated outline of a human toddler, like Mickey Mouse. We must not underrate the images that we have grown up with, or their potency. When teddy bears became fashionable in the nursery around 1902, they caught the public imagination so strongly that their popularity is still huge to this day. One line of storytelling led from teddy bears to Winnie-the-Pooh, a plus point for our civilisation. However, there was a slightly bad-tempered, monomaniacal creature at the antipodes, the koala or eucalyptus 'bear', which isn't a true bear but a marsupial, related to possums and kangaroos. This dumpy creature eats only eucalyptus leaves, is only marginally brighter than the trees it lives on, has a nasty bite – but looks like an idealised teddy bear. We believe it to be a cuddly, nursery kind of animal that we could leave our babies with because it resembles one of our mythic icons. The koala doesn't know about our cultural icons, and must be totally surprised at the amount of cuddling it gets from the dourest of zookeepers.

Similarly potent images of aliens litter the bar scene in *Star Wars*, an archetypal spaceman's 'dive' in the insalubrious back-streets of the spaceport of Mos Eisley, populated with a diversity of strange creatures. Some are playing jazz on weird instruments, others are just having a drink – though what's *in* the drink we dare not imagine. For many people, including both of us, that was exactly the scene we had imagined from 1930s and 1940s space opera pulp SF. In our imaginations, we had peopled the typical spaceport bar with just such creatures, too.

Note the word: 'peopled'. The *Star Wars* bar-scene aliens are all human cultural icons – snake, owl, fox – straight out of the Shepard nursery lexicon or the American tales of Br'er Rabbit. Perhaps we should say that these images were well chosen to elicit our western responses: 'wise' for the owly ones, 'sly' for the fox, and 'ugh' for the snake – with overtones of Biblical evil. However, as we remarked earlier, those responses are mainly characteristic of western cultures. Inuit folk from northern Canada would trust the fox rather than the owl, because their stereotypic fox is brave and fast, not sly or cunning at all: he is the open-hearted hero of their nursery stories, even though he is the villain in ours. (It's not important that we're talking different fox species – theirs is the Arctic Fox – because the point being made doesn't relate to real foxes at all, but to their iconic status in folk stories.) In Norwegian fairy stories, the fox is clever and secretive, even discreet: you go to him to get clever advice, and he keeps your secrets. *We* learnt 'sly' in our nursery years because it was what the fox, the Fox in the Waistcoat ('vest' for Americans) did in the nursery stories. Indeed, it is probably more useful, following Shepard, to say that in our culture 'sly is the Fox', not that 'the fox is sly,' because the *real* fox isn't. The Fox icon is what our culture – our extelligence – uses to impart the meaning and the social associations of 'sly' and 'cunning' into our children's burgeoning intelligences.

The same happens with all those images evoked by the *Star Wars* bar scene 'snake'. For several native American peoples, the snake represents a messenger from a Nature-Mother or Gaia-figure (though these can't really be compared across cultures – that was the underlying error in Frazer's *The Golden Bough*). For a woman of the Water People in Southern Nigeria, a snake is evidence of a curse, but for a man of that culture a snake has little importance. For Yoruba and Hausa, 'snake' has other associations: mostly, people from these cultures kill snakes almost as a reflex, because that's What Men Do – like cowboys in the movies. So a Sioux or a Hausa seeing that bar scene, or an Inuit or a Norwegian, or indeed anyone not brought up on the subset of nursery tales associated with the Brothers Grimm and Hans Christian Andersen, would receive a totally different message.

There are other, perhaps more deeply rooted, prejudices about animal shapes and animal body-plans, which show up as phobias. Some people, for instance, have an ingrained fear of spiders. They know they're harmless, but it doesn't help. Jack used to run an animal handling course at Birmingham, for students from a variety of local backgrounds: English and other cultures, especially Asian. The students wanted to learn how to deal with a variety of animals, but because they came from different backgrounds, they wanted different things from the course. A few suffered from phobias, which they wanted to learn to control. Some had jobs, or anticipated getting jobs, that would require them to have expertise in microscopy to detect pests in stored food, or to be able to find and recognise parasites attached to farm animals and domestic pets. Some were, or wanted to be, zookeepers or assistants in pet shops. Others were zoology students, vets, and school biology teachers.

Across this whole spectrum, there were surprising phobic regularities. Most of the students seemed to be 'spooked' by a small class of animals, even though they knew them to be completely harmless. Jack would start by allowing each person one phobia: if it was 'spiders' (and for about half the class it usually was), it could include spider-like insects such as cockroaches, but it never extended to relatives of spiders, like scorpions. If it was 'snakes' (and for another third to a quarter it was), it could be extended to monitor lizards with their snake-like heads, and even to big worms; but the so-called slowworm, a docile legless lizard, was acceptable to nearly all of the students. If the phobia was 'rats' or 'mice', and for two or three people each year it was, then voles and gerbils were included, but hamsters, squirrels and rabbits weren't. Phobias for 'birds', including vagrant feathers, were grouped together with 'moths' and 'bats', but ducks, swans, and ostriches were not usually included. Those were the majority feelings: unease or outright inability to be in the same room with a few creatures known, rationally, to be harmless.

Two or three people every year professed to having no animal phobias at all, and some may well have been right. But a few of these turned out to be spooked by frogs or millipedes, or, in a couple of cases, by the discovery of previously unseen tiny mites in food or bed-linen. When the students were asked where they had 'caught' their phobias, a few of them confessed to having been frightened by such animals when they were young. 'Bitten by a dog' was common, yet it didn't lead to a phobia, at least not in that group – it may have done in many people who wouldn't go to an animal handling course. Nor did getting a shock from an appliance, a wall-socket, or an electric wire lead to anything like that spider-phobia, that odd feeling that so many of us have, letting ourselves be made very uneasy by what we know to be a perfectly harmless creature.

When we say that these phobias are deep-seated, we do not mean that they are genetically programmed. Origins of human behaviour are always more complicated than that; the inheritance of abilities or predilections to behave in certain ways is never simply 'built-in' to the genome in the same way that blue eyes or blond hair can be. We mean that the phobias arise as a side effect of the structure of our brains and the cultural development of our minds. In *The Adapted Brain* John Tooby and Leda Cosmides offer extensive evidence that our brains, the vessels that hold and make possible our burgeoning minds, have many different 'organs', which are programmed in various ways as we develop. All of us smile in response to a smile from mother; even blind children smile on being cuddled or spoken to gently and affectionately.

Nearly all of us can learn a language. Phobias may be a side effect of a (hypothetical) 'brain organ' that lists creatures – presumably from the wild environment in which we evolved – that it is best to keep away from. That would have been a useful list for anyone growing up in an environment where many creatures may bite, sting, or make you smell awful.

Phobias may arise because in a protected, perhaps urban, environment, that brain organ has no chance to put together its list from *real* dangers. So instead it invents or collects its list of creatures to be feared by observing what other people react to. In this way, common phobias become self-replicating mental constructs – 'memes', to use the term invented by Dawkins. One reason for thinking this is that these phobias tend to 'run in families', but without any discernible pattern of genetic inheritance. So perhaps we copy anyone who reacts strongly in our vicinity: in the wild that would have been a useful trick, because it would have saved us from getting stung or poisoned by learning from the experiences of others. The process involved here could be like the way we become allergic to harmless pollen or upholstery mites in our protected urban environment. Our immune system, no longer needed to defend us against all the real parasites to which we were prone before the invention of urban hygiene, is desperately trying to find itself something to do.

What has this to do with our concepts of 'aliens' in film or book, or indeed of real aliens? We all seem to have high emotional loading on particular beasts, particular body-plans – like our nursery animal icons. Filmmakers have exploited this by giving us Giant Ants and their ilk, and by equipping their terror-eliciting images with dripping mucus (is our phobia here to avoid catching colds from others?), fang-like teeth, arthropod-ish claws and sometimes jaws. The jaws usually work up-and-down and have *teeth*, showing to the biologically discerning eye that their ancestors were early vertebrate fishes in Earth's ancient seas, which is probably not what the producer intended. Unlike the nursery icons, which we mostly share, these little phobias are much more individual. Nevertheless, the subgroup of westerners who go to the movies, like being terrified, and have a phobia for spiders, is sufficiently numerous for the producers of *Arachnophobia* to have been well rewarded. These phobias interact with our nursery stereotypes, and with our adult superstitions about bats, werewolves and all the other 'supernatural' beasts, so that seekers of phobic images to portray the 'alien' have plenty to choose from.

The Voyage of the Space Beagle (A.E. van Vogt 1950)

The *Space Beagle* is on a mission to explore the distant reaches of the galaxy, where it encounters a series of threatening lifeforms. The first is Coeurl, intelligent and deadly. It is prowling a landscape of jagged rock. The moonless, almost starless night gives way to the dawn of a pale red sun. Coeurl has huge forelegs, tipped with razor-sharp claws. Tentacles grow from his shoulders. His huge head resembles that of a cat. 'The hair-like tendrils that formed each ear vibrated frantically, testing every vagrant breeze, every throb in the ether.' He can normally detect 'organic id', creatures that form his food supply, but now the ether is still.

A metal machine appears, and lands. Two-legged creatures emerge. It is the *Space Beagle*, and the humans spot Coeurl. He immediately realises that they must be a scientific expedition from another star. They know that 'pussy' is intelligent, and they believe that he means them no harm. Pussy knows that scientists can be very stupid.

While the humans observe Coeurl, he observes them. The ship's Nexialist – a multidisciplinary generalist who specialises in making connections – becomes worried about the creature's potential for havoc. When they land on a planet, Coeurl goes missing from the ship, expecting it to be their homeworld. Lots of organic id *there*. But all he finds is the remains of a dead civilisation, a ruined city, the next stop on their voyage.

Then the body of one of the scientists is found. The killer has attacked him, 'possibly with the intention of eating him, and then discovered that his flesh was alien and inedible. Just like our big cat . . .'

Belatedly, they realise that this may not be coincidence. Pussy was right: scientists *can* be very stupid. Later, after they have finally managed to deal with Coeurl, they encounter Ixtl, who can survive in the vacuum of intergalactic space . . .

The existence of a zoo of supernatural entities (ghosts, ghouls, trolls, leprechauns) in the myths of different cultures permitted the elaboration of Edmund Spenser's fictional *Faerie Queene*, the ghost of

Hamlet's father, and Count Dracula, in forms that were immediately recognisable to readers. Cultural myths provided common elements between what the writer dreamt up, and what the reader interpreted. The agreed existence of extraterrestrial life has permitted the invention of 'alien' icons to fill similar roles in moral tales, with just a few that, like the very 'logical' Spock in the first *Star Trek* series, do not have a clear parallel in the supernatural spectrum.

Most aliens in film or story, however, are transmogrified cultural icons. As we said earlier, The Mother Alien is the same as *The Hobbit*'s Dragon Smaug, and both are the same as A.E. van Vogt's Coeurl in *The Voyage of the Space Beagle*. The Ewoks in *Return of the Jedi* are teddy bears (*Return of the Tedi?*), as are Gordon R. Dickson's Ruml in *The Alien Way* and his and Poul Anderson's charmingly overenthusiastic Hoka in *Earthman's Burden* and *Star Prince Charlie*. And ET is the same as later versions of Mickey Mouse and a human toddler. These are not attempts to guess or work out what real aliens would be like; they are the translation of classical story icons into 'alien' models.

The transplantation of 'alien' myth-constructs into the slots previously occupied by 'supernatural' myth-constructs is part of the takeover by 'alien' models of much of western culture. Many good – and many more awful – 'alien planet' pulp SF stories on station and airport bookstalls are really just cowboy stories transplanted. Others are transplants of 'spider', 'dragon', 'werewolf', 'ghoul'; so many of the uneasy-making cultural icons have been lifted into the 'alien' slot that it is difficult to think about alien life without running headlong into these emotional loadings.

Similarly, people who now 'see' aliens in the form of Greys, LGM (Little Green Men), or even BEM (Bug-Eyed Monsters), are locating their fear-images, or sexual-incompetence icons, in this part of the modern cultural repertoire. In the past they would have dredged up equally 'supernatural' incubi, succubi, witches, or ghouls, to imagine their experience with. While the provision of these new cultural 'alien' icons has proved very lucrative for a few people, and very enjoyable for many, it has obscured what real aliens might look like. It has tied this subject to some of our most emotive myths, and made it difficult for anyone to think rationally about real aliens.

Dragonflight (Anne McCaffrey 1968)

The star Rukbat has five planets, and the third has air that humans can breathe. It was colonised, and named Pern. Then the colony lost contact with its homeworld, and was left to fend for itself. The menace of Thread, which dropped from the skies every two hundred years, for a period of about fifty years, became apparent. Thread destroys virtually any living thing that comes into contact with it.

To combat Thread, the Pernese breed huge flying dragons from a native Pernese lifeform, the fire-lizard. The dragons communicate with their riders by telepathy, and can teleport themselves by going *between*. Young dragons imprint on their riders when their eggs hatch, forming a lifelong bond. The dragonriders are organised into a group called the Weyr, united under the Weyrleader.

The Pernese live under a feudal system, and live in administrative groupings known as Holds, each run by a single Lord. The Holds are centred on vast caverns, which create protection against Thread. Dragons eat firestone and then can breathe fire, which is used to burn the Thread as it falls, to protect the fields and prevent Thread overrunning the planet.

The Red Star appears in the sky. The old ways have been forgotten, but the Weyr has ancient records of previous threadfalls, and the Red Star is a harbinger of terrible times to come. Thread starts to fall and battle is joined. The dragonriders re-learn the old ways, but the fight against Thread is long and dangerous. Then it is discovered that dragons can go *between* in time, as well as space . . .

There is also, of course, the tendency to ride on myths that are already well established. Think about dragons. There are Welsh dragons with wings, and Chinese dragons without (there was a lovely little story in *Argosy* by Bradbury, which had these two kinds as male and female). With some imagination, some African and native American rock drawings look like dragons too. The dragon image is what people – especially artists – aggregate together to represent a generalised concept of 'terrifying'. If your alien looks like a dragon, you need much less filmic expertise, or actual explanatory footage, to explain that it's dangerously impervious to weapons and wants to kill you, if not eat you.

Some SF authors have ridden on the dragon myth to good purpose, too: Anne McCaffrey's Pern stories have constructed dragons to help with the menace of Thread falling from the skies. According to later books in the series, the dragons were constructed by people, who made them like dragons because that's what people think of when they want to design flying fighting creatures. The flaming breath is part of the package too. That they were constructed on the basis of local fire-lizards was McCaffrey's cue to the reader to expect the teleportation facility she'd built into those little indigenes. (Jack's cooperation on *Dragonsdawn*, the 'explanatory history' of Pern, came after Annie had already written six books.) Heinlein's *Glory Road* is a piece of romantic adventure, with tongue firmly in cheek, whose climax is a skirmish with slightly-realistic dragons: they blow fire out of the more likely end, for example. But Heinlein knew that he could rely on the reader's 'knowledge' of iconic dragons, and didn't have to explain that they're usually portrayed as having flaming *mouths*.

Pratchett's spoofs of innumerable SF and fantasy tropes rely heavily on the common cultural patterns, the iconic images, of dragons, werewolves, golems, witches, elves . . . You name it, he's got it in the Discworld series. Just as he's got Death with a scythe and a white charger for free, so that he could guy them ever so slightly (the white charger is called Binky). Death, in the real world, is not a thing but a concept; indeed it is a 'privative', the absence of a thing – namely, life. Once Death is available as a character, Pratchett could play with other privatives (like cold or dark or ignorance) and make them 'real' on Discworld. For example, light goes at 600 miles an hour (1000kph) in the strong magical field of the Disc, but we are never told the speed of dark – except that it must be faster than light, because the dark always manages to get out of the way before the light arrives. This joke wouldn't work if Pratchett had to explain to the readers about people who believe that Death is a skeleton figure who collects your soul when you die, and that like dark he doesn't really exist.

In a way, what's going on when authors make such associations is like the English language, with all its built-in overtones arising from the structure and history of the words we use. Mainstream novelists can assume that much of the cultural background is shared by the readers, so they don't need to explain that if a group of oddly-assorted folk are brought together by horrible weather in a seaside hotel, then one of them will be murdered by Monday. It is totally expected by the readers, who apparently are not put off going to seaside or country hotels by this practice. Equally, vampires can be expected to drink blood (or, in Discworld, to have sworn off it, or to be in the City Watch); elves can be fairies with little magic twigs on their heads or the more ancient evil Sidh, the Irish 7-foot-tall devilishly handsome folk – sorry, Folk – who can tell you to eat yourself . . . and you do. These are the 'givens', available for free, with all the appropriate associations, as common ground between author and reader. We should not therefore be surprised that exactly this happens in SF stories and films. But it does mean that we have to excavate your associations widely and deeply to allow you to think about the real biology of aliens.

We start with a well-known story that doesn't have aliens as such: *The Lord of the Rings*. This resonates strongly with ancient sagas – parts of the nearly forgotten *Kalevala* stories, *Beowulf*, things like that – and that's where it picks up so much of its authority. Many of us came into it via the super-cute *The Hobbit*, where the hero is called Bilbo Baggins, so near to English usage but not quite . . . then there's the hairy feet bit. *The Lord of the Rings* is much more of an odyssey, *The Hobbit* more a children's story. But they are full of invented, or borrowed, species. We can imagine hobbits, who are strange child-like people, easily enough. And goblins and wise old men, and elves, are givens of the kinds we've discussed. But Gollum? Ents? Did we have to be educated by Disney animation to imagine talking trees? After the Muppets are we all much happier to import anthropomorphic characters into the stories we tell our children, and they hear?

Jack told his children a long story about Marmaduke the Noble Otter of Elterwater, which had a great assortment of talking animals and many nice little sitcom episodes. Were his children more competent at picturing these elements because of Disney, because of Peter Rabbit, because talking animals are part of nursery furniture? All human cultures put their children through a series of tests – experiences that encourage them to become the kind of adult that will pass on the *same* cultural view to their own children – along with the same tests. The Jewish *bar mitzvah*, a religious initiation ceremony for thirteen-year olds, is an example; school examinations are another. Call this package of tests the culture's 'Make-a-Human Kit'. Are animal icons part of all our Make-a-Human Kits? Surely this must be so. Equally, we think, 'aliens' of the several species which have caught the public imagination are placeholders for ghouls (think 'The Body-Snatchers') or dragons (think Mother Alien) or gremlins (think Gremlins, portrayed as 'aliens').

As we look around at our collections of SF novels, the stories about iconic aliens seem to fall into familiar and ancient patterns. One such pattern is common to lots of other kinds of story – detective stories, erotica, doctor-and-nurse, true confessions . . . Evening classes in novel writing often begin by stating that 'there are really only seven basic stories' – the exact number is debatable but seven is comfortable – and go on to list them. An example is 'Romeo and Juliet' – boy meets girl, boy loses girl, both stop thinking straight, tragedy ensues. This tale, thinly disguised as *Casablanca*, was spoofed in an episode of the British TV humorous SF series *Red Dwarf*: 'Android meets blob, blob meets blob, android loses blob . . .' Supposedly, the ancient Greek playwrights employed all seven formats, but whatever the literary merits of such a contention, the key point here is that *Homo sapiens* is a storytelling animal (see *The Science of Discworld II: The Globe*). The list of genuinely different kinds of story is necessarily short, but the possibilities for exciting combinations and nuances are infinite.

Another way that mythic stories illuminate our thinking about aliens, factual or fictional, is to alert us to which icons they replace. As we've seen, our culture has many icons that fit into stories, and we are exposed to them from the nursery. They include the Hero, the Princess, the Waif, the Ogre, and so on. As we grow up, these icons become more complex, and we read and hear more subtle variations on the themes. Just as 'sly' is what the fox demonstrates in the nursery tale, so the ogre becomes 'landlord', 'boss', or 'policeman', and our nursery lexicon generates a naive kind of sociology. Classical tales had other icons. The Greeks had gods, the Elizabethans had innkeepers and faery queens, and the Victorians favoured the explicitness of Mrs Doasyouwouldbedoneby. Vampires and werewolves made a successful transition to the movies. The icons in these tales were always mysterious: Cupid and Puck, witches and warlocks, demons and angels. It was natural that as our scientific cosmology replaced Mount Olympus and Faeryland, the stories came to contain denizens of the new mysterious reality, so the icons transmuted into aliens of various kinds. Compare Bradbury's Martians to iconic angels, Burroughs's Barsoom to the *Arabian Nights*.

Because such fictional aliens fit so precisely into iconic niches in western cultures, Aldiss has proposed in *Nature* that humanity has simply replaced the old images of the bizarre and the foreign, the mysterious and the outlandish, with these newly minted 'scientific' images. He insists, further, that those who are trying to make up a new science of the alien have merely reified – given credibility and substance to – nursery myths and cultural icons, and they need to rein in their imaginations because aliens are no more real than Bre'r Rabbit or a skeleton Death with a scythe. The biology that suggests that there are lifeforms on other planets, says Aldiss, is simply wishful thinking; scientists should keep their noses to the ground and get on with BSE and foot-and-mouth disease. We agree with most of his thesis here, especially when it comes to 'astronauts took my baby' or magic alien pyramid-builders, examples that fit his contention perfectly. But we differ in one crucial aspect: we don't think that the myriad aliens living in other galactic solar systems will disappear because they resemble Terran myths. No doubt we resemble some of theirs.

However, there is a great swathe of cultural-icon 'aliens' in SF novels and, especially, films, which we disbelieve nearly as much as we disbelieve humanoid aliens. We don't find them credible because they are lifted straight out of our cultural myths. We have already observed that ET is an anthropomorphic image: it fits our brain-template of a three-year-old infant, so is as unlikely as Spock to be instantiated in the real world. (We don't object to the magic bits – it's a kids' story, after all.) *Alien*, in all four films, is nothing more than a brain-template terror, and so are the various demonic women of alien extraction who do nightmare things to men. The resemblance of these beings to werewolf or vampire icons is much more obvious than the filmmakers' attempts to portray extraterrestrial lifeforms. It *is* possible to portray alien images on film: apart from the terminally stupid plot, the *Independence Day* alien was quite clever, both in what it showed and what it didn't. You couldn't easily see what was biological, what was appurtenances, clothing or weapons. The aliens in *Starship Troopers* weren't too bad; they were a bit too insectile, but maybe exoskeletal shapes have to be like that.

No, what we have in mind here, which drives us to agree with Aldiss more than we otherwise might, is 'Ewokery'. People like bears. Bears are popular in zoos and in the nursery. It is futile to discuss which came first; it was probably bear-baiting anyway. Bears look intelligent and, unlike owls, they really *are* quite intelligent. They are furry and cuddly, and they are (mostly) not carnivores. They can be portrayed as mad about honey, and Pooh Bear is a delightful icon of this ilk; so is Paddington Bear, in somewhat the same vein. We can think of half a dozen bear-like SF aliens, several of which we've already mentioned, and we don't very much mind the authors freeloading on the iconic mental package that they get for free when they say 'bear-like'.

They should be conscious, however, that they may import too much. We think of Axel Munthe's story of the tame bear that kept escaping from his stake in a Black Forest village. His owner was forever tempting him back with honey, or dragging him back and administering a beating. One day he found his bear on a forest path, not far from the village, and tried to chastise him along the path back to his stake. He failed, but when he arrived back at the village everyone assured him that the tame bear had been there all day, good as gold. The owner had been chastising a wild bear. Our culture is replete with such bear stories, and George Lucas *needn't* have made his Ewoks look like teddy bears.

Pigs, too, can be cutesy. Harry Harrison made a grand joke of this when he invented the Men from P.I.G. and the Men from R.O.B.O.T. The agents from P.I.G. have intelligent boars to help them overcome the alien army, while those from R.O.B.O.T. are the reverse of intelligent. All good clean fun, no harm done. More upsetting are the piggy aliens in Robert J. Sawyer's Starplex series; the plots are quite exciting, but not enough attention has been paid to biological plausibility. The pigs have four eyes, four ears, and four nostrils, arranged in pairs, but apart from that they seem to have orthodox porcine anatomy. The parochials are doubled up, the universals are not considered. And the character of these aliens rides heavily on nursery pig tales, too.

Cats are responsible for the greatest offence in this regard, though. Cats clearly possess so much of the psyche of many of our female authors, that these writers are driven to personify their aliens as feline. Even McCaffrey fell into this trap with the Doona series, though she claims it to be a spoof. C.J. Cherryh wrote a series of exciting stories whose major flaw is that cats are the heroes. Gwyneth Jones's *Divine Endurance* features a cat alien that cannot possibly be extraterrestrial. Fritz Leiber employs a green cat as alien in *The Green Millennium*, but the story is designed for humour, not hard SF. Niven's Kzinti are far too big-cat in anatomy and supposed – actually mythic – psychology; but despite this minor flaw many authors have provided readers with enormous fun in the *Man-Kzin Wars* series. When we saw Chewbacca, the Wookie sidekick of Han Solo in the *Star Wars* series, we at first were tempted to blame Niven – but then we saw Jar Jar Binks in the 'first' episode, and realised that film producers can be terminally stupid without needing a model.

More forgivable are Clarke's demonic aliens, who come to supervise our rise to transcendence in *Childhood's End*. They are forgivable because there is an explicit suggestion in the story that humans possess that icon for demons because those very aliens were present at our earliest cultural steps. Similarly, several stories have portrayed 'angel' aliens as visitors in Biblical times, or even genies-in-bottles in ancient Arabia. Ghosts are plagiarised in several SF stories, most obviously on film – the temptation to do special effects based on the audience's 'knowledge' of ghosts is just too great. The Jawas in *Star Wars* (now *Episode 4, A New Hope*, but once the one and only *Star Wars*) with their cloaks and luminous eyes, are clearly ghosts, just as the beasts of burden are elephants with dinosaur-glaze.

There is another kind of cheat that we're ambivalent about. It is to have the alien represented by a machine, perhaps an anthropomorphic machine, or in the limit an android or even a constructed human. This last ploy too often relies on *Jurassic Park* type technology ('We have a sample of their DNA here, Splxxyll, can you construct a human for us by tomorrow afternoon when we arrive at Earth?'), but we agree that galaxy-spanning aliens will be able to do all kinds of things that seem magic to us. We just think it's a cheap trick for the author to use this kind of plot element.

The collective entity Jack&Ian has a little fantasy about iconic aliens. Imagine some *real* aliens arriving here (camouflaged: perhaps dressed up as circus animals or university professors). They study our iconic alien portrayals, and find out how to get into our good graces, by working out how those icons expose our weaknesses. What they would pick up, we believe, is that humans are 'narrative animals', suckers for a rollicking good yarn. (In collaboration with Terry Pratchett we developed this idea at length in *The Science of Discworld II: The Globe*.) We pick up relativity easily, because it has a story and an iconic hairstyle to go with it, but quantum theory has never achieved that kind of popular recognition because there is no comparable narrative. Now throw in a second ingredient: Jack, who loves western opera, is completely thrown by the plots of Chinese opera. You would think that Italian operatic plots are silly enough ... they are, but it's a silliness congruent with western culture, and the silliness of Chinese operas simply doesn't fit. The Chinese are narrative animals too, but they have very different narratives, more like TV soaps that go on interminably than stories with a beginning, a middle, and an end. Our fantasy now continues: how different would those opera plots, western and Chinese, look to a real alien? Quite conceivably, a real alien wouldn't understand 'narrative' at all, but would hang its extelligence on some quite different concept. And then it wouldn't be able to communicate with us at all.

How many *really* different ways are there to interact with the universe and make technology, do you think? Seven?

9

MODELLING ALIEN ECOSYSTEMS

A TOUR BUS IS *conveying a flock of virtual tourists, about a thousand of them, to one of the most productive ecosystems they will ever experience. A sign on the bus says 'Parasite Park'.*

'Here two alien-to-each-other biomes come into intimate contact,' says Cain, back at the base, as the scene is relayed to him. The tourists look up, towards the light. Gigantic green blobs, some compact, some feathery and fractal, are dividing at high speed. They arrange themselves in vast herds, packed closely together. 'Those beasts duplicate themselves as fast as they can,' Cain explains, 'in order to exploit the nutrients from below and the light from above.'

There are long squirmy things, too, which keep their heads in the oxygenated water and their tails in the reducing soup below. Suddenly the green blobs are attacked by a swarm of tiny creatures, like bees around a whale. 'Those,' says Cain, 'are parasites, for which this park is justly famous.' Before their eyes, a blob dies, collapsing into a vile green sludge. The squirmy things home in fast, like vultures, to mop up the remains.

A huge, slow-moving creature comes into view: one corner of it fills the windows of the bus, cutting out the light. It is a gigantic cone, made from material like concrete; its softer insides spill out from the rim. The improbable beast is grazing on the green blobs, chewing out a path through the massive herds. As it passes, the blobs at the sides of the path, which have avoided being eaten, divide to fill it in again.

Each grazer's digestive gland is full of different kinds of fluke larvae, chewing away at its cells. When the grazer is eaten by a flyer (and about a tenth of them are, every day) the larvae become flukes in the flyer's liver and gut, and use the flyer's energy to make billions of eggs, to scatter back into the water.

'More than half of the energy is used to power that incredibly productive set of parasite cycles,' says Cain.

Abel pulls all the tourists together again, and the bus hitches a ride under the feathers of one of the flyers – the locals call it a 'seagull' – as it flies off up the Thames Estuary towards Gravesend, replete on its diet of algae-eating mud-snails.

Many sciences, in their early stages, go through a collecting phase. Botanists collected flowers, entomologists collected insects, lepidopterists collected butterflies, palaeontologists collected fossils. Today's biologists collect DNA sequences. These collections are great fun, but they are only the first step towards scientific understanding. Long lists of butterfly specimens do not, of themselves, tell us much about butterflies. As more and more information flows in, it has to be organised and systematised. Moreover, this has to be done on more than one level. Butterflies cannot be understood on their own: you must also study their food plants, their predators, and their parasites. And then you have to study their predators' parasites and competitors, and their parasites and competitors . . . so that pretty soon, a large chunk of the ecosystem is involved. By this stage, you're no longer thinking about individual organisms, but whole *systems* of organisms. And if an organism is difficult to understand, a whole system of them is even harder.

Xenoscience has not even reached the collecting stage yet, unless you count the odd pile of moon rocks as constituting some kind of start. But already there is a need to think about aliens on the level of ecosystems, as well as individual organisms, and that's what we'll discuss now. There are several reasons why we might wish to carry out this exercise. One – the commonest – is to provide the background and context for an SF extraterrestrial plot. Another reason is suddenly becoming very practical, as we prepare to investigate Jupiter's moon Europa.

We need to know what equipment to take along.

It's a long way to Europa. If we get there, and suddenly discover that we need some gadget, a phase-contrast microscope perhaps, but didn't bring one, then it will take a long time – about two years – to send one. By the time the gadget arrives, the beasts in the water sample that it was supposed to look at will have died; and even if not, the ecosystem sample will certainly have degenerated. So some prior thought, imagining the possible ecosystems of Europa's ocean and what apparatus we'll need to sample it, will be time well spent.

There is a third, academic reason, too. It concerns what we mean by 'understanding' a system like terrestrial evolution. How do we put evolutionary events in scientific context?

In *The Science of Discworld* we answered this question by thinking about some actual historical event, for example the assassination of Abraham Lincoln. There are several ways to view this event. The most boring of history books will simply tell you what happened, in detail: the name of the assassin, when and where he was born and educated, the calibre of the bullet and make of gun, some history of the theatre, the name of the play – Mrs Lincoln's response probably won't be mentioned. More interesting accounts may embed the event in a better historical context: how many other attempts on his life, how many groups were disaffected at that time, was John Wilkes Booth really pro-slavery or just anti-Yankee, how was it that Lincoln died only the next morning when he had been shot through the head? Perhaps the most interesting accounts, to us at least, are those that put the death into a context of possible outcomes: what meetings did Lincoln have booked in his diary, what effect did his demise have on contemporary politics, how might today's world be different if it had not happened, or if he had survived? The exercise of imagination, of conjuring those worlds that didn't quite happen, the geography of the paths not taken, enable us to value the likelihood of the path that *was* taken, to estimate the extent of contingency or intent in politics. Lincoln is less interesting than the phase space to which his history belongs. What is the geography of Lincoln-space?

In the same way, Jack's reconstruction of an Earth that was not hit by the K/T meteorite, contributing to the background of Harrison's *West of Eden* trilogy, tests our understanding of the evolution of mammals after the dinosaurs' demise, and of the extent to which something like the human would appear. The convolutions of that history, getting some early primates to South America and then having them produce quasi-humans (with different genitalia – that is a big difference between African monkeys and apes and the South American forms – read *West of Eden* again with new eyes) was a complete cheat, of course. If the Earth's evolutionary history had taken the path without a K/T meteorite, then there could have been no scientifically credible route to human beings.

Nevertheless, the attempts to warp possible histories toward humans taught us a lot about the crucial events in humanity's African history. So now we know more about human history as a result of constructing the spurious *West of Eden* one. Similarly, a history of the US in which Lincoln was not assassinated (and there are many – for instance *The Lincoln Hunters* by Wilson Tucker) would teach the author a lot, and the readers more than a little, about the course and importance of subsequent events, and the extent to which they have influenced our present human world. In SF, stories of this kind are called 'alternat(iv)e worlds'. The good ones always lead to a deeper comprehension of our own world.

The construction of convincing, consistent alternative worlds is a good way to test our understanding of evolutionary trends and mechanisms. On the small scale, this kind of scientific testing of evolutionary theory could go like this. First, we test our knowledge of the evolutionary patterns and processes on Earth by taking a sequence, a little historical snippet, and see if our predictions agree with what actually happened. This is like testing out a financial theory, about the prediction of changes in currency exchange rates or prices of shares, by looking at a real historical chunk of sequence, one you've not seen before, and then seeing whether you can 'predict' what happens in *its* future. Its future is the past for you, so once you've recorded your prediction to avoid cheating, you can see if you got it right. This kind of 'postdiction' is a stringent test of your predictive ability, and so of your understanding of what makes the system tick.

But before you can perform such a stringent test, the first scientific requirement is to know what the system *might* do; you need to construct a phase space of imaginative possibilities. That's the difficult bit, the kind of thinking that scientists (should) get paid for – the actual stringent test can be done by the hired help, and usually is. That is what the Lincoln example above, and the *West of Eden* exercise, demonstrate. Imagination is the necessary element in scientific understanding.

To illustrate this process, we'll consider a question that is extremely relevant to the existence of genuine, factual alien life. *What might we find in Europa's ocean?* Will we need to take guns, as the Victorians would have done? Harpoon-guns? Nets and cameras that can be deployed through a half-metre wide ice hole twenty kilometres long? What pressure must the cameras stand? What lighting will illuminate without destroying – if any? What about pumps and sieves? This kind of assessment cannot be left to engineers alone, because there is an enormous literature – and much practical experience – that tells us that biology is genuinely fragile in unexpected ways. We have two sets of problems, then, which inform each other. The first set concerns the equipment that we might take to sample the Europan sea, and is constrained by weight, by materials and energy requirements, but particularly by the requirement that it should be kind to any life – or even any complicated chemistry – that we might find. The second is that we should make informed guesses about what that life will be like, in order to be fairly sure that we can handle, sample, or photograph it.

The things we have to take depend on what kinds of life we expect to find, so the two questions are inextricably linked. The most pessimistic anticipation is of very complex chemistry, like that of watery exudate lying on mine tailings on Earth – but with no lifeforms taking a tithe of the chemical energy as they do here. In this case, there would be chemical 'systems' on Europa, but their analysis would excite only chemists. We would be very surprised, however, if there were not overt organisms, explicitly living things, at least where the energy is, at the bottom ('top' in 'Monolith') of Europa's seas; later in this chapter we'll develop the arguments involved. The least interesting scenario would be if Europan organisms had similar chemistry to Earth archaeans or bacteria. That would *not* imply that such chemistry is the only way to make life (if only because we can already envisage numerous alternatives). A far more plausible explanation would be that there had been an exchange of early lifeforms among the planets of the inner solar system, and that the Earth had been infected from Europa or vice versa. In *Wheelers* we had one of the scientists find a piece of a diatom (a common marine alga with an ornamented silica shell) on Europa's ice. Bearing in mind the big splash about 60My ago, we don't think it's unlikely. However, the thick ice layer has probably kept Europa's ocean from being infected, so we expect to find a different experiment in the origin of life when we get there – even though that protective ice layer does break up from time to time.

Has Europa produced anything comparable to our diversity of life? We doubt it very much, but would be happy to be proved wrong. On the most optimistic assessment, the ocean would be full of bacteria, like Earth's, with many filter-feeders and grazers on the rocky seabed (the 'roof' in 'Monolith'). There *could* be 'fishy' forms, streamlined feeders on the grazers and on each other, with shapes similar to Earthly fish, because shape is determined by hydraulic swimming constraints. Probably not, though, for evolutionary reasons. It took Earth, with a much more energetic geology and sunshine, nearly three billion years to get any complex eukaryote life, and nearly another billion to get a Burgess Shale explosion of multicellular forms (the Cambrian Explosion). There is no reason to suppose Europa to have life that evolved faster, and many to suppose it evolved more slowly, so we probably won't find 'fish' – they are only shadowy potentials to be expected in Europa's far future.

A more likely possibility is enormous jellies, their lives interwoven with the bacterial ecology so that we would not be sure if they were symbionts, parasites or food (as in the various worms around our own black smokers, which have bacteria in their cells in all those capacities). Only in the last twenty years or so have we discovered that the middle depths of our own deep oceans possess a prolific and varied biota of this nature. The phylum *Coelenterata*, containing the corals, jellyfish and the wonderful siphonophores as well as the humble *Hydra*, is probably about to be expanded to include several more classes of creatures, the likes of which have never been seen directly by human eyes.

Television cameras, with very special lighting, have exposed these bizarre organisms to our surprised gaze. Surprised, because all of the nets we have used, which picked up all those strange gulper-fish that have eaten fishes bigger than themselves, little angler-fishes with lit-up lures on their foreheads, phosphorescent shrimps, and pteropod sea-snails, simply shredded the most common living creatures and left the bits behind.

Those creatures are desperately fragile, lace-like nets of jelly. Perhaps, like the siphonophores of the plankton, they are long washing-lines of different kinds of polyp, catching anything they touch by paralysing it with sticky nematocysts – cellular poison darts. But the likelihood is that they are totally new creatures to our eyes, probably coelenterate but possibly even remnants of the first multicellular life on this planet, the Ediacaran fauna. The fossil evidence for those baffling organisms comprises great flat symmetrical discs, found in late pre-Cambrian rocks; it is assumed that they have left no modern descendants, but we don't really know.

The lesson here, of course, is how little we know of some of the most common life on this planet, never mind Europa. For years our apparently very sophisticated sampling methods told oceanographers that the middle depths of the deep oceans are vast, empty, sterile wastes with a thin rain of debris, shells and bodies and faeces, from the lighted zone above. This view has now turned out to be totally wrong; our nets have simply been inadequate to sample the characteristic creatures of those depths, which are fragile jellies. How much less successful are we going to be, sampling the much deeper ocean of Europa though a tiny hole several miles long? The problem would be to find any organisms, perhaps even to see any. And that's where the question of apparatus makes its appearance.

Here is an example of the preliminary thinking that should go into deciding what to take. What sort of pump should we use to extract water, and the micro-organisms that might be contained in it, from Europa's ocean? It may sound like a purely engineering question, but there's more to it than that. Experience with the pumping of seawater in marine aquaria suggests that piston/valve pumps and regular rotary pumps tend to 'bruise' seawater. The water that used to be pumped from the sea for large public aquaria in seaside towns was, alas, a suspension of maimed, dead-and-dying plankton in a thin soup of their body fluids. In such a system even normal bacteria begin to look grossly abnormal, and the photosynthetic ones are killed after a so-called 'recovery period' of days in huge dark holding tanks. The holding tanks have a population of filter-feeders which clean up the worst of the mess, but how representative of the sea is the final effluent that is sent to the aquaria? It is not suitable for the maintenance of delicate marine organisms such as small squids, for sure. So we shouldn't simply pump up Europa's water with the usual kinds of pump, and hope to see what lifeforms used to be in it.

How will we sample Europa's sea, then? It seems that 'mesh pumps', where the liquid is carried between the teeth of cogwheels in volumes of about one cubic centimetre, are least damaging to planktonic sea creatures, because most of these organisms have less than a thousandth of that volume. These gentler pumps should therefore be standard. Except . . . what if Europan organisms are mostly long and stringy? Or spherical lumps of jelly? There is a problem even for those. Human eggs, recovered from ovarian follicles in their coats of cumulus cells some millimetres across, can be damaged rather nastily by being drawn up a tube which is a couple of millimetres wide, even though the egg itself is only about a tenth of a millimetre across. The damage occurs because the water forms rolling tori (the plural of 'torus', shaped like a doughnut with a central hole) against the walls of the tube as the jelly mass is drawn through it, and these tori alternate their direction of rolling, causing positive and negative pressure oscillations that shake the egg to pieces as it moves along the tube. This problem can be solved by drawing the egg slowly.

What metal should instruments be made from? In many people's minds, especially engineers', stainless steel is the epitome of cleanliness and sterility. Is stainless steel a good material for handling Europan lifeforms, then?

Absolutely not.

What you have to ask here is: *why* is stainless steel stainless? The answer is that it continuously releases heavy-metal ions (nickel and chromium at best, molybdenum and cadmium at worst) as a result of electrolytic effects on its surface. As a result, stainless steel is a very nasty material if it is used in saline solutions such as body fluids or seawater: the heavy metals damage biological organisms. Stainless steel forceps, for instance, are *toxic* to human eggs and embryos, and to much microscopic marine life. Including, very probably, any that lives on Europa. Polypropylenes and teflons are not toxic to our kinds of life, although some samples release nasty plasticisers or detergents introduced in the manufacturing process . . . but it will not be amusing to discover, too late, that they are just the polymers that Europan lifeforms have evolved as storage products or skin films.

Solaris (Stanislaw Lem 1961)

The planet Solaris orbits two suns, one red and one blue. At first no humans visit it, because of a widespread belief in the Gamow–Shapley theory that a double star's planets cannot harbour life. The reasoning is that the complex gravitational dance destabilises the planet's orbit, leading to wild fluctuations in temperature, which life cannot survive. Theoretical calculations confirm the instability, but observations show that Solaris's orbit is *stable*. The Ottenskjöld expedition is sent to make a study of this anomaly. It finds an ocean world dotted with low-lying islands. There is no oxygen in the atmosphere.

A second expedition discovers the secret of the planet's impossible stability: active movements of the ocean. The biologists analyse the ocean and find complex chemistry: they declare it to be a 'pre-biological' entity, in effect a single giant cell that surrounds the planet with a miles-thick colloidal envelope. The physicists propose that the ocean is a complex form of organisation, a 'plasmic mechanism' capable of modifying the planet's gravitational field.

Further investigations lead to the strong suspicion that the ocean is best considered as some kind of lifeform. Moreover, it does not affect the planet's gravity after all: instead, it controls the orbit directly by influencing the flow of *time*.

Then the researchers from Earth start to receive 'visitors' in superficially human form, but they are not fully convincing. Their clothes look right but the buttons don't actually *do* anything, their skin is too soft. The researchers realise that the ocean has made contact with them.

It is not only alive: it is sentient.

That's a tiny bit of the prospective biological thinking about the engineering. What about the biology itself? Here we have to start by questioning even the most basic of terrestrial life's generalities. For a start: has Europan chemistry taken the same origin-of-life paths as ours? In this case we expect the answer 'yes', in general terms, because the iron sulphides, heat, water, and clays involved are likely to be much the same on the sea floors of both worlds. Has Europan life formed discrete items, organisms, rather than a vast collective oceanic entity, such as that found in Lem's *Solaris*? The latter may seem outlandish, but that's another word for 'alien', so let's not dismiss it out of hand. This *is* xenoscience, right? On evolutionary grounds, it seems virtually impossible for a Solaris-style single-beast ocean to occur right at the beginning. Perhaps something like it could take over later, though. However, we are far more likely to encounter Europan life in its early stages, because those stages usually hang around for a very long time. Moreover, separation of competing entities would be likely to subvert a collective entity, by taking parts of it away to compete with other parts for substrates. So individual organisms make more sense. What would they be like? We must expect to find organisms with, say, tens of thousands of molecules organised into semi-cyclical chemical networks, as in terrestrial bacteria and archaeans. We would be very surprised, then, if there were not bacterial-equivalents on Europa, but we don't expect some oceanic analogue of Gaia.

Europan heredity is probably not mediated by DNA/RNA. Indeed, if the biochemistry turns out to be very close to our system, then the most likely explanation would have to be some kind of interplanetary infection, probably caused by impact splashes. We don't expect DNA because many other iron/carbon/sulphur compounds could play the same universal role, serving as long linear backbones for hereditary molecules. Moreover, elegant as our long-linear systems have turned out to be, there may well be other two- or even three-dimensional systems that can constrain synthesis of molecules in the next generation.

Will Europans have amino acids making proteins, sugars making polysaccharides, fats making membranes? Just as there are alternatives to DNA, there must surely be many alternatives to proteins. Polysaccharides are unlikely to evolve independently on Europa, for similar reasons. But Europans will probably have fats making membranes, because a hydrophobic membrane to separate different chemical compartments is such a useful – and apparently easy, hence universal – trick. Our lifeforms achieve this result in several different ways. It is, however, *possible* that amino acids and sugars are somehow basic: they turn up 'spontaneously' in primal soup experiments when energy is added to early Earth atmospheres. Alternatively, though, do we just know a lot about them, and look for them in origin-of-life simulations because *our* life has capitalised on them?

And what happens if everything occurs underwater, as it would have been on Europa? Our best guess is that there will be Europan bacterial forms, but their chemistry will probably be unlike ours in surprising ways. Their reproduction, on the other hand, will probably look much like the variety of prokaryote reproductive methods employed here: there will probably be little spheres, little rods, and little spirals – all busily multiplying by dividing, because that is what simple three-dimensional geometry dictates.

The energy sources for Europan lifeforms will be less potent than photosynthesis, but perhaps this will not be as restrictive as Earth's lifeforms, with their high-energy throughput, lead us to imagine. The major problem for terrestrial photosynthesisers on land is not to take up light, but to reflect the excess energy – most of them use only 2–4 per cent of direct sunlight, and even shade plants get 5 per cent at best. A possible comparison is with Earth's decay communities, such as the collections of organisms that break down dead plants in forests. These communities have produced many organisms more complicated than simple prokaryotes, which resemble real eukaryote cells, and even fungi. At least seven kinds of fungoid creatures have evolved on Earth as separate origins from much simpler forms, so we would expect several kinds of fungoid life to have evolved on Europa. But not fungi.

Europa's energy budget could well start with black smokers, where the hot rock meets the seawater, much as the vents in Earth's deep-ocean trenches provide a start for whole ecosystems. On Earth, most of the 'higher' creatures living at the vents – like fish and crustaceans – evolved for millions of years in an oxygen-rich environment before they were recruited to the dark hot sulphurous depths. However, a growing body of microbiologists believes that terrestrial life may initially have been powered by these hot sulphurous systems, and that the archaeans and bacteria – whence all of Earth's lifeforms – may have started there. So we should not expect Europan life to be especially 'slow' compared with ours, even though its energy base is so restricted compared with photosynthesis.

On the other hand, much of the vitality of Earth's modern smoker faunas is maintained by chemical energy derived from aerobic/anaerobic sulphurous interfaces, the interaction of highly-reducing hot water with cold oxygen-charged water – and the oxygen comes from photosynthesis up at the surface. So maybe we should expect life on

Europa to be less energetic, after all. But we should also expect to be surprised.

In early prokaryote history on Earth we find evidence of complex structures, stromatolites, built by (and from) filamentous lifeforms. Many bacterial families have some filamentous members, so we would expect that route to have been exploited on Europa too, as well as colonial forms. We would also be sure to find 'grazers', the creatures that live on bacteria. On Earth there are giant underwater bacterial towers, for example in the Baltic Sea. Here bacteria have formed immense structures, which benefit their metabolism by separating hot reducing seawater from cold oxygenated seawater. It's a kind of bacterial analogue of a termite's nest with air-conditioning vanes in the basement, and it is similarly a collective construct made by large numbers of simple organisms. We can't be sure whether such bacterial towers will be found on Europa, but it looks likely. Many of Earth's bacteria exploit such interfaces: in every old farm pond there's a grey film on the mud, whose basis is a kind of bacterial rod with one end down in the anaerobic black mud and the other up in the light and the oxygen. The famous *Amoeba proteus* is perhaps the best known grazer of these.

Let us suppose, then, that such interfaces will be exploited in a similar way on Europa, leading to the generation – evolution – of specialist 'amoebas' that earn their living by grazing these interfaces. These systems are probably not as energetic as Earth's black smokers, because there's not likely to be much free oxygen in Europa's ocean. But we wouldn't be at all surprised to find 'animal' organisms, pumped up to high energy using, say, sulphur chemistry. Reactions involving sulphur dioxide, sulphur trioxide, and sulphuric acid can channel energy reasonably efficiently: Earth's use of free oxygen is only twice as effective.

All this is conjecture, but within the next few decades, one or more of Earth's nations will send a probe to Europa, and we'll find out. The best first shot would probably be some kind of camera-probe, very well sterilised . . . but also with no *dead* bacteria on or in it. The probe would have to be able to detect micro-organisms and larger ones, with various kinds of illumination. The microscopic detector would need to be capable of dark-field microscopy, to make transparent organisms visible. The probe would have to be very mobile, yet very stable. We suggest a 'gentle' pump for jet propulsion, and some kind of gyro-stabilisation with an inertial navigation facility so that the probe could return to promising locales. The lights would have to be very low powered, illuminating only the regions that the optics are looking at, with absorbing plates to prevent the light escaping into the dark ocean around. Get the design right, and we'll gain our first glimpses of an alien ecosystem. Get the design wrong, and we'll see nothing, because our apparatus will have destroyed the things we want to observe. So, for entirely proper reasons of 'serious science', we *have* to speculate about Europan life. And we *have* to be imaginative, not conservative, when we do.

Legacy of Heorot (Larry Niven, Jerry Pournelle, & Steven Barnes 1987)

Rich, comfortable, but overcrowded Earth sends colonists to the fourth planet of the star Tau Ceti, which they name Avalon. The colony is based on an island, Camelot, 50 miles (80km) away from the main continental landmass. The ecology seems benign, with silver rivers full of fish-like 'samlon' and golden fields. The colony prospers, and all seems idyllic.

Only one colonist, Cadmann Weyland, worries about security, putting up a perimeter fence, laying a minefield and barbed wire. The remainder of the colonists see no reason to take such precautions, and Weyland is a figure of fun.

Then a dead calf is found, its bones neatly sheared off. The colonists tie up another calf as a lure, and wait in the darkness with infrared goggles, to see what comes to call. The interloper resembles a Komodo dragon, but with a thicker, spiked tail, like that of an ankylosaur, and a rounder head. It moves like a boneless crocodile, it *ripples* ... Then infrared light flares at the centre of its body, and the monster accelerates to an incredible velocity. It attacks and kills one of the observing colonists.

The colonists name the creature a *grendel*, after the monster in *Beowulf*. Grendels store a chemical, *speed*, which carries a large amount of oxygen, so it can react at a rapid rate to power their unbelievably fast movement.

The humans go on the offensive. They also realise that the island's ecology is drastically simplified – grendels, samlon, plants, pterodons high in the mountains, not much else. There are insects, all of which fly: no crawlers. It is like Earth must have been, soon after the K/T meteorite hit. They come to the conclusion that the grendels must have swum over from the main continent, or were carried across on driftwood.

The colonists pursue one of the grendels into an underwater cave. One of them is killed, but Weyland blows the monster's head off. When laid out, it is fourteen feet long.

'Is the nightmare over?' Weyland asks himself. 'Or just beginning?'

In order to illustrate other aspects of alien biology and ecology, and how they arise in xenoscience and SF, we will now take our speculations much further, and think about a full-blown alien ecology: the SF ecology of the main continental ecosystem of Tau Ceti 4, which appears in the second of Niven and Pournelle's Heorot books. Nature is not sentimental about parent/offspring relations, and neither should xenoscience or SF be. In the front of *The Legacy of Heorot*, Niven thanks Jack for the genesis of the McGuffin in the tale: *Xenopus*, the aforementioned 'frog with nasty habits', which uses its own tadpoles as a food store and eats them whenever it feels the urge. This terrestrial system works well, and is one of many such biological transformers.

A converse example is Alaskan salmon, where the parents die after mating and laying eggs, and their rotting carcasses fertilise the very pure water, run-off from the mountain snows. Their smell calls out midges from over-wintering pupae, and these lay eggs on the water surface. The hatchlings feed and grow on the bacterial bloom, and are big enough as larvae fifty days after the salmon-run to serve as food for the hatching salmon.

These are two of the ingredients that led to the genesis of the complex ecosystem of the Avalon continent, after Niven, Pournelle and Barnes had had fun inventing a reasonable – but not very deep – ecology for the island of Camelot, where the Earth colony had built its settlement. We don't want to give away the climax of *Legacy of Heorot*, but we have already told you that the 'samlon' are the grendel's larvae, which it normally eats. After the novel had proved a great success, the authors asked Jack over to California to craft the background to a forthcoming sequel, on the basis of what they had done with the first. Pournelle rang first: 'We want you to do a proper job on the ecology of the continent, which they've only made mapping and mining expeditions to.' Jack replied: 'Do you want me to do an ecology for us to go adventuring into, or do you want to start off with a skeleton plot, and then I'll build the ecology around it?' There was a long silence from Pournelle, then: 'You invented this speciality of yours. Come over prepared to do either or both.'

Jack's speciality of Alien Design functions in two contexts. Sometimes (Harrison's *Eden* and Gerrold's *Chtorr*) the authors want a novel ecosystem to adventure into. Others (McCaffrey's *Dragons*) want their plots, including (retrospectively) those of any existing books in the same series, rooted in a credible 'scientific/historical' background. In many ways *Heorot* was both, as Pournelle had seen: the authors had invented what an ecologist would call a depauperate (impoverished) ecology on the island. The intention was that the (brighter) second generation of Earth colonists would adventure on to the continent, where the first book had suggested plot lines. Niven, Pournelle and Barnes, spent a very tough week with Jack, taking hints from the first book and building them into the ecosystem on the continent. There was only one kind of grendel on the island (so far as we had been told), and there was evidence that as a successful carnivore it had pretty well cleaned out all the valleys; there were, however, little furry joeys (altogether too mammalian) on the upper slopes, and there were flowers.

Grendels are what reproductive biologists call r-strategists: many offspring, not finely tuned to any particular ecological niche. We decided that what kept the grendels from undergoing a population explosion on the continent was a great variety of other, specialist grendels. A successful way of life like that would have taken over as a great flowering, but then the major competition would have been other grendels, and evolutionary pressures would have specialised them not to compete for the same resources. That way of life would have been most subject to subversion by grendels that ate other grendels' babies instead of their own. So we invented several kinds of more 'K-strategist' grendel that made their own lakes, sometimes with their own bodies (dam grendels), and sometimes by building them (beaver grendels). K-

strategists are more highly specialised, and look after their babies. Some of our *K* grendels specialised in taking other's samlon from such lakes, some specialised in using those lakes for their own samlon, camouflaged as the lake guardian's own (nest parasites), and so on.

The Camelot-type 'general-purpose' or basic grendels were forced out of the lush forests, up the valleys to the sparser wildlife high in the hills, where their fierce competition honed the kind of violent competitive behaviour between females that occur in the first *Heorot* story. The specialised species would not adventure out to the island, but the basic type would. On the continent, there was a symbiosis with a parasite that made one kind of adult grendel rather more intelligent, and infected its samlon, which on average survived better. The relationship between the hero/villain Aaron and his grendel was set up with an intelligent grendel of this type.

The flowers presented another problem. There was no choice: they had to be pollinated by a flying colour-sensing animal, because they were native flowers, not ones imported from Earth. There were terrestrial bees on the island, but there had to have been a creature that did the job before humans arrived. We invented a whole phylum of flying creatures like bees but rather bigger, and with a different flight principle, like some beetles. They had a front pair of 'fixed' wings, like the 'elytrae' of beetles, but their rear wings could not begin to hold them up, as beetles' rear wings can. They were simply propulsive, like little propellers: airscrews, really. The Heorot bees were like small model aeroplanes in flight. They didn't have a tracheal system, like our insects, but a very efficient lungbook organ, like some terrestrial crabs, some species of which have more efficient breathing than small mammals. The 'bees' did the pollinating, and they had some odd predators, including a monkey-like creature that spins a web between trees.

Some bees had some nasty habits of their own, which we won't totally give away here because the plot hinges on them in several places. What we will give away is where Jack got the idea, though. There had been a paper in the *Biological Journal of the Linnaean Society* about the bees of the Amazon, some of which took the very nutritious juices of corpses back to the nest. They could do this because their ancestors, in the course of specialising on nectar-feeding, had developed antibiotics, mostly anti-yeasts, that kept the sugary solutions from becoming infected and alcoholic. These salivary antibiotics sterilised the corpse-juices, allowing the bees to store a new and much more nutritious diet. The Amazon bees have not, so far as we know, evolved as far as their analogues on Avalon, but if they do, even fictionally, it will be a much more frightening scenario than Giant Ants.

That was a taste of what went on in the planning of *Heorot #2*. Great fun, with very intelligent and knowledgeable people (Pournelle, particularly, would run with every idea until it started paying back plot lines), and the knowledge that Niven's professional writing would turn it into a pleasure for many readers. His Known Universe series of novels and short stories did exactly this, with a variety of imaginative aliens such as the cowardly three-legged Pierson's puppeteers and the great galleon-like bandersnatchi.

This leads on to a more general question: 'Where do you get your crazy ideas from?' This is the standard question asked of SF writers; it should also be asked of would-be xenoscientists.

We have talked as if the range of conceivable kinds of life is enormous, and complained about the lack of imagination in otherwise excellent books such as *Rare Earth*. But how much imagination do you need to invent a new and plausible alien scenario? Astrobiologists typically write and act as if science comes on tablets of stone. They treat biology and astrophysics as givens, not to be questioned or modified. They give the impression that every idea is the result of hundreds of person-years of tremendous thought and work, so that a truly novel idea strikes once in a lifetime, if you're lucky. We think that the exact opposite is the case: original ideas in this area require very little imagination. In order to demonstrate this, we set ourselves the task of inventing a new alien ecology from scratch in ten minutes, while having lunch in a pub. Also, it had to be 'plausible' – in the sense that its existence or non-existence would be a sensible xenoscientific problem.

There was an obvious way to get started. Terrestrial parochials are an endless source of ideas for new universals, hence new aliens. The idea that earthly life is all very much the same (because it uses the same molecular biology to shuffle its DNA) may seem evident to a modern biologist, but it wasn't to zoologists and botanists before their subject was overrun with molecules, and it isn't today if you look at the actual organisms. We saw just now that grendels were an invention modelled on the 'nasty frog' that eats its own offspring. Several of Jack's attempts at aliens have been validated by that terrestrial parochial, and we have discussed the basis of this kind of validation.

By now, it scarcely needs saying that simply transplanting a terrestrial parochial example into an 'alien' setting doesn't work. It results only in 'cowboys on Mars' scenarios, where the cowboys haven't been presented as interestingly different *because* they're on Mars. At worst this approach results in tall thin humans with big chests, because of the low gravity and the thin air. It doesn't exploit the interestingly counterintuitive ideas that come from even the most minimal engagement with possible evolutionary scenarios. As an example, it is obvious – once you've thought about it – that flight as an evolutionary strategy is *easier* to develop on a heavy-gravity planet than on a light-gravity one. Why? In heavy gravity, the atmosphere is so much denser at the sea surface, making it easier to 'swim' out of the no-gravity sea, rather like terrestrial flying fishes. Clarke missed this point when he made the winged Overlords in *Childhood's End* come from a low-gravity planet.

In pursuit of our self-imposed task, with its sharp deadline, we decided to start from the parochial example of Earth's salmon. Adult salmon migrate up-river to the place where they first came into existence, breed there, die, and their bodies provide food for their growing offspring. We interpreted this parochial adaptation as an instance of a universal: the migration of parents into a location that is free of nutrients, and therefore free of egg-predators, which is what makes it attractive in evolutionary terms. In this set-up, a local species acts as a transformer of the parents' bodily goodies into the offspring in the course of its own life history. But we challenged ourselves to invent a locale that had seldom, if ever, been used as context for an alien story before, or at least not often. Jack suggested the core/mantle discontinuity of an Earthlike planet, because it has an energy-exploitable boundary with continental drift up in the 'stratosphere'. The basis of the ecology was to be magnetism, not matter. SF *aficionados* will recognise some resonances with Fredric Brown's short story 'Placet is a Crazy Place', but that story was set on the surface, with subterranean 'birds' flying through the rocks below.

We chose our viewpoint to be that of the 'slamon', inhabitants of the lower mantle formed by patches of peculiar crystalline semi-magnetic materials. They move through the patchily-molten magma by leverage against local magnetic fields. These fields are concentrated as 'ley-ropes' of braided magnetic lines of force, forming an armature around the whole core. These braided ropes represent the primary production – plants – in our ecosystem. The core's rotation inside this armature both maintains its structure and produces the planet's magnetic field, and its Van Allen belts, by interaction with the pervasive charged solar wind. The slamon are formed of, and in, just-molten transition-metal complexes whose magnetic structure resembles 'spin-glasses': local magnetic fields that seem to point randomly, but which have structure resulting from local interactions, so that a change in any one is accompanied by a wave of re-orientation around it, spreading through the creature.

Unlike most creatures, their intelligence evolved at the same time as their extelligence; so, like most intelligent creatures in a technical culture, they were conscious mostly of their fellow-creatures, and much less conscious of their natural environment. The aspect of their lives in which we are most interested at this juncture is their reproduction, because it involves not only the transfer of parental energy and structure, but also of knowledge, through another species. Each slamon is a not quite bilaterally symmetric volume of magma about the size of a terrestrial automobile, not differing grossly in overall composition from its very-high-pressure-and-temperature environment, but with a very different fine structure (like you, dear reader). It has a slightly higher proportion of beryllium compounds in two nuclei near its front end. Internally it is in a great variety of resonant magnetic modes, and it has a rather simple pulsatile electric-current relationship with local rock, which becomes very powerful as it approaches any ley-rope.

Now the opening of a story, leading to further 'biological' details, presents itself. We observe three of these creatures approaching a confluence of the ley-ropes, and almost merging above it, with a very noisy peaking in their electromagnetic output. Other slamen avoid the trio, as if they are embarrassed by the shouting – or the sexual excitement . . . The core below the party *would* suddenly glow white hot, *if* there were any space at that pressure for it to glow in. Something, a spiral something, has swirled into being in the chromium-rich uppermost layer of the core, and we see threads of nickel-rich, molybdenum-rich, rare-earth-element-crystalline threads, winding themselves into a very well-contained magnetic bottle under the slamen. A complicated little dance ensues, in which the slamen are apparently destroyed, but not before they have scattered tiny, egg-sized pebbles, which accumulate at the metal/magma interface. As we watch the spiralling above, the neck of the molten-metal bottle draws some of the pebbles down into the core. The bottle tilts, and moves along under a ley-rope, whose braids separate as it cruises along collecting samples from several other patches of slamon eggs. Examination of these would show a high-beryllium surface being incorporated into the bottle's workings, while the organisation inside the eggs is, for the best of reasons, like a terrestrial computer's magnetic memory. When the bottle has accumulated enough beryllium, it fragments into several smaller tori, which seem to pass not their substance but their *organisation* up into the mantle, like smoke rings being blown from one room to the next. From the centre of each torus a tiny slamon appears, ready to undertake its journey into the upper mantle to collect beryllium and partake of the rich culture of its peers. At the surface of the core a slow swirl of nickel-rich threads withdraws into the depths, preparing for its grim metamorphosis into a Guardian of The Ever-Present Truth – or something . . .

We could go into the history of their slamonic civilisation, how the housewives' revolt several millennia ago forced a complete revision of their sexual mores and a disastrous collapse of the market in bismuth . . . But we hope the point has been made: parochials can be transferred in the abstract, not the concrete. There is a process of abstraction, of working out what aspect of the terrestrial adaptation can be said to be a universal trick, and then that trick is applied to the imagined alien context.

There is spin-off from this process in 'normal' science, too. Jack has found that some of his SF inventions can serve as counterexamples for his more orthodox professional biology. For example, he was asked to write an introductory chapter for a symposium on *Maternal Effects in Development*. In the course of thinking about this, about nests and the passage of information between mother and offspring as the nest's safety means that there can be trial and (non-fatal) error . . . he came to the firm conclusion that culture – we would now say extelligence – can arise only in so-called '*K*-strategists'. *K*-strategists look after a few offspring very well, they are the kind of animals that build nests (though many don't); they contrast with *r*-strategists, which produce millions and give little attention to each. Jack then used the *West of Eden* invention of the Yilané, to see if he could invent an *r*-strategy culture; he needed a system that would permit a few offspring to be valuable enough to the adults that a culture could be maintained despite loss of 99 per cent of the offspring. That seemed a credible enough system to a lot of people, including biologists; so Jack&Ian invented a somewhat wilder *r*-strategy for the Jovians in *Wheelers*, which we now describe.

Adult Jovians are 'blimps', octopus-like balloons that float in the upper cloud levels of Jupiter's atmosphere. Early stages of development take place two thousand miles further down, at the 'upper phase boundary' where a gaseous mixture of hydrogen and helium gave way to liquid hydrogen. There is no separation of genetic material into sperm and egg. Instead, when a blimp comes into season, it produced trillions of tiny packages of genetic material, 'nanogametes', each containing about 10 per cent of a complete blimp genome. These packages are small enough to blow away like spores of Earthly fungus or orchid, dense enough to sink, and compressible enough to continue sinking until they regain neutral buoyancy at the upper phase boundary. As the nanogametes descend they are winnowed by innumerable species of aeroplankton, so every organism's new generation begins its existence as food for everyone else's larvae.

At the phase boundary, molecular forces link the nanogametes into Fibonacci chains – first in twos; then in threes as single nanogametes attach themselves to a pair; then fives, as pairs and triples join; then chains of 8, 13, 21, 34, and finally the Sacred Number of the Lifesoul Giver, 55. Each completed chain of nanogametes closes into a ring, an act of fertilisation that creates a 'cyclozygote', which begins to grow and develop. Each cyclozygote incorporates parts of the genetic codes of 55 separate adults. The cyclozygote develops a double-membrane across the ring, and molecular machinery sucks liquid hydrogen from the surrounding ocean into the gap between the membranes. Chemical reactions heat the hydrogen, and the cyclozygote rises like a hot air balloon. A glimpse into the invisible book then occurs:

> How had such complex genetics evolved in the first place? Blimp scientists believed that back in the lost histories of Deep Time the process had been far simpler, with diminutive molecular machines competing for resources and driving each other into more sophisticated strategies. The evolution of the nanogametic ring, and much later its attendant double membrane, had sparked an explosion of diversity as the molecular gadgetry acquired the ability to explore the upper reaches of their planet's atmosphere. In a new environment, evolution played a new game . . . A game that had enormous implications for the old environment . . . A game that could change its own rules . . .

Blimp development continues with colonies of cyclozygotes merging together and growing into a variety of aeroplankton. The aeroplankton form a rich soup of competing molecules, which sustains higher forms of Jovian life:

> The blimps thus inhabited the converted bodies of countless trillions of their own dead children – a fact that they suspected, but did not know, and were completely unmoved by in any case. This is the way of the *r*-strategist, which has evolved not to care for its offspring . . . for there are too many.

Early pre-adolescents generate copious quantities of hydrogen and rise to the troposphere, just above Jupiter's clouds. There they extract energy from the sun, grow, and change. Then they descend to the waiting Cities, to be trimmed into proper adult form by surgeons . . .

There's a lot more, but you get the idea. Our Jovians violate almost every terrestrial parochial, and a great many developmental and evolutionary myths, yet they comply with the universals in an entirely consistent manner.

Now we get to the scientific spin-off. The invention of r-strategist aliens has led to genuine scientific insights. Jack now can't see why he originally believed that only K-strategists could develop extelligence; the alien 'simulations' were good, credible disproofs of the position that he initially had favoured. They were good science, in the sense that science is the best defence against believing what we want to.

10

Not As We Know It

*A*BEL DECIDES TO *take another look at the black smoker, for reasons of his own, so Cain rounds up a gang of interested tourists and they all take a bus to the ocean floor. Abel gets several of the tourists to discorporate and 'taste' the complex ecosystems around the sulphur bacteria, and several of them agree that there seems to be some kind of an overall organisation at work. But whether it controls the ecosystem, or arises from it, cannot be determined. It is part of its pattern.*

'OK, guys,' says Abel. 'Here's a challenge for you. Whichever entity gets the answer first – that's the right answer, you understand – wins free drinks at the bar for the whole of this evening. One drink each if you're a collective entity with more than five hundred members. Free lump, or whatever your cultural equivalent is, for those of you who don't drink, naturally.'

The announcement causes great excitement. Free drink (or cultural equivalent) is one of life's great universals.

'I want you to work out which organism is the commonest on the planet, counted by mass,' says Abel.

Several tourists calculate that the size of the stands of pogonophoran tubeworms, gutless as they are, make them very common animals indeed when multiplied up along all the undersea volcanic ridges. Abel doesn't actually know the answer to the question he has set, so he desperately tries to calculate the commonest animals in the sea, thinking they are little crustacean copepods in the plankton, grazers on the productive algae. Then he decides that these are probably outnumbered by all the slow-living jelly-animals in mid-waters.

Cain likes ants, and reckons that they outweigh humans at least twenty times. They start to argue, which amuses the tourists no end.

'Well,' says Abel, *'the viruses in seawater outweigh the whales at least twenty times, too. So what?'*

'If we find the answer, we can add it to the Earth Prospectus,*'* says Cain.

'We've already got enough for ten prospectuses,' says Abel.

Because of our current lack of evidence, our currently aqueous Earth *may* be the only place in the universe where life, of any kind, occurs – and ever will. If so, it must be a very strange universe: all that *stuff* and yet nothing more interesting than a few not-very-brainy apes on one tiny lump of rock. However, we not only have to be imaginative enough to contemplate alternatives to the familiar DNA-based life here; we must also be imaginative enough to contemplate that there might be *no* alternatives. If that is indeed the case, then most of what we're telling you is beside the point.

Most biologists would be happy with an intermediate position: that Earthlike life is likely to arise on Earthlike planets, provided they are sufficiently Earthlike and stay that way for sufficiently long. This, in essence, is the classic 'astrobiology' stance. Solid, though limited, science. But what's *really* interesting, and constitutes a central question for xenoscience, is whether aqueous planets need be the sole 'cradle of life'. What about the really alien: different theatres, different chemistries, and different contexts? What about life as we don't know it?

So far, we've mainly presented the cosy carbon-based argument, rooted in Earthly biology and our well-tested understanding of it. But now we must think seriously about the possibility of other kinds of life. Are they feasible or credible? And if they are, what differences would a different foundation for life produce in the ecology? There are four steps away from our life that we should consider:

- Different carbon-based chemistry in an Earthlike system, such as another nucleic acid conformation, or different linear chemistry altogether instead of DNA.
- Different metabolic circumstances around carbon-water-based life, such as sulphur instead of oxygen, different amino acids and proteins, or metals as support at temperatures above 300 °C.
- Different chemistry altogether, perhaps based in silicon and

silicones – or chemistry that we haven't thought of, for instance in Jupiter's atmosphere or core.

- Totally different recursive systems altogether, from reproducing tori in stellar atmospheres to complex systems of subatomic particles on the surfaces of neutron stars.

Later, we'll try to give you some feeling for what's known about such variations on the conventional RNA/DNA/protein chemistry of life on Earth. But first, we want to convince you that this kind of difference in 'materials' can occur without any essential changes to the abstract *processes* that they implement. Whatever the differences in the details, natural selection would still operate, and an ecology/evolutionary system would soon arise, producing diversity, especially if recombination (sex) could be exploited to shuffle the possibilities. Aliens in what to us seem exotic environments would evolve to succeed in those environments, and they would seem to any visitor to be exquisitely finely tuned to function in those environments. They would not only 'tolerate' conditions that we would consider extreme: they would thrive in them. 'Tolerate' is a very tempting word, but our habitual thinking here is very parochial indeed. Nobody gets excited about the remarkable ability of humans to 'tolerate' the terrible conditions on the beaches of the South of France – sleeting electromagnetic radiation, a corrosive oxygen atmosphere, and that terrible universal solvent, hydrogen monoxide, sloshing all over the place.

Any particular alien species (assuming 'species' is a concept that generalises successfully to aliens) will exhibit a huge suite of parochial features in its form, behaviour, and social organisation (if any). The main xenoscientific interest lies in the universals that those parochial features exemplify. We can use the universals to determine the general possibilities, and on that basis we can offer parochial instances to make the discussion more specific and more vivid. But we repeat: any particular alien that we describe will be at best a possibility that *could* be realised; we'd be astonished if anything exactly like it actually *was* realised.

What universals can help us here? Depending upon what energy source was being exploited, there would probably be synthesisers, which turn that energy into body-parts, and exploiters, which are predators on the synthesisers. On Earth, the synthesisers are plants and the exploiters are animals. Recursively, some plants exploit other plants, and even animals: consider the Venus flytrap. And some animals exploit other animals, often many stages along a complex food chain or feeding web. Alien habitats would evolve comparably diverse entities and strategies. Synthesisers would necessarily become very different from exploiters, because their ways of life must inevitably deviate from each other as time passes.

Simple, early alien creatures would initially be pretty much at the mercy of their environment, but as they evolved into more complex forms they would begin to buffer themselves against environmental changes. We would expect to see 'homeostasis', in which the creature would be able to maintain stable, constant conditions provided the environment remained within some appropriate range, just as warm-blooded organisms on Earth buffer themselves against changes in ambient temperature. (So do some cold-blooded organisms, too.) More subtly, we would expect to see 'homeorhesis', in which the development of each individual creature would be regulated so that it could function in varying conditions. For instance, many chemical reactions are temperature-dependent, and if those reactions are going on inside a developing creature, different parts of the body-plan may get out of step with others. On Earth, for example, frogs solve this problem by employing not just one enzyme to mediate certain developmental changes, but a whole suite of enzymes, from which development 'chooses' whichever enzyme keeps everything close to the normal timetable. There isn't a single developmental path from egg to tadpole: there is a set of contingency plans. The whole set-up is automatic, and has evolved over tens or even hundreds of millions of years. It evolved because (a) it is possible and (b) it works.

A big question, of course, is intelligence. Life – even exotic life made with different ingredients from ours – is one thing. Intelligent life is quite another. Our prejudice is that intelligence would be favoured even in such different living systems, that it really is a universal evolutionary strategy, but the topic is a big one, so we postpone the reasoning to its own chapter.

Assume that you are a terrestrial biologist, and you therefore know a lot about the kinds of life we have on this planet. Assume further that you are a very widely read biologist, with an enormous breadth of experience ranging from biochemistry and physiology to anatomy and natural history, from studies in behaviour to sophisticated population genetics. You have experience of working with bacteria and viruses, with fish farming, time-lapse filming of developing embryos and timber-felling with a gang of lumberjacks; you have been with whales and elephants, apes and raccoons. You have dived on coral reefs, spent many boring hours with dusty museum specimens, and farmed both cereals and livestock. We don't know any biologist who has such a wide experience, but we want you to imagine having done all those things, so that you see all the kinds of life on this planet as forming a seamless whole. If you have seen Attenborough's natural history TV series *Life on Earth*, *The Living Planet*, and *Trials of Life*, and you can imagine how he holds all that in his mental and emotional grasp, then you're coming close to what we want you to try to feel.

Now imagine that you are such a person watching the bar scene in the first *Star Wars* film. The creatures there would *not* seem alien to you, because their different aspects would all come from different parts of your experience. Plausible? Yes. But alien? Not at all. In the same way a very good book, called *Aliens and Alien Societies*, by Stan Schmidt, the editor of *Analog*, has a picture on the front cover. This depicts a couple of very rabbit-faced furry creatures, not only obviously vertebrate but obviously mammals with eyelids and external ears like a deer's, with some kind of artefact in their hands; there are odd–shaped mountains in the foreground (!) and a spaceship in the far distance to show us that it's an alien *place*, but most people would not be surprised to see those creatures in a London or Sydney zoo cage. Schmidt is also committed to terrestrial models in the text, more or less, with some honourable exceptions.

The Uplift War (David Brin 1987)

Startide Rising and *Sundiver*, the two previous novels of a trilogy, have set the scene for *The Uplift War*. There are intelligent aliens throughout the Five Galaxies, and they are all part of a vast interstellar society with its own rules and politics. All of them, save (apparently) humans, have been nurtured into intelligence by the process of *Uplift*. Here, a race that has the potential for intelligence is provided with an intelligent patron race, which develops the burgeoning intelligence (we would say 'extelligence') of its client race. At some time in the distant past, this recursive process of Uplift was started by the Progenitors, who are now represented only by their relics. These include an ancient battle fleet, lost long ago but rumoured still to exist.

Not only do humans seem not to have been Uplifted: they are busy Uplifting three of the other species of their planet: dolphins, chimpanzees, and gorillas. This is unprecedented, and distinctly annoying to more mature races of the Five Galaxies, who call the Terrans 'wolflings'.

The roost Masters of the Gubru meet in the Conclave Arena. They are descended from flightless birds, with highly coloured plumage, sharp beaks, downy breasts ... They are culturally rigid, and naturally cruel. There are Lords and bureaucrats, and a female conclave President. The Gubru were Uplifted long ago by the Gooksyu. They embark on a perilous invasion, to conquer the dying planet of Garth, in which Earth has a considerable interest. The Terrans assemble a small flotilla to defend the planet, and to the annoyance of the Gubru they do so 'with meticulous attention to the protocols of War'.

After that, it starts to get complicated.

David Brin's dolphins and chimps in his 'Uplift' stories, especially *The Uplift War*, are beautifully realised against a background of a great variety of extremely ancient alien civilisations, out there in the galaxy – indeed the Five Galaxies. The concept of 'uplift', by which older races drag immature races into the over-arching galactic culture by using them as slaves for many generations (oddly reminiscent of medical training) enables many kinds of creature to be exhibited to the reader. Some are very strange, like the mining organisms that live in the crusts of planets, but nearly all of them are insectile (with the obligatory six legs) or otherwise reminiscent of terrestrial forms. The Gubru, for example, are covered in feathers and – 'naturally' – are bipedal with claws on the feet and you can break their neck vertebrae in a fight.

Brin includes many nice touches, for example when a servant fails in a trivial duty and is sent to cut his head off, and as he leaves he expresses hope that his next head will do a better job. It is our strong impression that even the best alien depictions 'run the whole gamut from A to B', as Dorothy Parker said of the emotional range of a certain actress. Here A is an assortment of terrestrial, mostly vertebrate, characters in pick-and-mix mode, and B is the range of absurdities in *Men in Black*. Very few depictions get as far as, say, G – which might be our Jovians in *Wheelers*, or the walking trees on Epona. Most real aliens must be off the end of the alphabet altogether.

All very well, but why do we believe that life *has* a range that extends so much further than the enormous variety here on Earth? Could it indeed be that nearly everything that life can do is already being done somewhere in the range of organisms that have existed on Earth, so that other evolutionary histories can only repeat much the same stories? In that case, all these terrestriform aliens will have been well portrayed by SF novelists. However, there are two steps outward that show that this is not so: one rooted in observations, and the other founded in theoretical biology, the modern vision of what constitutes life. We *will* go on to extend our concepts of what life might include, so that we can take you, as that supremely competent terrestrial biologist, out into much wider considerations, into real xenoscience. But first, there is direct observational evidence that life, even 'life as we know it' is much broader even than that very competent biologist would realise. Among his competences we did *not* include palaeontology, but in life's history on this planet there have been forms more grotesque than nearly all invented aliens. And we're not (only) thinking of dinosaurs.

Perhaps the best documented kinds of life on this planet that do not resemble creatures with which we're familiar today are the creatures of the pre-Cambrian Vendian era, some 570 million years ago, and the creatures of the Burgess Shale, some 540 million years ago. The Vendian fauna, also known as the Ediacaran creatures, are mostly flat jelly-plates, but a few resemble planktonic worms like today's *Tomopteris*. There is a possibility that the jellies we have recently found to be very common in the mid-ocean depths are this unfamiliar kind of creature. Equally, in the Burgess Shale there are many fossil creatures that we can assign unambiguously to ancestral lineages of today's, or yesterday's, faunas; there is even a proto-chordate known as *Pikaia*, as well as proto-trilobites. But about half the creatures in that one little fossil site (and another one in China, with much the same fossils) have anatomies that were never seen again, body-plans that were simply lost. Their variety extended the range of body-plans beyond anything we have today (although one of them, *Hallucigenia*, turned out to be a fairly ordinary velvet-worm, but interpreted upside down).

These alien body-plans, like the later reef-forming sponges and bivalve molluscs, show that different body-plans, different ways of life, are indeed possible: corals form the basis of our reefs now, but it was not always so – ancient bivalves were making similar reefs when dinosaurs roamed the land. Today's biology is but a fraction of what Earth's biology has been, and presumably – if we give it a chance – will be. Dougal Dixon has written and illustrated *After Man,* an imaginative and well-reasoned book about the kinds of body-plan that could evolve on Earth over the next fifty million years or so. (*Could*, not will, for the usual 'parochial instances' reasons.) For example, he posits that the great whales die out, and that their successors will be enormous marine creatures descended from penguins. That's just on this one planet, with our style of life, in one possible future Earth.

We have already done the thought experiment 'If we re-ran the Earth what would we get?' to seek evidence that distinguishes universals from parochials. Now let us move further out into the space of possibilities surrounding our own real history. One step out is to suppose, for example, that the K/T meteorite didn't hit. As mentioned before, Harry Harrison's *West of Eden* invents the corresponding alternative present-day Earth, and in *Winter in Eden* he presents us with great browsing herbivores with babies hanging on tens of teats, living around the Arctic icecap. The browsers feed on the fringe of brown algae, and in that alternative Earth the holdfasts of the algae have the trick of warming their way a little distance into the ice, forming an anchor. That didn't happen on this Earth – but snowdrops and crocuses do melt their way up through a snow layer, so we have a comparable parochial adaptation to justify that invention. The vast browsers, like Dixon's penguin-whales in our hypothetical future, are a necessary ecological/evolutionary consequence of the fringe of brown algae around the ice cap.

Let's be clear what we're saying here. We're not claiming that all lifeforms on all planets, or even all carbon-based lifeforms on aqueous planets, must do what Earthly life has done, as universals. Anthropists would claim that this set of tricks must be unique, for many steps up the ladder of argument. Our view is different: the Earth is just one instance, though *variations* of it may be replicated many times in the Galaxy. We don't say that other systems are impossible, only that the system we have on Earth is not, so far as we can understand it, difficult to achieve, so it is probably an entirely 'ordinary' example. We consider its basic elements to be parochial instances of universals, but not universals in their own right, and this view enables us to extend the argument about terrestrial evolutionary innovations to hypothetical alien environments. We think that the default may be something like we've got on Earth now, if we 're-ran Earth' – another ordinary planet with a great variety of lifeforms. But possible differences begin to give us a geography of the imaginative phase space outside *Life on Earth*.

First, let's think about the possibilities for different hereditary material . . . but not *very* different. Laboratory experiments have shown that it is possible to ring the changes on virtually any feature of our DNA/protein system, even if most of that system is left unchanged. That is, although the roles that these molecules play may be universal, the actual molecules that we use here are *very* parochial.

Recall that DNA is a double helix, composed of two complementary strands, each a sequence of bases chosen from a standard set of four. The way the strands assemble leads to a 'backbone' of sugar molecules running along the core of the helix. Each triple of bases encodes one of twenty amino acids, or a 'stop' instruction. Proteins are linear chains of amino acids that fold into complex three-dimensional shapes, and these shapes do all the work.

Alternatives to DNA definitely exist. To date, only two have been studied in any depth, but it is clear that these are the tip of an iceberg. P.E. Nielson and colleagues have synthesised nucleic acids whose backbones are related to those of proteins. A. Eschenmoser's group has studied RNA analogues in which the sugar (ribose) forming the backbone is replaced by a different sugar, such as hexose. The resulting 'pyranosyl RNAs' (pRNAs) spontaneously pair off to form double helices, and could therefore, probably, carry genetic information and replicate. Strands of pRNA do not bind to RNA, so pRNA is not a plausible precursor to RNA on Earth. But pRNA could conceivably be the 'molecule of life' for aliens.

In 2000, Eschenmoser's group modified pRNA to match the geometry of conventional RNA, by synthesising a variety of so-called L-α-threofuranosyl oligonucleotides, or TNAs. These form stable Crick–Watson helices, *and* can also form hybrid helices in which one strand is TNA and the other is DNA or RNA. Eschenmoser has stated that an enormous variety of RNA-like molecules can be conceived, any of which might be a plausible RNA precursor; and, though he didn't say that, a plausible starting-point for alien genetic chemistry.

It is also possible to extend the range of bases in DNA – to add new 'letters' to the code. The resulting 'artificial DNA' could generate novel proteins. In 2000, Floyd Romesberg's group made twenty new, distinct, bases that combine with sugars just like the four 'natural' bases. 'The biggest surprise,' said Romesberg, 'was that there's nothing special about a natural base.' These new bases can be inserted into DNA, extending its alphabet from four letters to twenty-four. Some, at least, of the new bases can be replicated by the same polymerase molecules that replicate conventional DNA.

On the level of protein manufacture, new amino acids can be brought into play. The standard twenty used by DNA to make proteins are not the only amino acids in the molecular repertoire. In 1986 it was discovered that one of the 'stop' triples, UGA, could sometimes produce a twenty-first amino acid, selenocysteine. (Here 'U' stands for 'uracil', a molecule that replaces thymine in RNA.) In 2000 the mechanism behind this change was unravelled, first for *E. coli* and then for mammalian cells. It is related not just to the DNA sequence, but also to the geometry into which the molecule is folded, namely the occurrence of a hairpin-like 'stem-loop structure'. In *E. coli* this section of DNA is adjacent to the UGA triple concerned. Remarkably, in mammals it is some distance away along the DNA strand, and is more complicated. The only other known 'natural' interloper is formyl-methionine, which also substitutes for a 'stop' triple.

In 2001 two groups, one headed by Lei Wang and the other by Volker Döring, persuaded *E. coli* to operate with a new repertoire of amino acids. There are plenty of amino acids to try, such as L-ornithine and L-citrulline. These exist in all cells, but do not normally gain access to the genetic code. In order for them to do so, various of the cell's protection mechanisms must be overridden. The unnatural amino acid must penetrate the cell membrane and bypass the usual checking system that ensures accurate protein synthesis. Then the ribosome, the cell's DNA protein translation machine, must be able to use them as building blocks. Wang's group achieved this using an amino acid rejoicing in the name of O-methyl-L-tyrosine; Döring's group used aminobutyrate.

Clearly the specific constituents of our ubiquitous DNA/protein chemistry can be changed at almost any level of the process, without interfering with its ability to carry out the fundamental information-carrying and replicatory processes of life. There is nothing sacrosanct about DNA and the genetic code. They are parochials – very parochial parochials.

So alien biochemistry will, almost surely, differ in every detail from our own. Even if the biochemistry were exactly the same, however, Earth's lifeforms cannot possibly be duplicated elsewhere, even on another 'ordinary' Earth, for evolutionary reasons. Parochial phenotypes will not be repeated; even if they have the same biochemistry the context will be different. So there will be no extraterrestrial insects or dinosaurs or australopithecines, unless they were exported from Earth by visiting aliens. However, the initiation of life on Earth could well be a common, almost identical situation on all aqueous planets: being bombarded with asteroids, keeping the seas boiling and the chemistry in ferment, may make them all equivalent.

This may sound like part of the *Rare Earth* argument, but there's a difference. Our point is that whatever happened here 'worked', so anything sufficiently similar has a good chance of working too – although we'd expect the result to be different in detail, even if we re-ran the whole of evolutionary history again on *this* planet. The *Rare Earth* view boils down to 'whatever happened here is *essential* everywhere'. Logically, this confuses sufficiency with necessity; scientifically, it is not sensible to deduce from a single example that a successful trick must be the *only* trick. In fact, given what we now know about alternatives to DNA on Earth, it is completely ridiculous to assume that our specific DNA/protein system is the only way to make life.

*

Our expectation that there may be some underlying chemical universals is restricted to aqueous planets: these are likely to be fairly common, but they're not the only game in town. On aqueous planets, we expect some similarity in early chemistry to occur for two reasons. One is 'bottom-up', based on internal features of chemistry and physics; the other is 'top-down', based on contextual influences.

The bottom-up view stems from the observation that the array of amino acids made in Miller-type experiments has a fairly rigid pattern. Some amino acids, like glutamine and arginine, are 'easy' to make, while others, like tryptophane and cysteine, are 'difficult'. These require longer times, special starting mixtures and are rare in the final mix. The same distinction occurs with sugars, fats, and oils: all the 'starter molecules' that pave the way to life. The reasons why some molecules are easier to make than others have to do with various physical properties, which can in principle be predicted by experts. So we expect there to be a relatively common easy set of chemical compounds building up in the seas of most aqueous planets as the planet cools and the seas condense. Whether these are precipitated from the atmosphere, or are received from meteoroids and comets' tails, some amino-acids will be common and others rare. This provides bottom-up constraints on what can happen later.

There are some complexities, to be sure, that build on this bottom layer. We have seen that some simple peptides (amino acid chains too short to count as a protein) have turned out to be moderately effective catalysts to help other amino acids join together. Some sugars are very easily oxidised, and are destroyed by high concentrations of oxygen; as they are destroyed they reduce the level of oxygen, allowing their fellow molecules of sugar to survive more easily. This can lead to a repetitive cycle, like that in the BZ reaction, if there is a reservoir of sugar available. So the standard system of 'small but interesting' molecules complicates itself on all aqueous planets in not-altogether-predictable ways. The 'initial conditions' on a planet that is going to acquire seas, and life like ours, are immensely complicated – but they are not, in the important aspects, very variable.

We don't think that anyone will ever argue convincingly that kind-of-life-X will appear on 15 per cent of them, no life on 70 per cent, kind-of-life-Y on 5 per cent, and so on: this is more detail than any theory could safely predict. In the same circumstances on other aqueous planets, much the same rules will apply, leading to much the same complexities and complications, simply because a planet is so big and lasts so long, and because the space of the adjacent possible is so much bigger. Phase space for a planet has more dimensions.

The 'top-down' reason why the same complexities could appear in different places is more subtle. Perhaps some of the reaction-systems are dynamic 'attractors'. That is, many different starting-points will converge on to the same behaviour. Even if there is not much likelihood of any particular system getting started anywhere on such an early planet, once it *has* got started it is autocatalytic: it encourages the production of more systems of the same kind. If so, then that system could take over locally wherever – and however rarely – it appears. This is what spirals do in the Belousov–Zhabotinskii experiment: they are quite difficult to initiate, and appear only very rarely without some kind of intervention, but once a spiral appears it takes over everything else until the whole dish is a seething mass of 'spiral chaos'. One spiral will gobble up any number of target patterns and turn them into more spirals. This is interesting, because a spiral isn't a *thing* – you can't pick it up and lift it out of the dish. It is a kind of organisation, but it can act like a thing. Life is like that too, only hugely more complicated.

In an environment, such as an ocean, that contains several different autocatalytic chemical systems, there will be some systems that 'trump' and take over other systems. Many of these 'winning' systems, probably most, won't be very interesting: their effect will be to 'damp down' other systems, rather than to enhance them. Every so often, though, one of these winning systems might become a dominant process. Cairns-Smith might point to polymerisations, which can exploit the stability given to their intermediates by certain common clays; Wächtershäuser might consider the sulphur-iron possibilities to be so ubiquitous that the organics *have* to link up with them to become permanent features of the complex system. Because these are complex systems, school chemistry is not a sensible way to predict how they will progress – though it does give us a 'bottom-up' soup to start from. And convergence to a few stable complexities could easily be the way these things work, so that whatever the beginning states may be, the end states are much the same because they are the most robust.

We don't *know* that this is so, but to us and to chemists like Cairns-Smith it seems the most profitable way to think about these primordial 'self-complicating' chemical systems. We wouldn't be surprised if, unlike the later evolution of lifeforms, this early biochemistry is rather unoriginal in its invention. We think it will converge on to a few standard autocatalytic tricks that engage local chemical systems in cyclic transfers of energy, and regions containing such autocatalytic mixes of chemicals will compete with other local regions for rare molecules or for energy or for catalytic surfaces. This is 'pre-life' – chemistry with some of the surprising features that life exhibits, but not (yet) all of them.

There have been two kinds of experimental work in this area, mostly directed towards 'our' kinds of chemicals. In such work, success, identifying such pre-life, is *defined* as the achieving of one of *our* kinds of molecule. If other possible life-starting molecules appear they are not 'noticed'. As we saw earlier, it has been shown that there is much possible diversity, but there is also some convergence on to particular families of chemical reactions. Laboratories concerned to achieve synthesis of 'our' living chemistry by heroic methods (not by duplicating possible 'natural' early seas or substrates) have made various kinds of nucleic acids, but not yet the 3–5-phosphate-ribose or deoxyribose (RNA or DNA) that we know and love. A few of these 'failures', however, could make the same kinds of double helices that DNA has – though not using the same component molecules – and could in principle go on to be a hereditary message, different from the one we use. We don't yet know whether Earthly life's current DNA/RNA system is the only one that really works, or whether it is the *best* system so far that has won over all the other attempts at pre-life, or whether it is simply the system that by chance happened to appear on Earth and went on to be our heredity, even though many other systems *could* have done the same if they'd got their noses in front early on.

We may know more about this question, soon. Better analysis of what happened on Mars (assuming that anything is found to analyse) or Europa or any other of Jupiter's moons with water will give us examples of chemical systems that have self-complicated. They are cosmic laboratories practising origin-of-life experiments. If Mars, Europa, and the rest *also* have DNA and RNA, then this would appear to strengthen the argument that the DNA/RNA system is universal for watery worlds. Unfortunately the question can't be decided as easily as that, because – even if what's out there is DNA and RNA – we still won't know whether this is the only possibility or the best, or neither. The problem is that life on these nearby worlds might have the same origins as ours, splashed through space in the same way that Mars rocks have been splashed off the planet's surface by meteorite impacts, to land in Antarctica. But we predict that if it's just as possible to use other molecules *instead* of DNA and RNA, then Europa and Mars will probably have done just that. Indeed, the early Earth may well have done so too, but the evidence has been mopped up – 'eaten' – by the winners here. If Europan life uses different molecules from ours, that will be informative, but if the molecules are the same as ours, it won't be.

Amino acids are important because they can stick together ('polymerise') to form long chains: proteins. The vast array of catalysts, structural elements, chemical switches and chemical triggers that can be constructed from proteins is truly amazing, and could be effectively infinite (in part because some special kinds of protein, called 'chaperonins', can affect the shape taken up by others). This potential diversity has been thoroughly exploited by our kind of life: for example there are about 100,000 (some say 300,000) different proteins in the human body.

It looks as if the diversity can easily be accessed, because amino acids were certainly present in the early seas – being produced either by Miller-like reactions up in the sky, or from the tails of passing comets, carbonaceous meteorites, and other cosmic junk. Their polymerisation happens easily too, either on a clay substrate or once a few peptides have formed on other catalysts. Our major evidence that protein structure is easy to exploit is that the other kind of experimental pre-life laboratories, those trying to duplicate 'natural' primordial Earth-conditions (in contrast to 'heroic' systems), nearly always get at least as far as moderately long peptides, along with sugars, oils, and those ubiquitous resins and tars. On the other hand, these workers are *looking* for peptides and proteins. If we looked for other possibilities, who knows what we might find?

*

What would Earth life look like if it had rather different chemicals as its basic heredity, but still used much the same peptides, proteins, sugars, fats, and oils? Surely it would not look very different? The point about DNA is not that it is the only molecule that can programme our development, but that it is the molecule that our kind of life has chosen to fulfil that function. It sits behind life's functions without affecting their workings. Another kind of molecule could necessitate somewhat different mechanics of cell division, perhaps, but if it sits at the bottom of causal chains whose end product is a particular protein, or a fat or an energy-using system, then the process will be selected for, or not, independently of the detailed mechanism.

How different are these differences really? Natural selection is a universal, and it would lead to an evolving ecological system. Recombination, another universal, would cause the ecosystem to diversify rapidly. Depending upon what energy source was available, plant-like organisms (synthesisers) and animal-like organisms (exploiters) would necessarily deviate from each other as ways of life. We've already pointed out that as the bio-system evolved, organisms would develop homeostasis – regulatory mechanisms rendering them independent of environmental vagaries like changes in temperature. And we also said that the more subtle phenomenon of homeorhesis, which maintains not just some fixed state, but regulates the dynamic development of each, would arise in some lineages.

Both of these mechanisms make development more robust, more able to resist external disturbances, and in evolutionary terms that's a good thing. Whatever the genetic chemistry might be – indeed whatever the chemistries that underlie other biological functions – photosynthesis would still be a profitable enterprise for lifeforms, and there would be profit in the competition for the foods that this makes available. Carnivores would have their eyes in front and herbivores would have them at the sides – whether or not DNA, or even proteins, sugars, fats or alternative molecular families were used to make the eyes.

Let us be slightly more imaginative. Imagine that it is possible to produce a vast array of switches, catalysts, and structural molecules from sugars and polysaccharides instead of proteins, and that our life had done this. If you find this hard to imagine, it's worth pointing out that in fact Earthly life has done just that, but polysaccharides are not tied to our heredity as directly as proteins are. 'Genes for homosexuality' sell newspapers, but sugars don't. At any rate, in such circumstances, human chemists, trying to duplicate 'the' origin of life in simple primordial systems, would be testing them for those all-important polysaccharides. Polysaccharides are *more* complex than proteins: the molecular architecture of proteins is linear (though folded), whereas polysaccharides form treelike branching structures. So the chemists wouldn't pay much attention to mere linear peptides – those are clearly not going to have the versatility of the many-branched polysaccharides, and *obviously* anything that tries to use simple linear peptides to make complicated lifeforms won't get very far. The same goes for a hereditary story constructed out of fats and phospholipids – our lifeforms use these for the barriers between the watery components of our cells. Indeed there is a vast array of other possibilities there too – silicones, for example.

Today's biologists have homed in on proteins as the central characters in their terrestrial heredity-determined chemistry; but even here, that's just one way to tell a much more complex story. So even on this planet, there are reasonable alternatives to DNA and proteins. Other creatures might use very different molecules for their heredity, and exploit the vast possibilities of other kinds of carbon-based chemistry. Again, we could soon know, from the other bodies in our solar system, whether this has actually occurred.

If our lifeforms *had* specialised on, say, polysaccharides, polymerising sugars instead of amino acids, then we would expect the kinds of life that arose to look much the same as the ones that actually occurred, bearing in mind that such a different evolution could never repeat our insects, chordates, or mammals. But fingers with chitinous fingernails for climbing, or arthropod-like creatures shedding a keratin armour, would not make much difference to overall patterns of life: the opposite happens here on Earth, but the arrangement described would pose no great problems. Just as a bridge's function can be served by rope, wood, concrete or cast iron, life's functions can be served by a variety of chemical substrates. Whatever chemistry has been exploited, lifeforms will use it to exploit each other.

Our argument for parochials and universals would apply to such different systems on other aqueous planets, whatever chemical substrates they use. Photosynthesis (if there is starlight, which there would not be in, say, Europa's oceans), flight, sex, and intelligence will be favoured, even in such different living systems. The criticism that we have applied to aliens portrayed as terrestrial forms does not depend for its power on the likelihood that they have rather different chemical bases. Similar forms would appear despite different chemistry, because form is an adaptation to environmental context: our criticism is of the range of morphology, not of the existence or non-invention of chemical differences.

Given time and enough variety, a range of curious beasts moderately similar to those on Earth should appear, even if the chemical basis is quite different. Intelligence really is such a universal evolutionary strategy that creatures of very different kinds would be expected to originate nervous systems and diversify them. Extelligence is perhaps another matter. We are not persuaded that colonies of intelligent creatures would automatically become extelligent, even though real-life meerkats and invented 'Tines' (*A Fire Upon the Deep*) both make that development seem plausible.

A Fire Upon the Deep (Vernor Vinge 1992)

The galaxy is home to innumerable alien races, many extremely ancient. The older ones either go extinct, or Transcend and become Powers. The Powers know everything, have seen everything before . . . many times.

There is a galactic analogue of the internet, used for interstellar communication, much of it cryptic. On a dead planet a company from Straum has discovered an ancient archive, which has been lost to the nets for five billion years. They plan to skim the archive's surface, identify its origins, and extract a few secrets that will make Straumli realm rich beyond belief.

Something goes wrong. A human spacecraft fleeing from the Straumli realm crashes on to an unknown planet. Two children survive the crash. They discover that the world is inhabited by dog-like aliens. They are not mere animals; they wear clothing, green jackets. These are the Tines. They possess a medieval level of civilisation, and spend most of their time skirmishing with neighbouring domains.

The Tines are not exactly like dogs: their necks are long and slender and their heads are more like the head of a rat. They go about in small packs, six at most; each pack seems to be a social unit, but different packs stay well away from each other. The packs often behave like a single unit; the individuals almost *merge* . . . It turns out that each alien possess a tympanum, which transmits sound; they translate their thoughts directly into sound-waves, and broadcast them to the others in the pack. It seems almost a form of telepathy, yet really it is not much different from ordinary speech – just far more direct, and far more effective at conveying thoughts.

Something dreadful has happened to the Straumli realm: the Blight. A Class Two Perversion has manifested, probably a self-booting evil from an earlier time, accidentally activated by humans investigating the archive. The Blight begins to spread, threatening to take over large parts of the galaxy. A rescue mission races against time to find the children and recover a vital weapon against the Blight, which their ship is believed to have been carrying. It is necessary to enlist the aid of the Tines . . .

Now let's move further out, conceptually speaking. Aqueous planets may not be the sole cradle of life. What about the really alien: different theatres, different chemistries, and different contexts? What about life as we don't know it?

For a start, what about water and oxygen as a basis for life and its metabolism? Hydrogen and oxygen are very common elements. Even if alien life uses sulphur, for example, instead of oxygen in its metabolism (as we've seen that many archaeans and some bacteria do on Earth) then it will have to contend with lots of oxygen in its environment – just as our life has to contend with quite a lot of sulphur, which becomes obvious in anaerobic places like estuarine mud.

Life requires the ability to 'transduce' energy – to transfer it from one entity or process to another. There could be many energy-transducing systems, differing significantly from the phosphate/oxygen one that is ubiquitous in our lifeforms. That one uses adenosine triphosphate (ATP) as the sterling currency, demoting it to adenosine diphosphate (ADP) and then adenosine monophosphate (AMP) as it is spent. Phosphorus has just those properties that our system exploits; but there may be many other compounds, of other elements, that can link up with different kinds of photosynthesis and thereby power alien forms of biochemistry. There may be places where there is very little hydrogen, and oxygen/sulphur/transition-metal chemistry could be the fashion. Until recently it was thought that Mercury rotates in synchrony with its revolution round the Sun, so that one side was perpetually hot and the other extremely cold. Although we now know that Mercury is not like that, such a configuration is dynamically stable and could easily happen elsewhere. If such a planet is close to its star – but not too close – then its *hot* side would be ideal for that kind of exotic heat-resistant metabolic chemistry. (What price the 'habitable zone' then?) Of course, to the inhabitants of such a world, that kind of chemistry would be the 'obvious' way to do things, and our amazing low-temperature proteins would be exotic. At the high energy levels found in such environments, the rare earth elements that inspired Ward and Brownlee's punning title, or even fluorine compounds, could mediate the energetics of living systems.

It turns out that, contrary to what we were taught as students, there *are* proteins that can stand very high temperatures – some are used by a few very-high-temperature archaeans that live near undersea volcanic vents or in very hot springs. If these proteins could be synthesised without free water, then complex life might arise in a high-temperature environment. We don't know whether this is possible, but this is one of those cases where absence of evidence is not evidence of absence. We've not made theoretical or experimental studies of the chemistry of complex systems that are different from the ones that drug companies pay biochemists to investigate; in fact, we know virtually nothing about the universe of chemical possibilities that *our* kind of life exploits and invents, let alone potential alien alternatives.

For similar reasons, we don't yet know enough high-temperature chemistry to deduce what energetic autocatalytic systems could run on the hot side of a synchronously rotating world – say at the edges of barely molten pools of lead with interesting metals dissolved in them, with crystalline sulphides on the banks that transduce solar energy into electric potentials and currents. We don't yet know enough high-pressure chemistry to begin to make guesses about what's happening at the surface of the semi-solid hydrogen, nearly dense enough for fusion reactions, at Jupiter's core, but we know enough to suspect that surprisingly complex things might go on in such conditions. That is, although we don't know the details, the new ways of looking at complex systems that have been advanced by books like *Investigations* encourage us to make some general predictions.

For example, there will be lots and lots of ways to have autocatalytic systems – some, but by no means all, of which will exploit carbon compounds – on the hot side of a pseudo-Mercury, near Jupiter's core, in the methane/ammonia seas of Titan, or anywhere where there are energetic disparities and a variety of chemical substances and processes that react differently to these energy sources. So if we look for aliens, we should not restrict ourselves to aqueous media, and to carbon as a source of chemical variety. When we look for planetary atmospheres that are far from equilibrium, like our own oxygen/nitrogen atmosphere, as evidence of life, we should not restrict ourselves to warm blue planets. Indeed, it should be rewarding to re-examine the other planets and moons in our own solar system, looking for non-equilibrium chemistry. And we will soon, within a decade or so, be able to study the chemistry of planetary atmospheres in other solar systems, which raises the prospect of detecting the presence of alien life, without having much idea what it is. The main point here is that astrophysicists should not restrict themselves to looking for an environment that astrophysicists *like themselves* would tolerate. 'Alien' could mean rather more alien than astrobiologists have in mind.

Here's just one example of the neglected possibilities for autocatalytic chemistry. Like many of our examples, it happens on Earth. (That's a major reason why we deplore the use of 'folk biology', and with it 'folk biochemistry', in astrobiological speculations. The astrobiologists aren't even restricting their search to life-like systems that resemble those found *here*: they are restricting their search to systems that resemble their own oversimplified mental images of ones here.) The example that we have in mind is a little chemical system that Jack was told about by an industrial chemist at the end of one of the POLOOP lectures. One of the standard hypothetical alternatives to carbon-based life is silicon-based life: silicon occupies the same column as carbon in the periodic table of the elements, and it can also form quite complex molecules. In his lecture, however, Jack expressed scepticism about silicon-based life, because silicon is nowhere near as *good* as carbon at making big molecules. 'What about mortar rot, then?'

enquired this very respectable looking guy. 'What about it?' asked Jack, having no idea what mortar rot was.

It turns out that church walls in Wales have peculiar problems, and that these are transmissible. The complex silicate mixture-and-process that is mortar can 'catch' a kind of silicon virus, which grows at the expense of the substrate, so the mortar crumbles and loses adhesion. There is a variety of these virus-type 'rots', and they specialise in mortars of different kinds – some needing zinc, some needing sulphur compounds. Mortar rot is a terrestrial example of pre-life, which happens to be parasitic on the work that the carbon-based lifeforms have put into inventing Christianity and then building church walls. At the time, Jack's answer was that interesting as mortar rot was, and despite its potential as a life analogue, he didn't think there would be planets where carbon-based pre-life destroyed the church walls of silicon-based lifeforms. Silicones may be more interesting than mortar rot as bases for lifeforms, but they would probably have to be constructed – see below.

Over the last fifty years, scientists have learned that the material world is reluctant to bow to artificial distinctions between chemistry and physics. There are all kinds of very interesting properties of nearly pure crystals, for example, in which tiny amounts of other substances are diffused in patches. Indeed we are using just those properties to write this book on our word-processing computers: silicon chips won't work without all kinds of cunning 'impurities'. There are many compounds that show peculiar properties like superconductivity at (reasonably) high temperatures, and with all kinds of reactivity to magnetic fluxes. As we write there has been a flurry of excitement about superconductivity in magnesium diboride, a molecule so simple that nobody had bothered to test whether it could possibly *be* a superconductor. As it happens, it is – and it could become an extremely useful one. So it would be unreasonable to deny natural selection the capability of building really complex systems that exploit successful tricks in this area.

Standing back from the chemistry that we learned at school, and in parochial university classes – for example 'subsidiary chemistry for biologists' – and looking at the variety of properties of materials that humanity is just beginning to discover and exploit, there seems to be no reason why carbon chemistry should look a much better bet for making astrophysicists and biologists than semiconductor technology with a dash of superconductivity and some quantum photon-swapping. In *The Emperor's New Mind*, Roger Penrose argues that human consciousness stems from quantum phenomena in tubulin molecules – long tubes of protein units that could act as waveguides and thereby generate coherent quantum wave phenomena. In *Figments of Reality* we described the case against that view – for example, its all-too-easy equation of quantum indeterminacy on the small scale with human free will on the large scale. There are many technical problems, too – such as there being too much water in cellular tubulin for it to act as a waveguide at all. But it is possible that Penrose could be right for lifeforms based on superconductivity. *Their* consciousness could indeed depend on waveguide molecules.

We are not saying, 'We don't know it *can't* happen, so it probably does.' Lack of evidence against is not evidence *for*. We are saying that life is not a single system, adopted (or not) throughout the universe; instead it is a 'meta-system' like a bridge, a general scheme that can be realised in many ways. Bridges can be made of rope, or steel, or concrete. So living processes can implement that meta-system using a variety of substrates. These other substrates look, to our inexpert but widely read eyes, at least as interesting as carbon chemistry. Pre-life processes, such as autocatalysis, competition, synthesis, budding off separate systems, and incorporation of some systems into others, can and will happen on these other substrates. There is some evidence for this assertion: it has been demonstrated by Cairns-Smith, both theoretically and experimentally, for clay substrates on Earth.

Yet again we see that the possibilities on our *own* planet are known to be much broader than the astrobiologists permit themselves to envisage *anywhere*. The indirect evidence that autocatalytic processes can potentially utilise a huge range of physico-chemical substrates is extensive and convincing. It is therefore up to pessimists to demonstrate that such systems will *not* self-complicate into living systems. Perhaps the 'living systems' involved will be temporary, with evolutionary scales lasting minutes or centuries; perhaps they will be so slow that tens of billions of years of pre-life are still experimenting on some distant world – but if carbon on Earth has put up such a good show, we can't expel these other interesting properties of materials from the quasi-biological arena.

Our anthropist friend, of course, would espouse the opposite position: life is immensely hard to achieve, and only carbon-based chemistry can begin to show enough controlled complexity to do it – and then only very, very occasionally. But naturally, when – if – it does happen, the living creatures that result see all kinds of possibilities in the processes around them that *seem* to be capable of increasing their own complexity, and they therefore assume that life is both easy and ubiquitous. From that viewpoint, mortar rot and computer-life are irrelevant. Anthropism denies our claim that the early – in geological terms immediate – origin of life on Earth shows that life is 'easy'. In support, it advances the counter-claim that life can happen *only* in those first few boiling aeons, and then occasionally or possibly just once. Since we're around to enquire, we are the descendants of lifeforms that arose through this exceedingly improbable and rare process. We have already considered such arguments, and dismissed them because of independent confirmation that life 'ought to be' common; but it's worth bearing in mind that the anthropists *might* still be right.

There are many other energetic systems in the universe. As we said, life requires the ability to 'transduce' energy – to transfer it from one entity or process to another. All energy-transducing systems – from cosmic-level stars, Black Holes, tidal stresses in orbiting moons, magnetic fields induced in and by solar winds, to planetary-scale energies of weather and weathering, sun-powered transport of water up into mountain snows and thence to rivers and gravity-powered return to the ocean – are capable of generating parasitic recursive systems. Each of these causes and exploits some phenomenon with the property of 'hysteresis', which roughly speaking means that some locally irreversible change takes place, some kind of 'switch'. As a result, subsidiary systems are set up. Whirlpools in rivers are a familiar example. Smaller systems appear on the boundaries of such systems, or at least on the boundaries of the recursive, cybernetically stable systems – those that last long enough to be counted as genuine systems.

All of these systems are rich enough to have the potential to complicate themselves into lifeforms. The further away those system are from carbon chemistry, the more 'alien' those lifeforms will be. For example, earthly whirlpools are vortices in a liquid medium, but in stars there could be magnetic vortices. A single vortex lacks the complexity even to count as pre-life, but an interlocked collection of vortices possesses tremendous topological complexity – which, being topological, is very stable. 'Creatures' made from magnetic vortices in stellar plasmas could be at least as complex as any lifeform on Earth, with topological linkages of vortices taking the place of the linear topology of DNA. And they would not only survive inside a star: that's the best place for them to live. (Even more so, what price 'habitable zones'?) The plasmoids in *Wheelers* were aliens of this kind.

This kind of argument needs to be pursued with caution. It has led to an interesting misunderstanding about Lovelock's Gaia hypothesis, in which 'life-like' is extended to intelligence and consciousness. The correct part of the argument is this. Any such complicated energy-transducing system as the early Earth must have had billions of recursive subsystems. These will have ranged widely in scale: examples include the melting/crystallising energetics involved in establishing the core/crust boundary, the interaction of a spinning planet with the solar wind and the generation of a magnetic dynamo system, and the molecular-level interactions between clay and organics that resulted in both layered geology and lifeforms. Some of these processes lasted only until they had used up their substrates, or otherwise burnt themselves out, or some sub-process hit the zero-line. Those were very temporary history. But there would have been others that had some self-reinforcing circularities, and negative feedback loops to keep them within bounds.

Those systems would have lasted longer, and would have become tied in with other similar systems that formed part of their context. This build-up of symbiosis between systems, and evolution of the larger systems they contributed to, gave us a planet with many of its large-scale systems overtly in various kinds of dynamic balance. Those that couldn't fit or last are not around any more, or are not systems: their processes have been subverted into other, more robust processes. So, even without life, the Earth can be understood as a complex of complex systems, most of which have sub-processes that keep them within a viable range of variation.

This is Lovelock's concept of Gaia: the Earth as a self-regulating 'organism'. In the above sense, the Earth is indeed life-like, but some enthusiasts go much further and attribute consciousness to it. Here is where we and Lovelock disagree with the enthusiasts. Instead, we hold that the existence of Gaia-like properties in the Earth demonstrates how complex systems that are not carbon-based can become life-like.

Dawkins has pointed out a neat gloss on this question: to what extent can Gaia be considered an organism? While conceding that the processes involved in the maintenance of various balances and continuing complex processes are rather like simple lifeforms reproducing in an ecology, he denies that the whole Earth is genuinely like an organism. This is because the whole Earth, unlike those sub-processes, has not been competing with anything for resources, has not been *selected* for existing – it just is. Whatever the sub-processes do by way of interaction and competition, the Earth does not 'do' better or worse; there is no scale on which the Earth prefers sub-process A to sub-process B.

We disagree with this argument, though not with Dawkins's overall position on Gaia. We believe that there *can* be top-down control, starting from the most mechanical constraints. Some sub-processes are very compatible with each other and with the structure of the Earth, while others are not so compatible. It is not only the other sub-processes that constrain any system, but also the amounts of substrate available, and the temperature relations, both within the Earth and in Earth's relation to the Sun. So the complexity of Gaia at any moment is both a function of Gaia overall and of those sub-systems that are compatible with 'Her' and with each other.

And with cosmic accidents. The life subsystems were anaerobic marine, then aerobic; fishes, then dinosaurs. But the dinosaurs met their end because of a meteorite that just happened to occupy the same bit of space as the Earth. Some SF authors have suggested, not very seriously, that this was Gaia's way of getting rid of the dinosaurs – who clearly could not protect the Gaia system from big meteorites – and replacing them with another set of big terrestrial vertebrates, one of whom might be able to. In this view, if humans ever get round to deploying cometary defences, then this will simply be another layer of Gaia. Gaia didn't think it up, it was implicit in the nature of materials and processes – but Gaia will be protected from meteorites in the future *as if* She had wanted to be.

It happened to be carbon chemistry that did the big trick here on Earth. That's where we look for life now. But it genuinely would not surprise us to find that there are plasmoids in the Sun's outer layers, or that the universe is much older than the twelve billion years or so that physicists currently estimate, from present-day data attributed to the Big Bang. The data might be misleading, because 'living' processes could have been parasitic on, for example, the inflation phase of the universe. Or the remnants of such 'lifeforms' have led us to *assume* that there had to have been such an inflation phase. We used both of these ideas, semi-seriously, in *Wheelers* – because they were a good way to persuade our readers that life will appear anywhere it can . . . and also anywhere it can't.

Several fictional aliens have been used to make the same point. In Fred Hoyle's *Black Cloud,* particles interconnected into a complex system by radio form a sentient network. Stanislaw Lem's *Solaris* envisages a sentient ocean with mystical and mystifying properties. Robert L. Forward's *Dragon's Egg* is a neutron star inhabited by the quasi-crystalline cheela, believable alien lifeforms of exotic provenance. In many ways the cheela show the complete puzzle laid out, the existence of genuine lifeforms with a wholly different basis from organic chemistry. They do not even share our kind of matter. Yet they inhabit a world that is entirely consistent with our current understanding of physics, they exhibit biology and sociology that are reasonable, and finally they evolve – believably – far beyond our capabilities.

We have learned so much about Earthly life, so quickly, and with such amazing results, that we forget how special it all might be. Ask a biologist what life is, and ten to one you'll get an answer containing 'DNA'. On this world, that's a good answer, in that it states a crucial feature of all (well, most) life here. However, it is an answer that addresses a parochial realisation of a process that is probably far more universal. 'Life' is a more general phenomenon, capable of many distinct realisations, employing many different types of organised matter – and even non-matter.

Part of the evidence for this proposition, which forms a key distinction between astrobiology and xenoscience, is the rapidly growing area of research known as 'artificial life'. This is not about making Earthlike lifeforms in the laboratory, but about hypothetical complex systems whose properties resemble those that make lifeforms interesting. The scientists who study artificial life, usually through computer simulations, have constructed systems that by all the usual criteria, except that of being based in organic chemistry, conform to our ideas of 'living'.

Computer viruses are an all too familiar example. Several computer scientists have invented little evolutionary ecologies in computer systems. We are not very impressed by the ones, like Dawkins's 'biomorphs' in *The Blind Watchmaker*, that are refined using artificial selection *by a human*. We do see that more sophisticated programming could eliminate the human element, but this has to be *done*, not just asserted. We are much more impressed by Tom Ray's Tierra system. Tierra plays out its evolutionary drama in the memory of a computer, and its 'organisms' are strings of binary digits. Ray discovered that its evolutionary lineages diverged into parasites, symbionts, efficient organisms that did well in the short-term, and more robust systems that outlasted them. A form of sexual reproduction evolved surprisingly quickly, as did rudimentary social behaviour. None of these possibilities was 'built in' explicitly, or even cryptically, beforehand. They emerged naturally as high-level patterns of the system, unpredictable consequences of the simple bit-level rules.

If there are advanced aliens out there, they will surely have the equivalents of our computer viruses, and probably of our computer games. (Play is probably a universal.) They may well have made, invented, initiated self-replicating machines for mining and manufacture. Our nearly-automated factories are not far from that – imagine an automated factory set up to make basic programmable automated factories . . . Come to that, there could well *be* something like that working on Earth *now*, given the speed with which robotics and nanotechnology are advancing.

Many thinkers, following this road, have argued (in some ways like the anthropist we've been so unkind to) that there are *no* extelligent technological aliens out there because, if there had been:

1. The aliens would have built automated mining machines and suchlike.
2. Improvement of these would have led to their becoming self-constructing and self-repairing.
3. The aliens would doubtless have seeded them on to the other bodies of their solar system.
4. Some of these self-replicating machines would have escaped that solar system – they would have spread.
5. Even at very slow spread speeds – say a walking-pace – they would have spread throughout the galaxy by now.

However, we haven't found any such machines in our solar system. Ergo, there aren't any aliens out there.

We don't buy this as a reverse-explanation of the Fermi Paradox, 'If intelligent aliens exist, why aren't they here?' The whole argument is incredibly fragile, and assumes that the aliens will be like us in all important ways. But we should not be surprised to find that aliens – or humans – with a level of technology very little above ours have made self-replicating and self-modifying machines. We're close that doing that ourselves.

There are many arguments for supposing that any 'aliens' that we meet will not be the organic beings themselves – who like us will not be well-adapted for lengthy interstellar travels – but their sophisticated machines. *We* will soon be sending such robotic probes out, after all, and although aliens will doubtless be different from us in many more ways than we can imagine, this does seem an obvious strategy. Which makes it a potential universal. It is the obvious way (to us anyway) of combining the short lifetimes and fragility of organic living creatures that originated on planets, with the durability of boredom-proof machines – exploratory robots. So we would be surprised to meet organic beings across stellar distances, but we would not be surprised to meet their – non-replicating – robots.

However, don't restrict your expectations to a picture of a greeting robot – for example the Tralfamadorian in Kurt Vonnegut's *The Sirens of Titan*, who went right across the galaxy in order to carry the message 'greetings'. The monolith in *2001* is just such a mechanism, too. And the aliens may well be on Earth, all over the place, sending their messages back to the home planet, or at least to relay stations on the Moon. An advanced technology (say a million years old, or five hundred million, compared to the two hundred year history of 'advanced' human technology) will be able to make, distribute, and hide artefacts that we cannot begin to imagine. That may well be the most likely scenario: Fermi's Paradox doesn't apply because the aliens are here, but we simply don't recognise them.

Hi, guys.

Other possibilities for exotic kinds of alien include symbioses and cyborgs ('cybernetic organisms', mixtures of organics and machinery). A major lesson from our own history is that few organisms consist simply of only one lineage. You have many kinds of bacteria living in your gut, and quite a variety on your skin, that would be necessary to your well-being without processed foods and without soap. But in a very deep sense, all of us eukaryotes on this planet have several lineages combined into each of our cells. The mitochondria that power those cells still have remnants of their ancient bacterial DNA sequence. Many chromosomal mechanisms are performed by proteins whose genes can be traced right back to the archaean organism that accommodated the original free-living oxygen-using mitochondria. Centrioles, tiny organelles that construct the apparatus for sharing chromosomes among daughter cells during cell division, and possibly flagella and cilia too, are remnants of other originally free-living organisms that have had most or all of their heredity taken over by the cell nucleus. This looks like another universal, so we would expect aliens to have piecemeal lineages too. Much complexity and versatility in 'higher' organisms on Earth is due to a symbiosis, different organisms sharing their pool of different abilities.

Would aliens have special 'powers'? We don't believe the most popular suppositions about alien abilities: telepathy and other 'paranormal' phenomena. These are no more than hopeful transpositions in myth from the world of ghoulies and ghosties. There isn't a physical basis for how they work. The alleged 'evidence' that some humans have rudimentary paranormal abilities is seriously flawed, and not credible.

To be sure, Jack had to invent a plausible scenario to make teleportation and telepathy possible in McCaffrey's dragons – more specifically, in the fire-lizards from which they evolved. To achieve this, he appealed to the well known and widely accepted trope of FTL (faster-than-light) travel: readers would go along with the story that a few creatures had been found previously, on different planets, that exploited hypothetical FTL physics. Jack invented a scenario in which the fire-lizards disappeared from under the hands of those who were about to investigate them. The human characters took this in their stride, so he and McCaffrey hoped readers would do the same: 'Oh, they're like the so-and-so's on planet X, we need a mu-metal mesh to hold them down, I guess.' Telepathy was not much more difficult to nail down to the real world – of that story. A few people among the colonists were descendants of an experimental group of people who had been infected with 'mentasynth', a device used by ancient aliens to assist in – possibly-radio-mediated – communication without external machinery. The experiment had been a failure, however, and the few people that McCaffrey needed to have the 'Power' were unfortunate descendants of these mistakes.

By making the ability a downgrade rather than an upgrade we thought that the story could carry conviction even to those – like us – who didn't believe in natural mind-to-mind communication. Aliens will certainly have abilities that we don't expect, including some that we won't understand and that will seem like magic – but not ones that are physically impossible . . . to their technology, remember, not to our primitive one. However, certain types of alien – like Vinge's Tines – could have *apparent* telepathy that is actually just analogous to human speech, but realised in a different way.

Extelligent aliens, like us, will supplement their abilities with mechanical aids, and these will rapidly attain, as ours have in some cases, magical status. We were impressed by the brief view we got of the alien in the *Independence Day* film; you couldn't tell which bits were him, and which were clothing, weapons, or prosthetics. In the course of the evolution of extelligence, we would expect increasing sophistication of the prosthetics for the intelligent organic creatures. We have spectacles, hearing aids, special energy-conserving rubber on the soles of running shoes . . . and we will have much more and better as our technology improves. Sockets for direct neural interfacing with computers are not far off now, and 'intelligent' prostheses will be close behind.

From our experience of higher-level technical civilisation, we would expect more variety in the intelligences, the organisms that run it and are run by it. In today's human society we increasingly find people specialising for different jobs. Today, even the infantry has a different speciality for each soldier: no longer are they generalist 'grunts' whose main job is to die. So our technical cultures are increasingly symbioses of differently specialised intelligences, with some great branches of our extelligence rapidly becoming different 'species': musicians, artists, computer technicians, mathematicians. Our cities *need* integrated communities of these different species of extelligence, just as they are beginning to thrive on multicultures where there are twenty kinds of restaurant and ten kinds of visual entertainment. This trend seems evident in all affluent societies, but we cannot be certain of the extent to which it will be a universal. However, we'll stick our necks out and guess that alien extelligence will diversify similarly, and that it will generate increasingly sophisticated software and hardware – perhaps also wetware, brains with enhanced abilities – to do it with.

The prospects for our own species were given horrific exposure in Bernard Wolfe's *Limbo–90*. In this story, young men often volunteer to become 'Immobs': their limbs are amputated and replaced with much better nuclear-powered mechanical ones, their eyes are replaced by much more sensitive, much wider-ranging and more acute sensors, and so on. The story revolves around two primitives, who have not had this done and therefore still have their natural organs – and only their natural organs. There is a terrifying chase scene, in which they are hunted down by improved humans who can run at 50 miles an hour (80kph) forever, who can see in the dark or see warmth by infra-red, and who have other enhanced abilities like built-in 'telepathic' communication (a bit better than mobile phones, that is to say). Both aspects of this plot have attractions for the reader. We can understand why the primitives don't want to lose their legs, eyes, and hands, yet we can also sympathise with their pursuers who genuinely want to help them. It's not like being hunted by inquisitors who want to save your soul by (if necessary, that is, if you disagree) destroying your body. In this case it really will be better for the primitives if they are caught. What makes this novel so enormously potent is that readers know that in less than a hundred years their grandchildren will be in just that position. And this tendency will increase the further we proceed into the future.

So far, with our false teeth, reading-glasses, hearing aids, and wooden legs, we've not gone very far along that path. However, we appreciate that as prostheses improve on nature, there will come a time when everyone is a cyborg. So when we meet aliens who are ahead of us technologically – as very nearly all of them will be – their physical form will mostly be an assembly of technical tricks, and they may have left their organic history behind completely. Or they may have some other, alien, strategy whose pattern we have not begun to guess. *Limbo–90* puts our minds into a new state, where we would expect any advanced civilisation – one with 20,000 years of technology, say, instead of 200 – to be composed of individuals who are enhanced beyond our dreams. (Or possibly no longer individuals at all, but let's not pursue that development here.) Considering how far out, from our parochial organic history, any aliens that we find will be, leads inexorably to the conclusion that we will find them incomprehensible: not only as minds, but also as structures. Their biology won't be relevant, because whatever biology they still retain will have been altered beyond recognition, even by a primitive of their own species.

In such circumstance much of this book – the parts that argue for differences in the biology of alien lineages – will be irrelevant to our real meeting with real aliens if that event is too long delayed. By then they, and probably humanity too, will have left biology behind. But our speculations will still apply to the organic stage of development that brought them into existence, and that will govern how they acquired the technology to rebuild themselves.

11

THE SENSUAL TRIBBLE

ONE OF THE *Kleptosporidians has caught a virus, and there is a heated argument until its parent insists that it has enough pets already and must put the poor thing back where it found it. Abel is inspired to make a search of the younger tourists' baggage and confiscates several further items of contraband. Their protests are distracted by the announcement that the final trip for this bunch of tourists is a party.*

Cain and Abel have decided to show off the really high-technology abilities of their St Albans base, and the party is an excuse. The tourists all get processed for the event, and Cain stacks their various bodies in all the proper systems. Then he collects Abel, some of whose particles now have tourist ego-flyers attached.

The communication system allows some chat between the various lifeforms as they head towards the bus. Then, suddenly, they are all caught up in the most frenzied bit of living they've yet experienced. All around them are organisms stretched to their limits, adjusting to new conditions from moment to moment, changing the milieu for all the others. Their sizes range from thousands of times bigger than the little flyers, down to a millionth of that size.

There are organic chemicals of all kinds, from long paraffins and proteins to sugars and strange forms never seen in chemistry laboratories. The aliens have a fine time tasting the chemicals – and, it must be said, the inhabitants – with fine abandon. 'It is like being present at the origins of the planet's life,' they all say. They agree that it has a freshness that the other trips have lacked.

As usual on this trip, when Cain tries to have a roll call, several have gone missing; they will have to be reconstituted from previous memory-

recordings, but they won't have missed much. Abel slowly lifts his load of
egos out of the party, so that they can see where they have been. It is a little
pool of dirty water, with an oil smear over the top of it, in one of the bus's
tyre-tracks. It has provided one of the most exciting experiences of Earth's
lifeforms of the whole trip.

Do aliens have sex?

More specifically: is sex a universal or a parochial? If it is a universal,
then we will expect many alien species to have evolved *some* kind of
sexual reproduction . . . by which we mean that the offspring acquire
characteristics from more than one parent. Whether they acquire them
genetically – that is, by mixing up 'coded' representations of parts of the
developmental process – or by some other mechanism, doesn't matter
(yet). Whether they have two parents, or a trillion, doesn't matter (yet).
Heavy breathing and/or emotional excitement may or may not
accompany the process: that doesn't matter (yet) either.

It is a matter of record in terrestrial biology that sex has been
repeatedly lost and regained, or reinvented, in many lineages. Would
we expect this in alien lifeforms too, or is it a peculiarity of our almost-
panterrestrial DNA-recombinational system? Unfortunately, sexual
processes are rarely fossilised, but from the vast variety of fusions and
recombination mechanisms, it is clear that there is great pressure to
keep sex, or to reinvent it, and that argues in favour of some kind of
universality. If sex offers generalised evolutionary advantages, then we
would have theoretical reasons, as well as empirical ones, to expect it to
be a universal. So one important question about Earthly biology is:
what are the advantages of sex?

There are many other questions about sex, some of them less
terrestrially oriented. If aliens do have sex, will we find instances with
three or seven sexes instead of our conventional two? What are the
rules, the constraints, for them as compared to us? And as compared to
fictional aliens? These are very different kinds of question, which
require equally different styles of answer. If we are to remain scientific,
then we must first locate the theories and the data in the areas of
knowledge from which the answers must come. Then we must attempt
to disentangle what in other books we have called the 'lies-to-children'
aspects of the theories. That is, the oversimplifications that are always

needed when teaching (and learning) complicated ideas for the first time. After that, we must sort among the rest for those that deal most rationally with problems in the area, bearing in mind that any of these theories could look pretty silly in a hundred years' time.

The main point of this chapter is that sex *on this planet, now*, is far more complicated, and its variants are far more bizarre, than most possibilities found in SF stories. So alien sex has an awful lot of Earth parochials to parallel, even before it gets to anything radically new.

Before we can discuss alien sex, we need to consider the replication process itself, because all else rests on that. Is replication of hereditary material absolutely basic? Is it fundamental to reproduction? Today we hardly ever ask this question, because 'genes' have been consigned to the deep cultural background; we assume that *of course* hereditary material is fundamental, but we seldom ask why.

In fact, the necessity for heredity is not so clear-cut. Doubt appears as soon as we start abstracting the process: flames, after all, do reproduce, but they don't need to replicate their hereditary material – they don't have any. How, then, can they reproduce? The answer is that each new flame accesses the same regular behaviour of flammable material, the same rules (such as 'warm air goes up') for making the flame its customary shape and drawing new oxygen into the base to fuel heat-producing chemistry.

Many other physical systems also show multiplying behaviour without possessing an overt heredity. A cute example, with the possibility of becoming a neat alien invention, concerns the reproduction of tori. Doughnuts don't usually multiply, but a smoke ring is a more dynamic torus, a vortex in the air marked by the accompanying smoke, and that begins to have interesting properties. 'Smoke rings' in liquids can exhibit reproduction, if the trick is done properly. Find an old-fashioned ink pen with a device (a pump or a bladder) that will permit you to release just one drop of ink at a time from the nib. Stand a pint glass – a tall transparent cylinder like those used to store spaghetti is better – for an hour or more to allow the movement of the water to die down to essentially zero – don't jog the table. Then release an ink drop from about half an inch – a generous centimetre – above the surface. The splash will generate a torus that will slowly fall through the liquid ... then, quite suddenly, about two inches below the surface, it will generate between three and eight baby tori and these will fall through the water. These will again multiply, and with the right ink and a tall cylinder you can get six generations.

Larry Niven was very taken with this phenomenon, and he and Jack used it to invent a new and original alien. First, you need to know about the so-called Bussard ramjet spaceship, a standard feature of Niven's 'known space' novels, which collects hydrogen from space using a magnetic trap and fuses it at the focus, producing ample thrust energy by using the hot products for rocket propulsion. The pilot of such a ship is investigating a new dust cloud, like the Horsehead Nebula in Orion, where stars are condensing all around. To his surprise, he sees another ramjet ahead – but he knows that he's the only one from his home system. This must be an alien starship. Wow! First Contact with an extelligent alien species! He tries all his radio tricks, but can get only 'noise'. However, he knows that sophisticated messages are always encrypted, so (Lachmann's theorem) they *do* sound like noise – so he keeps trying to contact the other ship. Meanwhile he gets on its tail and begins to catch it up; that doesn't work, of course, because he can't burn the other ship's exhaust. So he runs around to ahead of the other ship – which suddenly disappears. 'Oh my God,' he thinks, 'the aliens have FTL travel – that ship's gone into hyperspace.' Soon he finds another spaceship, chases that and the same thing happens. The plot gimmick is that these are not alien technology, but *aliens*: 'natural' Bussard ramjets, reproducing lifeforms created by an unusual – but by no means impossible – collapse of a proto-star into a torus instead of a sphere. The torus didn't disappear because it suddenly went FTL: it disappeared because it died, unable to 'eat' the ramjet's exhaust.

These simple reproductions without heredity can just about support a story, but the problem is to imagine them *evolving*. We pointed out earlier that a credible alien lifeform must not only 'work' as an organism: there has to be a plausible route for its evolution. We said then that it is difficult to imagine an organism that works but does not possess a reasonable evolutionary ancestry, but the Bussard tori are so closely linked to specific physics that finding an evolutionary route that could lead to them is distinctly challenging. It is probably necessary to have mutable heredity in order to produce an interesting evolutionary development, but Bussard tori reproduce like flames, without heredity, and therefore they can't evolve in any very interesting ways.

Earthly lifeforms do have flexible heredity, which is presumably why very complex organisms have been able to evolve. The chemical basis of that heredity is DNA and its associated molecular assistants. The machinations of DNA and RNA are the only workable system of this kind that humanity has observed. Are other systems available? To approach this question, we have to extract an underlying universal from the terrestrial DNA parochial; then there is a good chance that an alien species might realise that universal using some *other* parochial instance. In other words, we must generalise the hereditary system that we use before we try to consider alien versions of 'it'.

In the late 1940s the mathematician John von Neumann was thinking about self-copying machines, although his ideas were not published until 1966. At that time, many people thought that there was a philosophical obstacle to the existence of such machines – the 'homunculus problem'. This goes back to an old debate about human reproduction. Sperms and eggs somehow 'contain', if only potentially, the complicated human being into which they grow. Including *its* sperms or eggs, which must also contain the complicated human being into which *they* grow. And so on . . . for ever. Your own body somehow 'contains' an infinite sequence of your future descendants, like nested Russian dolls. One theory took this image literally, and held that every sperm contains a tiny human being, called a homunculus – complete with its own sub-homunculus, and so on. But atoms set a lower limit to the size of material objects, so this idea makes no sense. Living creatures must reproduce by some other method – but what? There was a feeling that whatever that method was, it involved special, mystical properties of 'life', and was thus impossible for a machine.

Von Neumann exploded this view by devising a completely abstract mathematical system capable of making copies of itself. It was a 'cellular automaton', an array of cells like a gigantic chessboard. Each cell could be in one of a number of internal states, and at each tick of a clock, the state of a cell would change depending on what the states of the surrounding cells were. He set up suitable rules for those changes, and came up with a 'machine' – a configuration of states – that could copy itself. Its cells had twenty-nine states, and there were 50,000–200,000 cells (the literature is contradictory on this point). In 1968 E.F. Codd reduced the number of states from twenty-nine to eight, at the expense of having 100 million cells – later reduced to 100,000 by J. Devore in the 1970s, and published in 1992 by Devore and R. Hightower.

Von Neumann's automaton avoided the homunculus problem in a very clever way. It came in three pieces: a 'blueprint' which specified in coded form how to build the entire machine, a 'constructor' that could make *any* machine given its blueprint, and a 'copier' that could copy blueprints. The blueprint specifies the structure of both constructor and copier. The machine reproduces by using its constructor to read the blueprint and make a new constructor and a new copier. Then it uses its copier to make a new blueprint. Its constructor, still obeying the original blueprint, now joins the three new pieces together: done.

The homunculus problem is avoided here because the blueprint has a dual interpretation. To the constructor, it is a list of instructions to be obeyed; it results in actions by the machine. But to the copier, it is just an inert list: to be copied, but not to be obeyed. At first sight it looks as if there is another trick that makes the process work: the constructor and copier are *universal* devices. The constructor can turn any blueprint into a machine; the copier can copy any blueprint.

Strictly speaking, it's not necessary to go that far: all we need is a constructor capable of obeying the blueprints for a constructor and for a copier, and a copier with similar limitations. Chris Langton invented a specialist replicator in 1984, a P-shaped loop of eighty-six cells in a two-dimensional eight-state cellular automaton. However, universality of the constructor and copier has a fascinating side effect: the machine can evolve. Changes to the blueprint cause changes in the resulting machine. Random changes probably lead to a machine that doesn't work, but there might be changes that lead to a machine that can still copy and construct. If so, it is a 'mutant' self-reproducing machine.

Von Neumann invented this system well before Francis Crick and James Watson worked out the structure of DNA, but he didn't publish it until later. Our DNA-based organisms use the same generic trick. The blueprint is the DNA sequence. The copier is the cell with its battery of molecular tricks to pull DNA strands apart, form matching strands, and reassemble them. The constructor is the organism; the construction process is the events leading up to cell-division, or in more complex lifeforms the entire process of biological development. The Von Neumann three-part system is a universal, available for alien implementation in innumerable parochial forms. Earth's DNA chemistry is merely one such instance.

Now we're ready to think about alternatives to the usual reproductive set-up. Recall the key features of DNA: it is a linear coding system that can be transcribed and translated into proteins, which have their own complex catalytic structures. That is how DNA is 'turned into' organisms. Furthermore, DNA has a clever replication system. It consists of two complementary strands paired together: unzip them and arrange for each to build another complementary sequence to itself, and lo: two *pairs* of sequences, identical to the first *pair*! This replication mechanism is a so-called template system. It is (conceptually) simple, and numerous variants are possible. For instance, we could make a two-dimensional one with CDs: the master disc from which they are pressed, and a pressing from it, form a pair – which could, in the pressings factory, generate more identical pairs. So could a photographic negative and the print from it. Note how different these pairs are from a single letter that can be replicated on a photocopier, or a computer floppy whose message can be copied to another one in a computer.

We are not simply considering copying here, where God puts a photocopier in the corner of the office, or Ray puts a code-copier available to the denizens of his universe Tierra, which resides in the memory of his PC, and that generates evolutionary ecosystems of computer-codes. Each member of the template pair constrains the message on the other, whereas a succession of photocopies from a letter, each being copies from the one before, degenerates rapidly into noise. In principle, a correction system can be applied to template pairs, each copy having the other half to check itself against. So template systems are a good trick for alien heredity, and as such are likely to be universal; but there's no reason for alien reproduction to be restricted to DNA, to a one-dimensional message, to carbon chemistry. And, as always, 'universal' does not imply 'unique', as we see a little later.

There are more abstract lessons to be picked up from Earth's tricks with template replication. In principle, the templates can be replicated alternately, allowing a phase shift in alternate 'generations'. Aliens might alternate their generations in this way. The proposal may seem absurdly complicated, and it probably would to an astrobiologist, but the truth is more prosaic. It is not generally realised, even by biologists, that *most terrestrial organisms do this*.

Viruses show how the two aspects of their reproduction, their infectivity and the replication of their heredity, can be separated. When a virus particle infects a cell, its protein coat does the work: the coat interacts with receptor molecules on the cell's surface, and the virus bullies or wheedles its way inside. The hereditary material is usually DNA, and when the virus has got into the cell or bacterium this DNA links in to the cell's own mechanisms for replication. For a virus, the photocopier *is* indeed sitting in the corner of the office – think of the protein coat as the key to the office. The viral DNA is often incorporated into the host's own DNA, and copies of the next generation of DNA are generated from it. But the cell has to make messenger RNA: an RNA 'photocopy' of the DNA sequence in the nucleus is part of its normal processes, and this step lets the virus use the cell's protein-synthetic machinery to make viral protein-coats. The viral particles assemble, DNA packed inside the infective protein coat, before the cell bursts and releases them . . . and each has *two* generations in its structure. The protein coat is that of the parental DNA-transcription/RNA-translation; but the DNA-'copies', inside those particles, are the *next* generation, *progeny* of the original strands. So a virus's protein coat is translated from DNA of generation n; its DNA contents are at least generation $n+1$, and may even be generation $n+7$, depending on how the DNA is replicated in the host cell.

This 'phase-shift' of generations also happens for more complicated terrestrial organisms. In fact, it happens pretty well everywhere in terrestrial biology. When a bacterium divides after replicating its DNA circle, the two new circles sit in *mother's* biochemical machinery, are enclosed by *mother* bacterium's membranes. The same is true of the slipper-animalcule *Paramecium*, but these little eukaryote creatures have a complicated coat of cilia, and an anomaly in the pattern of these cilia (perhaps a scar from an accident, or the result of shaking the culture vigorously) can be passed through many generations. This implies that the coat of cilia has a separate replication mechanism from the *Paramecium*'s normal DNA heredity. A chick embryo also has two generations in its make-up: the egg structure, food material and biochemical machinery were maternally provided (generation n), but the little bird's own nuclei are generation $n+1$.

Reproduction, then, is not as straightforward as we usually think. The complexities and alternatives are crucial to xenoscience: when nature makes an alien, it will use whatever is available and suitable, not whatever human beings feel comfortable about. The same point arises when we think about alien sex. We must be aware of the peculiarities of Earth's own organisms, because those help to expand our mental horizons and stop us placing unnecessary limits on what aliens might do. We must also be aware that 'peculiarities' is the wrong word: behaviour that seems peculiar to us is perfectly normal to whichever organisms are doing it.

Von Neumann's strategy for replication is a clever way to overcome the homunculus problem, but in principle we can envisage radically different alternatives. Indeed, the homunculus problem may not be a problem at all. In *Wheelers* we invented aliens without heredity, which look like robotic machines (some have wheels, hence the title) but are actually excreted by the organic balloonist Jovians to get rid of metals. The wheelers' heredity resides in the Jovians, not in the wheelers themselves. The idea is far-fetched, and we introduced it deliberately to show that there is no clear dividing line between life and non-life; in a sense the closest terrestrial parallel is with viruses, which employ the replicative machinery of a host bacterium to reproduce – but viruses contain genetic material of their own to direct that process, whereas wheelers don't.

Here's another possibility. Suppose machinery could exist that can 'scan' any object at the atomic level, and build a copy by depositing atoms in a similar arrangement. A sort of 3D photocopier with *very* high resolution. Copiers of this kind could replicate by scanning themselves (or each other). Any mistakes would be faithfully preserved, which is the feature of Von Neumann machines that renders evolution possible. In such creatures, the genotype is effectively identical to the phenotype. So you could argue that they do have heredity – but they don't play the Von Neumann trick of having a code with a dual interpretation.

Atomic-level scanning is rather ambitious, but there is a variant on this idea in which the creatures are built from standard components, like Lego™ blocks. The copier-creature disassembles the one that it is going to copy, recording the arrangement of components; then it rebuilds the original plus one or more copies.

It's a reverse engineer.

Having dealt with reproduction, we can return to the main topic: sex. What is the fundamental issue of sexuality? According to sociologists, the key issue is gender roles, but if we are looking at biology far outside the human condition then we are driven to the genetics of the reproductive cells: the spores, eggs or sperms that pass genetic heredity into the next generation. In conventional human-style sex, the genome of the offspring carries a combination of alleles, some coming from one parent, some from the other. The key feature here, a good candidate for a universal, is not the details, but the mix-and-match genetics. Essentially, sex occurs when an organism carries the heredity of several simultaneous ancestors, and is not just a clone of one ancestral organism. On an alien world, Von Neumann machines could mix-and-match their blueprints by whatever process is appropriate for their parochial realisation of this great universal. Lego™-copiers could combine different assemblies of components to make a new one. On today's Earth, computer hackers refer to SEx – Software Exchange – and warn of SExually transmitted viruses. Inside those computers important problems are often solved using 'genetic algorithms', which interbreed potential solution methods and select those that perform better, trying to *evolve* the best method they can find.

The Lovers (Philip José Farmer 1961)

The novel is an expansion of a short story published in *Startling Stories* in 1952, and it was one of the earliest SF tales to explore sexual issues. Hal Yarrow is a joat (Jack Of All Trades). His marriage is on the rocks. The Sturch (State-Church) believes in unlimited procreation; already Earth is heavily overpopulated and humans occupy their world in shifts, twenty-four hours a day.

The Sturch has found a planet, Ozagen, that is suitable for human habitation. The *Gabriel* is being sent there, male crew only; any married man who volunteers will be given an automatic divorce. Yarrow promptly volunteers. Ozagen is already inhabited, by sentient aliens; the Sturch's solution is simple: extermination.

The Ozagenians are humanoid, with round bodies and skinny limbs; they have mouths like two shallow V's, with four lips. In place of teeth they have serrated jawbones. Their skin is pale green: they use copper, not iron, to carry oxygen in their blood.

While researching the language of Ozagen's chief nation Siddo, Yarrow meets Jeanette Rastignac, an entirely human-looking woman who tells him that her parents arrived on Ozagen many years earlier. Against the teachings of the Sturch, he has sex with her without the intention of procreation, and they become lovers. But he secretly subverts the precautions that she insists they take, and she becomes pregnant.

Only then does he learn the truth: she is not human, but a *lalitha*, a mimetic parasite. The *lalitha* are intelligent, and female. They co-evolved with humanity from an ancient parasite; they are insect-like pseudo-arthropods. And when a *lalitha* reproduces, her larvae grow inside her body and eat her from the inside.

Surgical intervention fails. Yarrow's grief leads him to revolt against the Sturch, which triggers an Ozagenian insurrection. The Sturch's genocidal plans are wrecked, but Yarrow still has to live with his past.

Among Earth's organisms, sex happens in many very different ways. A simple 'primitive' sexual strategy is the almost-random swapping of hereditary material that occurs among bacteria and archaeans, mediated by special sexual tubes joining the participants, or by virus-like particles, or by just picking up DNA that has been left lying around. As far as universals go, their almost haphazard disposition of hereditary material should be interpreted as sex, too. There are many creatures that seem, like the classical *Amoeba*, never to have been sexual – every individual is a clone, has only one ancestor in *any* previous generation (back until the eukaryotic ancestor separates into separate ancestral prokaryote stem lines, that probably *did* swap genes). Other creatures, like the insects, seem always to have been sexual – like us, their ancestors increase in number as you go back: there is no convergence to an Adam and Eve. You have two parents, four grandparents, seven great-grandparents (or thereabouts . . . not always eight because there is often a cousin marriage in the family tree by that stage) and so on.

The problem of explaining sex is made more acute because many asexual animals are descended from sexual ancestors. We know that because they produce eggs in the sexual way and have the right anatomy to get them fertilised, and they occasionally produce functional males. The tropical dock cockroach *Pycnoscelis* and the Indian stick insect *Carausius* have simply lost the need for fertilisation – females produce females from eggs that don't need sperms. If an SF writer had invented aliens whose children emerge *alive* from the adult body when it is accidentally killed, you wouldn't believe it, but in *Pycnoscelis* the eggs start developing in the female's ovary. It is very dramatic to step on a *Pycnoscelis* and have its babies run out from the burst body all over your shoes. And a surprisingly large number of common plant weeds, even those with very dramatic flowers like dandelions and daisies, mostly multiply asexually – only very occasional seeds have been fertilised.

Many other asexual forms are said to be amazonogenetic, which means that the females use males for other reproductive purposes, but seldom, if ever, do any of the males' genes pass to the resulting offspring. The term comes from the legendary Amazon tribe of women, who raided local villages for their men, and correspondingly, amazono-genetic females still need males and sperms to trigger the development of their eggs. But they use the males of closely related species, and the genes of the males are only rarely incorporated into the offspring. These amazon species are usually generated by what seems a very unlikely event, the conjoining of three hereditary systems by the mating of a female 'mule' hybrid between species A and B, with a male 'mule' between B and C; female offspring are occasionally produced from such a miscegenation, with an ABC heredity.

Most terrestrial sexual forms are diploid: that is to say, each individual has, in each of its cell nuclei, two sets of pretty-well-identical chromosomes, one from the father and one from the mother. In the production of sperms these chromosomes are scrambled together, and each sperm receives just one set, with contributions from both parents. Each egg reduces its diploid set to a scrambled single set too, and the two single sets of egg and sperm fuse to form the nuclei of the new individual. But in amazon species, most cells are triploid (having three sets, one each from the three species A, B, and C), and in order to make triploid eggs they usually go hexaploid (six sets) and then undergo an orthodox reduction to half as many, triploid again. The A, B, or C males oblige by producing sperms, whose entry into the eggs starts development.

Sounds unlikely? Not at all: several of the most common earthworms do this, and the most common frog on the European mainland, *Rana esculenta*, is an amazon. Four of the most common North American lizards are amazons. The amazon molly, a viviparous fish closely related to guppies, swordtails, and platies, and sometimes found with them in tropical-fish shops, and in most Florida streams and ponds, was the first example of amazonogenesis to be explained by biologists. There are also grasshoppers, crickets, stick insects, and crabs, whose *usual* representative is an amazon.

Having established some of the variety of terrestrial sexual strategies, we have a general context in which to consider whether sex is universal or parochial. First, the theoretical test: does sex offer a generic evolutionary advantage? Strangely, there is no general agreement among scientists about the advantages of a sexual reproductive strategy. The simplest view, popular in textbooks until the seventies, is that there is *no* advantage: on the contrary, asexual females should outbreed sexual females because they generate two producers for each one – plus an unproductive male – that would be produced by the sexual form. Males 'cost'. Indeed, many asexual forms have arisen in sexual species, and they are all female – as the above argument would lead us to expect. Moreover, they do indeed form a majority in many ecosystems. However, they seem usually to be temporary, ousted by another asexual form newly arisen from the sexual form.

What about the observational test: did it happen independently more than once, *here*? It is difficult to find out what terrestrial sex was like in the past, because sexual processes are rarely fossilised, but from the vast variety of fusions and recombination mechanisms to be found on today's Earth, it is clear that there has been great evolutionary pressure to keep sex or to reinvent it here. So presumably sex has universal evolutionary advantages – it is a successful strategy against universal survival problems – so it is likely to be a universal.

However, this argument falls a bit flat unless we can work out what those advantages *are*. A crucial question is therefore: What is sex *for*, in an evolutionary sense? At the moment there are several good suggestions, but there is no scientific consensus that favours just one of these. So we can argue that sex is a universal, like fur or wings, but we must be careful when extrapolating sex to other planets because its evolutionary utility may not be so obvious as in those suggestions.

First, we must acknowledge the existence of arguments indicating that sex is actually a *bad* idea, evolutionarily speaking. Many biologists have argued that if heredity and natural selection are about passing on as many of your genes as possible, then wasting effort on producing males doesn't make sense. Being a parthenogenetic female (with no need of fertilisation) and producing only females exactly like yourself is obviously a far more efficient way of getting your genes into the future. This argument has become very confused, however. Many organisms have indeed produced races that are all parthenogenetic or amazono-genetic.

A commonly cited example used to be the parthenogenetic (so in particular asexual) Indian stick insect, which was viewed as a dead-end organism specialised to a tiny, very precarious niche. If you made your living by resembling a stick, and then you evolved so that you ceased to look exactly like a stick, then a bird would eat you. So it was important that the organism should not vary much, in order to retain the wonderful camouflage that its ancestors had achieved. But it turns out that in the real, variable world out there, sex is necessary to keep those adaptations honed. Just as antelopes have to run *faster*, not just fast, and cheetahs have to get even faster to catch them, or go hungry – and then extinct – so, presumably, stick insects have to look *more* like sticks, as their potential predators become more discriminating.

In 1995 Robert Dunbrack and colleagues carried out experiments using the flour beetle *Tribolium castaneum*, and demonstrated that whatever the reasons may be, the advantages of sex can outweigh the 'cost of males' – the 50 per cent of the population that cannot give birth directly. If males are replaced by females, the population should be able to grow at twice the rate. They bred the beetles in two jars, each containing low concentrations of the insecticide malathion. In jar one, they removed all offspring and replaced each by three adults from the original population (some were kept separately for that purpose). This tripled the reproductive rate – *more* than the doubling that would occur if all males were replaced by females – and in effect made their genetics 'asexual'. Jar two was left alone to pursue its sexual course. Twice a week the two jars were fed flour, in proportion to the number of beetles, simulating competition. At first, jar one won hands down: it would contain 1000 beetles while jar two had only ten. The cost of males seemed high. But after a mere five generations, after twenty weeks, in every experiment, jar two suddenly bounced back, outcompeted jar one, and rapidly eliminated the 'asexual' competition. So not only does sex pay off: it does so very quickly. The question is: why? The experiments don't tell us in detail, but the evolution of resistance to malathion must be a major reason in this case.

The most obvious potential advantage of sex is that sexual recombination explores the adjacent possible, and what it finds gives some of your offspring, and *most* of their offspring, an edge. This explanation of sexuality was widely taught from the 1950s to the 1980s. But at the beginning of the 1980s Graham Bell wrote a book called *Masterpiece of Nature* (because that's the phrase that Darwin had used for sexuality). Bell's book disproved this simple-minded view. He pointed out that if that view was right, then asexual forms would mostly be found in stable, highly adapted niches, whereas sexual forms would be going out to conquer the variable world outside. This, indeed, is what stick insects do, and it was the standard textbook view. But Bell did a survey, and he found the exact opposite. The vast majority of world-conquerors, from rotifers in rain-gutters and lizards in croplands to cockroaches on tropical wharves, from grasses to dandelions and daisies to fishes in semi-tropical streams, were those that had *lost* their sexuality. In contrast, the sexual stocks from which they had presumably arisen were in tiny restricted areas, like little pocket ecosystems in the Atlas Mountains. There were many sexual successes, of course, but where both forms existed there were about ten times as many cases of the asexual form going out to conquer the world as there were of the sexual form doing it. The textbooks were teaching the opposite of what organisms actually did.

So what is going on, then? *Masterpiece of Nature* finishes with a rather weak attempt to bring back Darwin's argument that sexual reproduction is useful because it generates variability. Think of a very variable environment like a British hedgerow, and ask what are the chances that some plant's seed, which lands under a big thistle, will be able to compete with the thistles. If the plant is sexual, then any individual will have recombined its genetics from two parents. The chances of that specific individual possessing the right combination to thrive are very small. However, there are many different individuals, each with a different combination of alleles, so the chance that *at least one* will thrive is a lot bigger than it would be for an asexual plant, all of whose seeds have exactly the same genetics. Out of a thousand different seeds, perhaps ten will land in different, compatible places; but if all the seeds have the same genetics, then nearly all of the possible places will be 'stony ground' for them.

Darwin's proposal seems very sensible, but it doesn't fit very well with actual observations. Are there more convincing reasons for sex to offer an advantage? Ones that have been checked against what organisms do, and found to match reality? Probably the most popular justification of sexual systems, one that also allows for long but temporary success of asexual offshoots, is the 'Red Queen' hypothesis. In Lewis Carroll's *Through the Looking-Glass*, when Alice met the Red Queen again, they ran and ran – but got exactly nowhere. The Red Queen said, 'Now *here*, you see, it takes all the running *you* can do to keep in the same place. If you want to get somewhere else, you have to run twice as fast as that!' Running fast in order to keep in the same place is the basis of the Red Queen hypothesis, which emphasises the existence of parasites in the wild.

Most authors who have considered the biology of aliens have not equipped them with the variety of parasites found in all Earthly ecosystems. From the marine bacteriophages (viruses that attack bacteria), which are daily killing about a twentieth of the tiny photosynthetic plankton that account for about 15 per cent of Earth's photosynthesis, to the average of twelve kinds of blood parasites (like malaria, leishmania and sleeping-sickness) in tropical mammals subject to blood-sucking insects, to the fungal rusts, blights and mildews that attack higher plants – parasites are a vitally important part of any ecosystem. They are very common in both macroscopic animals and plants, and nearly always have many more species than their hosts. The parasites are the major selective elements in most ecosystems. They determine which of the hosts get to breed successfully. Not only predators, but parasites benefit from the 'Profligacy of Nature'.

Jack used to teach zoology, and one of his favourite practical exercises for his students was to separate the gross parasites from sticklebacks, the commonest little freshwater fishes in ponds and sluggish streams. These parasites were mostly nematodes and various stages of flatworms (tapeworm and fluke). The students were provided with sensitive balances, to weigh the parasites against the rest of the fish body. The many protozoan blood parasites, viruses, even the fluke stages in eye-lenses and in the brain, weren't weighed, for practical reasons. The fish were normal, healthy ones, yet every year, for about 3 per cent of them, the parasites outweighed the rest of the fish. It's worth recognising that these were the fish that the students themselves had caught, with little nets: possibly the healthiest fish had escaped, creating a bias towards those carrying the heaviest parasite load. Even so, it's a dramatic result, and it makes the point that parasites are extraordinarily widespread. This is only to be expected, because parasitism offers a clear, general evolutionary advantage: let someone else do all the work, and enjoy the fruits of their labours without taxing your own abilities. So parasitism would seem to be a universal.

For similar reasons, resistance to parasites is a universal, too. The hosts have evolved tricks to prevent or avoid parasite success, like the enormous variety of the molecules – lock and key systems – in their immune systems, which protect them by enabling each of them to recognise and destroy most variants of most parasites. If they were clones, the Red Queen argument goes, then the parasites would be able to 'catch up' by evolving a form that was invisible to the host's immune system, or possessed some other effective strategy for parasitising all of them. But by being sexual, having a large number of variant lock-and-key systems, a creature's progeny will be varied enough so that only *some* of them will be attacked by any plague. So by running as hard as they can – producing numerous sexual variants – they can stay just ahead of the parasite who's out to get them.

Parasites always have a shorter life history than their hosts – if it was longer, they'd be in trouble – and usually produce very many offspring, at the host's expense. So parasite genetics is constantly diversifying to counter the host's defences. Therefore the host has no choice but to diversify its own genetics in response to this. The result is a runaway 'arms race' in which each of the host and parasite 'tries' to outdo the other. Sex makes it far easier to generate a range of similar organisms that, nonetheless, differ in their reactions to various parasites. So surviving to breed is the result of a complicity between evolving parasites and evolving hosts.

The terrestrial trick of retaining two alleles at each genetic locus, but using only one (or, sometimes, both, or none), has another advantage in this context. The species as a whole can keep alleles that once were useful as a defence against some ancient parasite, one that is no longer around. Should that parasite, or one like it, reappear, the relevant alleles will once more be available. The trick works even if those alleles put their owner at a slight disadvantage when the parasite is not present. A well-known example in humans is the genetic defence against malaria, which has sickle-cell anaemia as an associated effect. In tropical countries where malaria is common, it is better to possess this allele than not; but if there's no malaria to defend against, you're better off without it. This kind of trade-off is quite common. But with two copies of everything, you can keep such an allele in the gene-pool without doing much harm, provided that it is the copy that in those circumstances is not normally used.

One piece of evidence in favour of this theory – that sex exists because it helps prevent disease – occurs in humans. Human tissue-types are very different in each individual (except for identical siblings). The reason for this is not to prevent you accepting somebody else's skin graft or kidney graft: it can't be, the system evolved long before grafts came on the scene. Tissue-types are different in very many ways: this difference ensures that we each have a different spectrum of susceptibility to diseases, and since this feature offers an evolutionary advantage, it persists. We have each labelled our own cells with a special code, matched by our own immune system's repertoire of what not to attack. In this way, when an influenza epidemic or bubonic plague hits a population, only some members will be susceptible. Some will be immune from the last occasion when that disease struck, and some will be naturally immune because they have a particular combination of genetic markers that permits identification of the parasites by the host's defences.

Many mammals choose mates with very different tissue-types, and this behaviour has been selected because it means that their joint progeny will be more varied, therefore less likely to be wiped out completely by an epidemic. Most human lineages have four major genetic hotspots, with hundreds of common variants (alleles) at three of these special genetic loci, and tens at the fourth, plus many minor ones. This arrangement pretty much *guarantees* survivors of most common epidemics. The biologists who specialise in this so-called histo-compatibility system (because it was discovered in connection with skin grafts, not with susceptibility to parasites) can identify the effects of previous epidemics, such as medieval bubonic plague or the 1919 influenza epidemic, by the spectrum of alleles now present in the progeny of the survivors.

How can the Red Queen hypothesis live with the existence of asexual variants, which seem very successful, if sexual variation is the major defence against infection? Sex and recombination are an effective way to keep that variation simmering. This makes it a major puzzle that so many creatures have lost sex, and seem to be survivors *despite* that. Why haven't plagues killed them off as their parasites, particularly viruses, have adapted to infect the whole population? Isn't that a big criticism of the Red Queen theory?

One answer is that parasites don't 'want' to harm their hosts: nearly all parasites would be happier in healthier hosts. So natural selection on the parasites favours them losing the nastier aspects of infection. Most of these nastier symptoms are produced by the host as part of its defence system (think what happens when you have a cold), so some of the parasite selection favours biochemical resemblances to molecules found in the host. Then the parasite can't be recognised as such, and reacted to. And that is easier if all of its hosts have the same genetics. So it could be expected that asexual, clone, animals and plants would carry several parasites that have 'learned' to live with them without causing any pathological symptoms.

The populations of some organisms seem to vary in fairly regular cycles: classical examples include four-year plagues of lemmings, and perhaps of locusts. Very often, this kind of periodic change in population structure is determined by epidemics. First, the population density increases to the point at which transmission becomes very common; then the parasite spreads rapidly; finally, the parasite causes the population to collapse, which renders the parasite rare again. Certainly there will be parasites in any evolutionary tree; there are even parasitic sequences that evolve in artificial-life computer memories. There is a school of computer experts who believe that some computer 'viruses' (quotes because it is, as things stand, a metaphor) would have arisen in the course of our ordinary computer work without people maliciously designing them. So in fact, the word 'virus' might be entirely appropriate after all.

Another answer, to explain the successful loss of sex in so many widespread species, is geographic. It applies to the sexless cockroaches, Canadian pondweed, and daisies that have 'conquered the world' and left their original habitat – with its parasites – back home with their original sexual stocks. By going out to conquer the world, the asexual form is actually running away from the parental environment where the arms race is proceeding.

Earth's colonial powers in the eighteenth and nineteenth centuries knew this philosophy well. In attempting to grow coffee, bananas, tea, opium, and cocoa, the English took stocks from the native country – where parasites were rife – and grew them at Kew Gardens for three or four generations. When the pests had been left behind, cured by good husbandry, they generated stocks to take to other tropical countries where there were no parasites. It was not simply bloody-mindedness that brought opium to China, tobacco to Virginia, cocoa to Africa – it was very good agricultural practice, especially as the crops always started from a very small population whose variation was deplored and bred out.

Sometimes, though, the chance effects that are common in all evolutionary stories take a hand, so that for example the parasitic medfly returns to its original host, citrus plants ... but now in California.

A final suggestion asks what the problem *is* about getting parasitised. *All* wild animals carry a parasite load, and most of the problems, the pathologies, come from the *host's* attempts to get rid of the parasites. But the best-adapted parasites don't harm the host very much, and if the host can avoid mounting a defence involving high temperatures, inflammation, pain, and so on both can perhaps live together happily –

provided that the host does *not* have sexual variation that would ruin the relationship, and that the host pathology isn't required for the parasite to get transmitted. One heredity of a cloned host population might be able to come to an arrangement whereby the parasite is selected for not being greedy or nasty or toxic – its interests are served by the host living longer, after all. So the successful asexual population might be parasitised by variants of the parasites selected to be benign.

All of the above suggests that however much aliens may differ from Earth's DNA/RNA/protein-based life, parasites – and therefore 'genetic' variation of a sexual kind – will enter the story. Will alien lifeforms, then, have two (or more) sexual forms, bisexuals, hermaphrodites, or systems not seen here at all?

The evidence from Earth's history is that sex is about the most universal of all the adaptations of organisms, which implies that we will find it even in such unorthodox ecosystems as atomic-nuclear creatures living on neutron stars. Similarly, alien evolutions will doubtless have parasites – even Ray's computer-based evolutionary system Tierra produces parasites – so the Red Queen model suggests that we should find such arms races in any ecosystem. It challenges us to imagine a credible evolutionary scheme that *won't* have the parasite problem, and therefore won't need (or find) sex as the solution.

Titan (John Varley 1979)

Titan is the first in a trilogy, the others being *Wizard* and *Demon*. A NASA probe looking for the twelfth moon of Saturn finds a huge, wheel-shaped structure. The wheel is hollow. The astronauts are attacked and taken inside. There, they find a remarkable world of bizarre creatures – one-eyed mudfish the size of hippos, giant floating blimps. And Titanides, which are centaurs – half human, half horse. Each half is equipped with a full set of appropriate genitalia. They lay clear green eggs.

The creatures, as well as the wheel, are constructs – built by the self-proclaimed goddess, Gaea. Gaea is a Titan, and she has sisters throughout the galaxy: they, too, are goddesses. She is the *same* as the wheel and its inhabitants, but she can manifest herself in other forms. She appears to astronaut Cirocco Jones as a dumpy middle-aged woman.

The plot of the trilogy is complex, and irrelevant here. What is relevant is the sexual possibilities of the centaurs, which can couple in many different ways, not always keeping the human parts separate from the equine ones. 'Couple' is the wrong word, for the Titanides are heavily into group sex. An appendix to *Wizard* charts some of their favourite configurations in graphical form, from the Aeolian Solo to the Phrygian Quartet. Proper attention is paid to the path of implantation of the egg, and both of its two fertilisations (one for the human component, one for the equine). There are foremothers and hindmothers, forefathers and hindfathers.

The roles often overlap.

Are *two* sexes universal? Very likely, because two is the smallest number that works, and meetings of two organisms are probably easier to arrange than meetings of three or more. Though the *serial* collection of genetic contributions, through encounters with first one of the complementary sexes, then another, might work. But, as we've said, 'universal' does not mean 'ubiquitous', and even on Earth, two sexes is not the only solution to the problem of diversifying one's genetics. Most of our sexual forms do indeed have two variants, but some forms like the social insects – and some birds – have 'carer' as a specialised reproductive role, in addition. Perhaps an alien would classify these as a third, necessary sex, the viewpoint adopted in 'Monolith'.

Three sexes is not the limit: on Earth there are fungi with as many as twenty sexual variants, each of which is self-sterile but can swap – recombine – genes with a spectrum of others. Many organisms have both male and female sexual roles combined in each individual, but expressed successively. These organisms usually start as males and become females later when they grow larger, because bigger females lay more eggs. However, in cases where the males compete for females, or guard territories, the small animals may all be female. There are several reef fishes that live in little shoals with just one male, the largest; if he dies, the next largest female becomes a male, sometimes in as short a period as twelve hours. About half of flowering plants are bisexual in various ways, with most being designed to avoid self-fertilisation – but to permit it if extreme conditions make it necessary.

The Red Queen hypothesis does make sex look like a good idea, but it may not be the last word. Choice of a sexual mate, from organisms that are related to you – but not too closely – seems consistent with the Red Queen, and that's what many mammals and birds do. But many don't. Gerbils, and many of the small rodents called lemmings or voles or hamsters, nearly always mate with their brothers or sisters in the wild, and there are several species that *only* inbreed like this. Lots of plant species, too, *only* self-fertilise. So there are many reasons to believe that the Red Queen cannot be the whole story.

We must therefore refine our view about the occurrence of sex in alien evolutionary histories. Probably we should judge its arising to be a universal – we'd be surprised not to find it – but then, perhaps, we should expect it to be lost again all over the place, because that *also* seems to be a universal. This is science, after all; we are ignorant, we don't *know* what advantage sex gives that can also be dispensed with to advantage as circumstances dictate. Biology offers several similar examples of evolutionary innovations that get 'built in' to the system, which is changed as a result; *then* it can become a good idea to *lose* that innovation. Snakes had four-legged ancestors that swam and climbed and got really good strong muscles, so good that by *losing* their limbs snakes became better swimmers and climbers than many of their four-legged cousins! The theoretical biology of sex is full of question marks and paradoxes. And why not? Terrestrial biology is full of evolutionary tricks that are great in one context, deadly in another, so that they have been gained and lost repeatedly. It would be highly suspicious if alien biologies weren't equally paradoxical.

There is another important reason why modern biologists see recombination, and therefore sexuality, as a major evolutionary adaptation – particularly for complex organisms that undergo development. If alien evolutionary stories include flexible development, so that the same heredity can produce different phenotypes in different circumstances, probably different 'environments', then it is a good idea to combine innovations from two or more ancestors.

Waddington, the great developmental geneticist, was very impressed by some strange examples where the development of an organism apparently anticipates, in detail, problems that it will face later. For example, when a baby ostrich hatches it already has a callus, a thickened patch of skin, where it will rest its breastbone in the stony nest. It would in any case develop a callus later from the rough stones, but – puzzlingly – it anticipates the need. Waddington did a series of experiments with fruit flies, whose results demonstrated what he called 'genetic assimilation'. For the baby ostrich, his mechanism would be like this. There must be genes that produce a callus when the ostrich is exposed to rough stones, and there must be a threshold of roughness. If you keep selecting for a lower and lower threshold, then eventually you evolve birds with threshold zero. They then produce the callus without there being any stones at all. Genetic assimilation shows that the wild-type genome is very diverse, and dramatically different phenotypes can be selected from it over a number of generations. The new types mostly do not have any new mutations; instead, they have the old alleles in new combinations. Evolution of new tricks can be genetically assimilated by selection, instead of waiting for the 'right' gene allele to turn up by chance.

The modern picture of evolutionary genetics, of which genetic assimilation is just one component, affects our view of alien evolutions. Alien evolutionary stories must be more subtle than those based on the sudden occurrence of a favourable mutant. Oddly enough, this liberates alien biology, making many more things possible. To demonstrate the point, and at the same time give you a feel for the connection between terrestrial evolution and credible alien scenarios, we'll take a new look at one of the classic *Star Trek* aliens – the tribble. This cute but deadly creature appeared in the episode 'The trouble with tribbles', written by David Gerrold. To quote James Blish's adaptation for book publication, in *Star Trek 3*:

> A tribble looks like a cross between an angora cat and a beanbag. It has no arms or legs, no eyes, and in fact no face – only a mouth. It moves by rolling, by stretching and flexing like an inchworm, or by a peculiar throbbing, which moves it along slowly but smoothly, rather like a snail. It does, however, have long fur, which comes in a variety of colours – beige, deep chocolate, gold, white, gold-green, auburn, cinnamon, and dusky yellow. Tribbles are harmless. Absolutely, totally, completely, categorically, inarguably, utterly, one hundred per cent harmless . . .

In these post-*Gremlins* days, of course, we know better. Back in the late 1960s, it was a new thought. The plot through which tribbles charmed their way into the SF consciousness was otherwise standard *Star Trek* fare. A cargo of grain, quadrotriticale, is urgently needed on Sherman's Planet, to bolster humanity's claim to that world and keep it from falling into the hands of their bitter enemy, which at that stage in the *Star Trek* saga was the Klingons. Lieutenant Uhura picks up a tribble from a peddler, a down-on-his-luck system locater called Cyrano Jones, to keep as a pet. Tribbles are cute and furry, and they purr like a contented cat. They brim over with *love* – or what their owners interpret as love.

They also – it rapidly emerges – breed at an unprecedented rate, and they overrun both starship *Enterprise* and Space Station K–7. They feed on almost anything, including quadrotriticale . . . A major setback to Captain Kirk's career is averted when the tribbles start to die, leading to the discovery that the quadrotriticale has been infected with a virus by a Klingon agent. The *Enterprise* is freed of tribbles by beaming them into the engine-room of a Klingon battle cruiser; Jones avoids prosecution by accepting the job of clearing them out of the Space Station, a task that Mr Spock calculates will last for 17.9 years.

As this plot unfolds, we learn snippets about the tribbles, though it is worth bearing in mind that many of them are told by Jones, a noted liar. Tribbles come from the far reaches of the galaxy. They are bisexual, devoting half of their metabolism to reproduction. Their purrs have a tranquillising effect on the human nervous system, but they hiss at Klingons. They are born pregnant, and produce a litter of ten every twelve hours. It all sounds like delightful nonsense, and the plot has several inconsistencies. The tribbles could have been removed by beaming them elsewhere, right at the start. And why can't Jones do the same in ten minutes instead of taking 17.9 years?

Leaving these errors, typical of media 'sci-fi', aside, we will now try to convince you that tribbles are an entirely plausible type of alien: more so, indeed, than most other *Star Trek* aliens, especially the all-too-humanoid Klingons. The big scientific problem is that tribbles have evolved to be pets, but on Earth no creature has done that. There hasn't been time. We *use* cats and dogs as pets, but they didn't evolve to fill that role. Nevertheless, tribbles are deeply rooted in a variety of evolutionary universals. As an exercise in alien design for SF, and xenoscientific imagination, here is how they could have evolved . . .

On a distant world – let us call it Tribblehome – there is a great prairie. Grass or its equivalent is not necessary, just a mass of small plants with several herbivores, large and small, and carnivores at the top of the food chain. Season well with a few plant parasites, animal parasites, detritus feeders, and other creatures. Particularly note a small burrower, a herbivore like our prairie dog or marmot, which lives in social groups and makes its own burrow. The terrestrial marmot has guests, among them the prairie owl and the rattlesnake; therefore it is reasonable for our alien 'exomarmot' to have guests too. These include a sluglike detritus feeder, whose ancestors escaped predators by crawling down holes: this unlikely beast is our prototribble.

Those that occupied exomarmot-holes did better than those that used chance crevices, so had more offspring, because the exomarmot holes contained debris, old nuts and roots, dead baby exomarmots, and excreta. The prototribble, being a detritus-eater, thrived on this diet. Soon a symbiosis developed. Exomarmots that put up with prototribbles had cleaner burrows and fewer parasitic diseases, therefore more healthy offspring, whereas those that ate or otherwise molested the prototribbles were worse off. So the symbiosis prospered. Many terrestrial detritus-feeders, such as worms and slugs, are bisexual; so was the prototribble. However, there was room for only one of them in a burrow, so on most moonlit nights they came out to mate.

Two processes could then occur, and we'll assume they both did because that makes the story more interesting. Some exomarmots, up in the hills, developed a special device, perhaps a stone across the door, to keep out other intruders; it has the side effect of keeping the prototribbles in. The exomarmots on the plains, like our marmots, became adapted to the presence of several house guests, and this gave their prototribbles a grave problem: they had to adapt so that none of the resident organisms found them tasty, or repulsive, or alarming, or unexpected. Both groups of prototribbles became more rounded in contour. Within a few hundred generations the hill prototribbles lost their eyes and ears and became able to fertilise their own eggs if the burrow wasn't opened at the right time for them to meet others of their species. Their round shape was an adaptation to cold nights, and was also less of an irritant to the exomarmots in their cramped quarters.

The prairie prototribbles, on the other hand, came more and more to resemble baby exomarmots. All of the symbiotic guests came to tolerate them, and they dutifully ate any organic material that wasn't living and animal. Soon all of the flat land of the continent was pockmarked with burrows, each with its assortment of occupants and its prototribble.

This idyll didn't last. A long dry period, during which the communities diverged in many ways, was followed by an extended glaciation. The prairie prototribble nearly died out, mostly because there were no more warm moist nights for its sexual adventures. The rest of the symbionts became a motley rabble. Dirty burrows allowed new parasites to appear, with stages in their life cycle spent in each of the exomarmot's guests; soon solitary living became the rule again. The symbiotic adventure seemed to be over; the few prototribbles left on the prairies made their way to caves in the foothills, and occasionally hybridised with their hill-living cousins.

In the hills, the solitary exomarmots had begun a stranger story. Like our own rodents, they had never been averse to the odd bit of animal protein if they could lay paws on it. When hard times came, one group became omnivorous, supplementing a vegetable diet with alien equivalents of insects and land molluscs. When the climate turned cold, they grew larger and shaggier, and so did their prototribbles, who now advanced rapidly to full tribblehood. The mixture of the two races gave much variation and hybrid vigour. They became 'furry' (though not with Earth-type fur, which is a parochial) and they became larger. The social life of the exomarmots, now more like badgers, became more complex; the tribble's sexual and social adaptations therefore did the same.

Because of its way of life, the prototribble frequently had to have recourse to self-fertilisation. All were male when young, and made all of their sperms then. When they matured, they made eggs, which they fertilised from their sperm store. Usually they had exchanged this with another female's at some time, but even if not, a few generations of inbreeding did little harm. (Female mammals make all of their eggs when they are embryos; as we saw, many Terran organisms are first male, then female; and the limpet *Crepidula fornicata* can use the sperms that it made as a male when it becomes a female – and so can tapeworms.) The prototribbles had been viviparous before they were symbiotic. The gestation period became longer until a new kind of sexual phenomenon became possible. The embryos were of course all male, with all sperms made before birth (like Terran female mammals' eggs). Some of these new embryos leaked sperms into their environment, which was the mother's ovary (guppy eggs are retained in the ovary for development, not in a special womb). Such a 'male' was of course more potent, particularly if his children behaved the same way. The arrangement did have some 'inbreeding' disadvantages, though, putting a brake on evolution.

The host exomarmots were evolving very fast at this time, as the climate warmed after the glacial. They brought tribbles out of the home burrows to play with, a game that was especially popular with the young, who were delighted by the remains of the tribbles' old mating behaviour. If two tribbles found themselves in a darkish, moist spot, then any contact between them started a whole series of sexual attacks. As both individuals lacked eyes and ears, indeed any external sexual apparatus, contact resulted in frantic attempts to mount each other until by chance both undersides were opposed. At this time the tribbles produced their most ecstatic noise, and of course formed perfect beach balls. The young exomarmots rolled them about, to their own great pleasure; this also caused some exchanges of sperms between the two females. If one tribble was pregnant and the other not, then the embryos might 'hatch' precociously in the rough-and-tumble, often into the genital aperture of the partner, who they would of course fertilise. Genetic assimilation would reinforce this behaviour, with the result that Junior fertilised Auntie.

Exomarmots were now chipping stone tools; tribbles were totally part of their lives. Odd colours, which turned up occasionally, were treasured and bred from. Also, the tribbles were consistently selected for their size. This resulted in a dissociation of their senility and growth factors (as in many of our dogs and cattle) until several different strains emerged. The most prized was effectively immortal by a weird tactic made possible only by its strange sex life. Only one embryo usually matured in the ovary at any one time. As it matured, it produced sperms into the environment, and some entered its own genital tract and ovary, to fertilise an egg there. As the mother (later, grandmother) grew and aged, so the embryo within grew too, and the one inside that, and . . . When her skin wore through, she gave birth and was promptly eaten by her offspring. When cross-fertilised, these peculiar creatures produced several young instead of only one. (If you think this is unlikely, consider *Gyrodactylus*, an external parasite of fishes. All are bisexual, and in each uterus is another *Gyrodactylus*, in whose uterus is, usually, a third. They are born pregnant, just like tribbles.) This odd variant did not survive into historic time, but the size-selected stock from which it had been produced did. They had about forty young at a time, all born pregnant (having, of course, been males as embryos). The exomarmots selected them for this trait, and also managed to select for a slight responsive purring when stroked.

The tribbles' nervous system had not been selected for sophistication. Beyond detecting dead organic matter to feed on, the purring, and basic sexual responses, nothing more was needed. As the exomarmots developed tribal cultures, though, they began to select tribbles to respond to a single individual. A few tribbles were found that hissed when strangers touched them; later ones of the same breed made excellent watch-tribbles; the most sophisticated became sensitive to 'fear' odours of intruders.

Meanwhile, back on the plains, in the region around a large river delta, some of the old denizens of the exomarmot burrow were still living on other small herbivores. Some had become pack carnivores, 'dawgs' like our own hunting dogs, and others had taken up solitary fishing like our otters. One of the two flying predators maintained an uneasy partnership with some of the dawgs. When the flyer spotted a prey, a large herbivore, it would hover over it; the dawgs had learned to associate this behaviour with food, and would find and kill the beast. The flyer usually got some of the meal, but as often as not the partnership failed. Sometimes the prey was wounded but not killed, and escaped, tracked by the flyer until killed by a solitary carnivore. Or the dawgs kept the flyer away from the meat.

Nevertheless, this partnership was a true relic of the old burrow-sharing days, so many of the signals that turned off aggression were recognised by the other partner. But the possibilities inherent in this symbiosis would probably have been lost, had the budding agri-culturalist exomarmots up in the hills not turned their minds to protecting their crops. Young exomarmots were always bringing strange creatures back to the cave, as they had done in the burrow, and eventually the caves were expanded and linked by tunnels, until each family of exomarmots had a large sett like a badger's. The sett contained the resident family, its many children and tribbles, and a variety of pets that the kids had brought home.

Down at the bottom of the slopes there were a few puppy-dawgs and some young flyers. Only with the coming of agriculture did these creatures endear themselves to the adults – until then they had just been a bloody nuisance, useful for keeping the kids out of mischief but not worth the food they consumed. All that changed with the advent of agriculture. The flyers were territorial, and one of them flying over the fields kept the big herbivores away; if any did blunder in, then the dawgs would drive them away again. All were rewarded by this accidental arrangement. It differed little in basics from the man/dog/cattle/sheep symbiosis that has long characterised much human agriculture, but still there was one significant difference.

Flyers and dawgs possessed behavioural traits that enabled them to work with one another, as well as with the descendants of their original hosts, the exomarmots. Tribbles, purring as of old, prevented aggression perfectly. They were the oil in the mechanism of the agricultural symbiosis, and they grew fat on its success. They evolved as a symbiotic, general-purpose pet. They were adapted to be symbiotic with three very different species, so they had reduced their form to the least aggressive: no head, no eyes. They never did have limbs. They responded to touch with a reaction that provoked curiosity: purring. They responded to organic odours, and to beings that produced them, by hissing: the old line of watch-tribbles coming to the fore. They ate by covering organic matter with their bodies and absorbing it, avoiding competing with the other dwellers in the household.

Their sexuality was highly non-aggressive. One tribble was usually a grandmother, containing mothers and their sons; even if no sperm had been swapped in the previous three generations, these 'descendants' were still genetically different and their mating reproduced new hereditary combinations. But the tribbles swapped sperms whenever they got the chance, and to do so they stuck together to make a furry ball. Thus the tribble became 'pre-adapted' as a spacefaring pet. The difficulties it faced in adapting to three distinct species led it to discard all possible threats, becoming a generalised, curiosity-provoking baby-toy. When space travellers arrived on the tribble world, they were captivated, and tribbles soon found their way on to human vessels.

On Tribblehome, the population did not explode, because it was regulated in a very simple way. To explain this, we must describe the tribbles' home solar system. Its sun was similar to ours. There was an inner Mercury-like planet; then the next one out, about 100 million miles away, was Tribblehome. Without other planets in the system this would have been a stable hydrogen/methane/ammonia/water world, just like Earth before chlorophyll was invented and oxygen appeared. But there *were* other planets in the system, and the sixth world, 800 million miles out, was a super-Jovian gas giant. Its effect on the tribble world was erratic, changing the temperature in a rough twenty-year cycle. Longer cycles were superposed (probably the dynamic was really chaotic) and at one period there was a million-year hiatus during which the only liquid water on the planet was a small brine lake about a mile across.

Life persisted, nonetheless, but all lifeforms evolved a highly cold-resistant stage. These stages had to be frozen and dried out before they could hatch by warmth and moisture. (Terrestrial plant seeds do this, and several fish that live in temporary ponds in the tropics have eggs that must be cooled and dried before moisture will cause them to hatch.) If tribbles were cooled below −10°C for longer than forty-eight hours, their reproductive behaviour changed dramatically. The fertilised eggs walled themselves in and came to resemble golf balls. Selection by the exomarmots caused the final hatching phase to drop out of the reproductive system altogether. The early agriculturalists froze nearly all of their young tribbles early, by leaving them outside on cold nights; after that, all the eggs they produced were non-viable. Only a few were kept out of the cold for breeding stock, so the tribble populations remained stable.

The non-hatching 'winter eggs' were used as ornamentation in ceremonial caves, but that was eventually stopped because occasionally an imperfectly frozen one would hatch, and a malformed prototribble would drop down the neck of a worshipper. Their use as projectiles, fired by methane pressure from a long tube of bamboo-like cane, doubtless accounted for the legendary successes of exomarmot warriors, and set the pattern of social life on the tribble world to this day. However, those tribbles taken off-planet were never frozen; instead, they were constantly cuddled by warm homoiotherms of various species . . .

The rest, as they say, is history.

This story of tribble evolution, of 'trials and tribble-ations', incorporates several abstract features of Earthly evolution, while avoiding mutations completely. Other abstractions from processes of reproductive replication on this planet can similarly be transferred to alien ecosystems. Reproduction here is apparently extremely wasteful, for example, with nearly all of the products failing to contribute to the future. And organisms seem to take care, through all the complications of evolutionary time, to keep their replicating heredity out of the way of their day-to-day machinery for living: they segregate their heredity from their 'ordinary' functions. This segregation is not very clear in viruses and bacteria, but the timing of their replicative events usually keeps them clear of other biochemistry. But the more complex eukaryote cells keep their heredity locked up in the cell nucleus except when the apportionment of chromosomes to the daughter cells is being carried out. The actual chromosome-doubling is done while the DNA is safely behind bars. Animals (and more subtly plants) keep their hereditary DNA in special cells, the germ cells, which don't get involved with growth and development – they become available as eggs and sperms only when the growth is nearly finished. Colonial animals, from siphonophores and bees to wolves and baboons, breed from just a special few reproductive animals, the others being devoted to getting food and protecting the breeders.

So if we're inventing an alien society for an SF novel, or if we're investigating an alien ecosystem that our xenoscientists have just discovered, we should bear in mind that the aliens' replicating hereditary material might be linear but possibly won't be, that there's a good chance that it will be a template system (which rules out such triple systems as have been suggested by SF authors – for example, in

Joan Slonczewski's *The Children Star*), that it will very probably waste nearly all of its products as far as reproduction is concerned, and that there's a high likelihood that each generation will have its structure programmed by previous generations.

There will also be systematic patterns in many alien evolutions, even though they may be based in different chemistry and take place deep in Jupiter's atmosphere or on the surface of a neutron star. For example, carnivores will get big teeth and eyes on the front, and animals will gain and lose sexuality, as needed in the ecological/evolutionary context. And aliens, just like terrestrial organisms, won't be waiting around for the 'right' mutation to turn up. Instead, they will be randomly exploring the phase space of possibilities that can be achieved by recombination from the existing gene pool, and if anything that they find offers a selective advantage, it will be selected, propelling them into the adjacent possible with a rapidity that would delight any resident Kauffalien.

Except . . .

In biology, there are *always* exceptions. Perhaps, sometimes and somewhere, there *will* be whole evolutionary histories that only do the mutation trick, waiting for the 'right' variant and capitalising on it. A very successful terrestrial group of animals, the nematode worms, have a prescribed number of cells in each of their organs, and can't repair wounds or regenerate lost parts. They do very well in their own way, but they don't 'learn' from other organisms and they contribute very considerably to the diversity of life on Earth. They are atypical of Earthly life – and they are a typical example of atypical exceptions.

12

THE UNIVERSALITY OF EXTELLIGENCE

*A*BEL IS RELAXING *in a liquefaction chamber, having his solid bits massaged and his gaseous parts frozen to refresh vital dynamic modes; Cain is counting his legs to make sure none of the tourists have taken any by mistake. Or on purpose.*

The ansible rings, and Cain takes the call. More accurately, the call takes itself to him, and turns out to be irritatingly argumentative. He bangs on the wall of Abel's chamber until his currently liquid friend siphons out, warms up, and becomes his usual gaseous-solid self. 'I was having a really nice bath,' Abel protests. 'Why did you have to interrup—'

'Because—'

'See, you're doing it ag—'

'Shut up. We've got trouble. The Prospectus.*'*

'It's been a failure?'

'Worse. It's been a huge success.'

'Oh, no.'

'Headquarters wants us to do a series. Ten million volumes.'

Abel dissolves into gibbering turbulent vortices, then pulls himself together. 'We don't have time to visit ten million habitats, Cain. Let alone describe them.'

Cain has coiled into a morose helix, legs tucked away inside to stop himself chewing them in distress. (He is now convinced that three have gone astray . . . probably that annoying little Fomalhautese.) Suddenly, Abel expands to three times his normal volume. 'I know! We can recycle!' Cain remains unenlightened and Abel explains his idea. 'We keep the same structure for all of the Prospecti, *and just change the names. There are only so many different ways for lifeforms to work, after all.'*

'A lot more ways than ten million, Abel. Think again.'

'No, I don't mean parochial differences, we can steal those from public sources, rip off a few macros from the newscasters . . . easy. What I mean is, if we write one master version in terms of generalities, then there has to be a way to specialise it to a mere ten million instances.' He rushes off to try the idea out.

Cain is left wondering how the Earth Prospectus *can be modified to deal with magnetic creatures in the photospheres of stars. But he suspects that Abel is right: there has to be a way.*

It would be wonderful to find solid evidence of alien life, even if it was just at the level of bacteria. We would learn a lot about our place in the universe, and about our own likely origins. It would be far more wonderful, though, to make contact with intelligent aliens. Unless, of course, they are like those in *Independence Day*, who equate contact with conquest. Intelligence *alone*, however, is not the true Holy Grail here. Many creatures on this planet are intelligent. The really interesting developments arise when an intelligent species starts to become extelligent: when it can store its 'cultural capital' independently of individual minds, and that store becomes available for all (or, at least, many) individuals to use and to contribute to. It is extelligence that has given humanity its literature, philosophy, economics, science, and technology. We invented the term in *Figments of Reality*, and its use has spread. It is *not* the same as culture.

So, if we want to bandy ideas about the nature of the universe or the price of CDs with an alien, then we'd better choose an extelligent alien. If extelligence is a parochial, that could be difficult: we could be the sole extelligent species in the galaxy. But if extelligence is a universal, there *must* be extelligent aliens out there, somewhere. Which brings us to the central question of xenoscience: is extelligence a universal or a parochial? Until now, we have merely asserted that extelligence is a universal, and assumed that it is without further discussion, but now we need to justify that assertion.

We've faced similar problems before: let's review how we handled them. We used a particular technique to establish the kind of biology that we are likely to find in other evolutionary systems. We identified certain innovations in terrestrial evolution, and we showed that these could be considered as universals if they had happened many times independently; we labelled them as parochials if we found them only once. We also devised a more general theoretical test for universality: 'a generic trick offering a clear evolutionary advantage'. Universals are things that we would expect to see in any re-run of Earth's evolution, but they would be associated with differently structured organisms the second time round. We would also expect to find them in alien evolutions – especially those sharing much commonality with our own, such as carbon-based life on an aqueous planet. In contrast, we would not expect to see any of our parochials in alien evolutionary stories.

Nevertheless, we can appeal to parochials to give some validity to alien invention, provided those parochials are instances of universals. So we used *Xenopus*, the 'frog with nasty habits', to validate Niven and Pournelle's grendels, and we used the male egg-pouch of the seahorse to validate part of the Yilané's reproductive strategy. In that sense we expanded the parochial example into a generalisation, which opened up a phase space of possibilities, from which we chose – invented – an alien example. *Xenopus* generalised to a 'biological transformer that transduces offspring into parent with no intermediate'. Earlier, we expanded the Earth-salmon example into the generalisation 'biological transformer (the other way round) transducing through another species', and then invented the 'slamen' at the core-mantle boundary. Universals tell us what kinds of innovation to expect – photosynthesis, penises, skeletons, antibiotic secretion (and therefore antibiotic resistance) – but we have to be imaginative to deduce anything about aliens from Earthly parochials.

When it comes to extelligence, then, we can certainly *invent* plausible alien parallels, but first we have to overcome one difficulty: as far as we know, human beings constitute the sole instance of extelligence on this planet. So our positive test for universality – 'Did it occur independently more than once?' – fails. However, that doesn't prove that extelligence is a parochial. *Every* universal will fail that particular test the first time it appears. We faced the same problem earlier with the development of an oxygen atmosphere, and we concluded on theoretical grounds that this development should nevertheless be a universal for aqueous planets. So our justification for the universality of extelligence will also have to be theoretical.

We'll start by playing Devil's advocate, and acknowledging that there are plausible arguments why extelligence might be a parochial. The question is illuminated on this planet by the problems facing manifestly intelligent dolphins if they were to evolve extelligence: underwater is not a good place to develop technology, and flippers instead of fingers don't help. Today, human extelligence can supply underwater technology (having evolved it on dry land) and in principle we could attempt to assist in the development of dolphin extelligence. On evolutionary timescales, however, such a process would take millions of years. It might have happened here, but the land-based creatures got there first. In a long-term view, the land-based creatures evolved from marine ones, so maybe that step in evolution was the marine world's way of heading towards extelligence; but even so, there were many other steps too, taking a lot of time, and today's dolphins are still back there in the water. If, in a few million years or so, dolphin descendants have evolved extelligence, they won't look much like dolphins.

Despite such counter-arguments, we are confident that extelligence is a universal, for theoretical reasons and because of our experience of evolution on Earth. Extelligence is a cultural phenomenon: it combines the abilities of many minds, making them accessible to all. A plausible universal here is that each individual in a society can use other individuals as social instruments; the different roles in society are, in this view, like tools for different tasks. And for *this* universal we do have other terrestrial examples. At the simple end, colonial insects have the queen using workers as foraging and nursing instruments, while the queen lays eggs for all of them. In simple mammal societies like those of terrestrial baboons or meerkats, there is negotiation of social roles: meerkats in captivity invent new 'begging' roles, and young animals grow up to implement these. In western societies we use lawyers as tools to implement the law for us.

These examples provide small extra bits of observational evidence that extelligence is a universal. Granted that, the same principle we used for biology, generalising and then inventing other examples, can take our imagination into realistic alien possibilities. However, as with the invention of parochials for alien evolutions, indeed as with arguing from universals on Earth to instances of universals elsewhere, we know that we won't actually find that *particular* invention. It stands as a placeholder for all of the possibilities inherent in the new phase space, whose geography can be investigated only when it has been made explicit.

One route to extelligence – perhaps not the only one – lies through intelligence and technology. There is no doubt that intelligence is a universal: it passes the observational-evidence test 'Did it occur independently more than once?' Increased intelligence (*whatever* we reasonably mean by 'intelligence') has cropped up many times in Earth's evolutionary tree, on several different branches. Mantis shrimps are clever, versatile creatures, and they can be taught tricks and can initiate little games with people. Jack has kept two mantis shrimps very rewardingly – the second was to impress Ian with how clever they were, and it worked. Some larger insects are similarly impressive problem-solvers. Octopuses and cuttlefish are very intelligent, much more so than their relatives the clams and the snails, but we don't know about previous cephalopods, the ammonites, belemnites, and the like. Jack kept a small coral octopus for more than a year, and it could recognise Nick Taylor, the graduate student who brought it crabs: it would reach out an arm for him and ignore everybody else. We're talking rat/dog levels of intelligence here, you understand, not Einstein.

Among non-mammalian vertebrates, parrots are surprisingly intelligent, if taught in the right way. Irene Pepperberg has taught the African Grey parrot Alex, star of many TV shows and her book *The Alex Studies*, to reach a startlingly high level of linguistic ability. Owls, alas, are not in the top league despite their iconic status as the wisest of birds, from Athene to Disney. To be frank: owls are very stupid. Some mammals have attained intelligence, especially primates, but don't forget dolphins and other toothed whales, and the larger carnivores.

Perhaps, if we took an 'outside' (dare we say 'alien'?) view of the incidence of intelligence among terrestrial creatures, we would count all of the mammals as just one instance, with minor variations. So, if mammals were the only intelligent creatures, it would be possible to argue that intelligence might be just a parochial. However, the previous discussion shows that there are plenty of clever non-mammals scattered about in different clades of the terrestrial evolutionary tree. So, if you accept our argument for the universality of oxygen atmospheres (excreted by creatures using sunlight to power their metabolisms) and of flying creatures (probably carrying the heredity of several prokaryote-grade ancestors, then becoming cellular and multicellular), then like us you will expect to find several intelligent species on any planet that has a great variety of creatures comparable in complexity to what we find here. There will be lots of different creatures with brains of various kinds . . . lots of rat, parrot, mantis shrimp, and doggy intelligence . . . probably many kinds that are intelligent and versatile in ways that we can't imagine. We can imagine that we can't imagine it, but we can't imagine what 'it' is.

So much for the universality of intelligence. What about technology? It is not hard to construct a similar argument for technology, which also has to be a universal. Evolutionary evidence runs from rotifers using tiny whirlpools to bring them their bacterial food to spiders building webs or throwing sticky nets at prey, from termite mounds having air-conditioning to birds' nests of complex shapes, from cellular pumps in *Paramecium* to the highly technical mammalian immune system, from alarm calls to human speech.

The last example is of course the critical one. Our hominid ancestor *Homo erectus* apparently didn't improve the design and manufacture of stone axes throughout a period of half a million years, so our ancestors' brains were interacting with a technology for a long time without any usable, improving extelligence appearing. We can find lots of similar examples of extelligence not happening, from net-throwing spiders to cephalopods and dolphins. We can find excuses for all three: spiders have well-known limitations on size, therefore on brain-size; cephalopods are nearly all annual, breeding and dying at a year old, so they have no chance to build up cultural capital by passing what they've learned across the generations; and dolphins are aquatic and do not have manipulatory limbs.

As we've seen, though, that needn't have stopped them in the long run, so we don't really know why dolphins have so far attained only the merely intelligent stage. There are quite a lot of creatures that look as if they might make it all the way to extelligence, yet to date have not: chimpanzees, of course, particularly bonobos; gorillas; several kinds of baboon; wild dog packs with their special languages; several other carnivores, especially communal ones like mongooses; and what about parrots, exemplified by Alex? But we don't know how many creatures have *had* extelligence on Earth.

In *The Science of Discworld* we fantasise about crab civilisations taking the 'Great Leap Sideways' and various promising dinosaurs, and while these ideas are surely nonsense, no evidence would remain today if in fact they were correct. Little evidence of our own civilisation would still be visible fifty million years into the future: see Dixon's *After Man*. Perhaps some dinosaurs, maybe even some ammonites, had the technology trick? As we've seen, absence of evidence is not evidence of absence . . . but we had better discuss extelligence as if we were the only example to appear on this planet, because anything else is far too speculative.

The origins of our cultural capital probably occurred in Africa, before the modern human *Homo sapiens* left that continent to follow *Homo erectus* into Europe and Asia. *Homo sapiens* had a succession of tool forms, and was probably putting together language, tool-use and social myths into a burgeoning extelligence some forty thousand years ago. If you like anthropic reasoning, you will argue that we cannot know whether the conflation of technology with intelligence is likely to generate culture. It did for us, but we cannot be sure that such an interaction is necessary, because we *are* here. So it looks as though we can assess how likely our presence is only through outside evidence. However, there is an alternative to evidence, namely theory. Technology is a generic trick, and extelligence opens the path to technology by making acquired know-how accessible to succeeding generations. The whole process looks suspiciously like a universal. The parochials would then be the precise choices of technology. For instance, Terran forks and chopsticks are two distinct parochial realisations of technological aids for consuming food. Food of our kind may well be a parochial, of course, but energy-acquisition is a universal. Communication is a universal, but the use of radio, especially simply modulated signals that have not been encoded to reduce errors, is a parochial.

The Mote in God's Eye (Larry Niven & Jerry Pournelle 1974)

A laser-propelled light sail from distant space, which is destroyed and falls into a star, alerts humanity to the existence of intelligent aliens near the red supergiant star known as Murcheson's Eye. These creatures have clearly not discovered the faster-than-light Alderson drive, a puzzle that is solved when it turns out that the appropriate Jump point is within the gaseous envelope of the red supergiant. A military expedition, two vessels, is dispatched to investigate; the ships' Langston Fields can easily cope with the relatively low temperature at the Jump point.

The expedition finds a thriving civilisation, which has occupied asteroid clusters at the Trojan points in the orbit of a gas giant planet, sixty degrees either side. The alien 'moties' come in numerous forms, some with three hands (which later stimulates a Terran saying, 'on the one hand . . . on the other hand . . . and on the gripping hand').

The moties are super-engineers, with an innate instinct for machinery. They routinely rebuild their ships as they travel between planets. They rebuild the Terrans' coffee-machine, and the result is far superior to the original. Belatedly the humans discover that the ship *MacArthur* has become infested by motie 'miniatures' that still possess the engineering urge, and these are rapidly demolishing the battle cruiser to turn it into something more suitable for themselves. The humans are forced to evacuate the vessel and destroy it.

Further plot machinations reveal the moties' dreadful secret. Their population growth is uncontrollable because of their own biological peculiarities. Denied access to the rest of the universe by the location of the Jump point and ignorance of the existence of Langston Field technology, they solve the problem of overpopulation by an endless cycle of war. This cycle has been going on for so long that animals have evolved to specialise in living in ruined cities.

The humans realise that the moties pose a horrifying threat: if they ever escape from their own system, they will overrun the galaxy and wipe us all out . . .

The evolution of technology is not a straightforward matter, even so. One issue for such evolutionary innovation questions concerns a kind of threshold effect: the innovation becomes recognisable only when it gets big enough, or significant enough. Darwin understood this point, and naive anti-evolutionists are grappling with it when they ask, 'What use is half an eye?' Because evolutionary innovation is based in a previous way of life, which didn't possess the trick concerned, it is reasonable to ask, 'How did it get started?'

Several satisfactory general answers to this question have been given, notably Gould and Elizabeth Vrba's invention of 'exaptation', change of use of a structure. Insect wings probably were used as ornamental fringes, then perhaps as heat exchangers, then as sails for insects floating on ponds, before flapping them became an option. Part of the evidence is that the flight muscles do not attach to the wings as such, but 'click' the thorax between two stable shapes; subtle flexures and changes in shape of upstrokes and downstrokes are modulated by thickenings and tensions in the insect's body, and by the shape of the wings. If wings had evolved *in order to* let the insect fly, the muscles would have operated the wings more directly.

Sometimes a series of improvements results in a complex organ whose function and effectiveness are completely different from the original rudiment. The original was a pre-adaptation. In 1994 Daniel Nilsson and Susanne Pelger showed that the vertebrate eye, contrary to the 'what use is half an eye?' criticism of Darwin, probably developed through a series of slowly improving proto-eyes from an original light-sensitive spot. There was never half an eye: instead, there was a half-as-effective *complete* eye. The compound eye of insects took a very different route (equally explicable as a series of improving steps), and cephalopods (ammonites, octopuses, squids) arrived at a camera eye, like the vertebrate one, but by a very different route. And once an organism has image-forming eyes it inhabits a different universe, with new rules.

The same goes, with even greater force, for the innovation that we have called extelligence. Extelligence may well have started as a series of pre-adaptations. For example (we invent) marking a rock with mud or dung to indicate territorial rights could have been one tiny step towards cave-painting and writing. A limited repertoire of grunts, used to scare away interlopers, could have headed us towards spoken language. There is a prediction here, though it may not be easy to test it. If an alien had sufficiently great intelligence, then the relevant store of know-how would *not* be beyond the capacity of any individual, and extelligence would be unnecessary.

Godlike aliens are probably solitary.

Tool using – for example, swords – has often been seen as the first step to extelligence. So we want to know if there are tool-using aliens out there, not just bright octopuses and dolphins. We must formulate the notion of 'tool' carefully here, though, because nearly all organisms exploit part of their environment to their own ends. For instance, birds make nests, sea turtles pump their blood into a more dilute state by using sodium ions in their lachrymal glands, and *Hydra* uses the glutamine released from water-fleas to arm its weapons. So we must narrow the question down to the intelligent use of tools. That is, the creature must 'know' that it needs the tool for some purpose. It must *plan* to use the tool at some unspecified stage in the future. Chimpanzees behave as if they do this: they take tools up at the beginning of a journey. According to some accounts, the burying of bones by dogs is also a planning action. Mantis shrimps build elaborate tunnel systems, and decorate the entrances with brightly coloured fish fins: does this count?

How universal are tools and weapons? In this area of knowledge, again, we are hobbled in trying to think clearly because we have filled the subject with metaphor. We speak of the weapons of a poisonous snake including its venom, of the weapons of the wasp being its powerful jaws and its sting. We speak of the long third finger of the aye-aye as a tool for winkling insects from under bark, and we speak of the complicated genitalia of insects in lock-and-key terms. We speak of chimpanzees using pebbles as tools to crack nuts, and using de-leafed sticks as processed tools to fish out soldier termites. We are so committed to the kinds of tools and weapons that *we* use that it is easy to see them as a likely option for other extelligent creatures. But this could easily be pure human prejudice, the too-easy reification of a metaphor.

Given this difficulty, we will have to start somewhere quite different and sneak up on tools and weapons from a new angle. Let us think of mathematics as a tool. (Language would serve equally well, here, but mathematics is a less familiar example and deserves attention in its own right.) Children learn a lot of mathematics at school, far more than, say, early *Homo erectus* did 200,000 years ago. We have a much bigger brain than they had, of course. But what were the selective forces on our ancestors, that led to the development of a brain that could do mathematics if pressed?

The usual answers are very diverse, but they all fall into one category: we needed our big brains (and ever bigger brains as time went on) for language, or for gossip about our social group, or for planning the hunt, or for recognising different roots and learning which to plant when, or for learning to lie to our colleagues, or for all of the above. The common thread is this: we started with specific problems, and solving those gave us a bigger brain. As a side effect, we can now use that clever big brain to attack much more abstract problems – like those of mathematics – and get real-life solutions.

We can see this kind of thing happening with computers: competition between PC manufacturers has driven companies to invent faster systems with much more memory, and with more user-friendly interfaces that exploit the speed and memory. So PCs can now do tasks that only the largest computers could do twenty years ago. And the assumption is that these tasks were sitting there in some ethereal space, waiting – like mathematics for the big-brained primate – till the computers were clever enough to tackle them.

There is a hidden assumption in that kind of thinking, and it's a subtle one that relates to the universal/parochial distinction. Are eyes, or wings, discovered or invented? Is mathematics discovered or invented? Are swords discovered or invented?

Try the middle question as a lever to pry open the other two. We are used to the idea that when we meet aliens, we can start to communicate by using the hydrogen atom, the periodic table of chemical elements, or the solar system and Newtonian geometry, or the series of prime numbers and more advanced number theory, or logical rules and Russell/Whitehead paradoxes. (Bertrand Russell and Alfred North Whitehead attempted to codify a perfectly logical mathematical system, resolving all its paradoxes by the use of systematic argument. Such attempts were shown to be in vain by Kurt Godel in the 1930s, and this suggested our logic might be parochial.) This is a useful trick for authors, anyway. We used it to introduce the 'alien' Zarathustrans in *The Collapse of Chaos*. Sagan used the sequence of primes in *Contact*. And NASA put a plaque on its Pioneer space probe along just these lines. It's a pleasant image, but far too naive. It assumes that aliens will have mathematics, chemistry, and physics that are similar enough to ours for meaningful comparisons to be made. Unfortunately, they won't, so communication with aliens can't really be initiated like this. Plaques on space-probes are based on the assumption that evolving alien brains will carve up the universe in the same way that ours do, and order what they find into the same patterns and logical schemes that we use. Most people seem to think that there is only one way to carve up the universe, and that our brains have homed in on the One True Picture.

This belief is probably false.

The question of how our brains became capable of doing tasks that are so much more general than the capabilities that honed them during evolution is phrased the wrong way round. Where does mathematics come from? The pragmatic view is that we make it all up as we go along; the mystical one is that mathematics is already in existence, waiting for us to discover it. The second philosophy is generally called Platonism, after the ancient Greek philosopher Plato who had some rather way-out ideas about 'ideal forms'; the first lacks a snappy name. Platonism in its most extreme form holds that mathematical concepts actually exist in some weird kind of ideal reality just off the edge of the universe that we all know and love. A circle is not just an idea: it is an ideal. We imperfect creatures may aspire to that ideal, but we can never achieve it, if only because pencil points are too thick. But our imperfect attempts are approximations to a perfect circle that really exists, somewhere 'out there'. Though no Platonist ever says *where* . . .

It's very tempting indeed to see our everyday world as a pale shadow of a more perfect, ordered, mathematically exact one, and to assume that alien intelligences would therefore mirror the *same* ideal in a similar manner to our intelligences. The evidence for this viewpoint, at first, seems compelling. The more deeply physicists delve into the 'fundamental' nature of the universe, the more mathematical everything seems to get. The ghostly world of the quantum cannot even be expressed without mathematics: if you try to describe it in everyday language, it makes no sense. It is tempting to conclude that the universe is in some sense built from tiny bits of mathematics, and many people, philosophers and scientists, have seen mathematics as the basis of the universe. Plato wrote that 'god ever geometrises'. The physicist James Jeans declared that God was a mathematician. Paul Dirac, one of the inventors of quantum mechanics, went further, opining that He was a pure mathematician. This is powerful, heady stuff, and it is highly appealing to mathematicians. Mathematical patterns permeate all areas of science; moreover, they have a 'universal' feel to them, rather as though God thumbed His way through some kind of mathematical wallpaper catalogue when He was trying to work out how to decorate His universe.

However, it is equally conceivable that all of this apparently 'fundamental' mathematics is in the eye of the beholder – or more accurately, in the mind. Human beings do not experience the universe raw: they experience the universe through their senses and interpret the results using their minds. So it is not obvious to what extent we select particular kinds of experience and deem those to be important, or to what extent we are picking up things that really are important in the actual workings of the universe.

There is certainly something about the human mind that causes it to seek out, and give especial weight to, mathematical patterns – whether or not they are actually significant. Just listen to any discussion about this week's lottery numbers. Ironically, what mathematics tells us about the lottery is that any patterns we think we see are illusions, but that doesn't alter the fact that human beings consciously and deliberately look for simple numerical and geometric patterns when they are trying to understand a complex world. This tendency has led to Newton's law of gravity and the equations of quantum mechanics – but it has also led to astrology and an obsession with the measurements of the Great Pyramid.

Is the universe based on mathematics? Do we select those parts that are, and ignore those that are not? Or is the presence of mathematical pattern an illusion, resulting from inhabiting prejudiced minds? The problem is difficult because mathematical 'things' lead a virtual existence, not a real one: they reside in minds, not in physical hardware. But unlike, say, poetry, that virtual world obeys rather rigid rules, and those rules are pretty much the same in every (human) mathematical mind, because the rules are held in a common extelligence. The world of mathematical ideas is a kind of virtual collective, very like Jung's famous 'collective unconscious' – the idea that all human minds include vast, evolutionarily ancient, subconscious structures and processes, which govern much of our behaviour. These things are 'collective' in the sense that they are much the same in all of us. The crucial distinction here is that between a single collective unconscious, into which we all dip, and a large number of distinct but very similar unconsciousnesses, one for each of us. A single unconscious mind for all of humanity is a mystical and rather silly concept that leads in the direction of telepathy, whereas a collection of more or less identical individual subconsciousness – rendered similar by their common social context, their extelligence – is more prosaic but a great deal more sensible.

The same point lies at the heart of how we view mathematics. By introducing a single word for the virtual collective, we make it sound like Jung's mystical telepathic unconscious. This makes us think of mathematics as a single thing, into which all mathematicians – human or alien – dip, and this is a very difficult concept to capture. Where is that thing? What is it made of? How does it grow? What is it? Instead, think of mathematics as being distributed throughout the minds of the world's mathematicians. Each has his or her own 'mathematics' inside his or her head. Moreover, those individual systems are extremely similar to each other. Not in the sense that each head contains the whole of mathematics. One contains dynamical systems, another contains analysis, and a third contains algebra . . . but all three are logically consistent with each other, because of how mathematicians are trained, and how they communicate their ideas. The more established an area of mathematics becomes, the more strongly it feels as if there is some kind of fixed logical landscape, which mathematicians collectively explore.

All very well – but why do the abstractions of mathematics match reality? Indeed, do they, or is it all an illusion? Nowadays it has become fashionable, especially in academic arts departments, to embrace the idea of cultural relativism. This leads in particular to the idea that science is a social construct, and therefore it can be anything scientists want it to be. Yes, science is a social construct, and scientists who claim that it is not are making the same mistake as those who think that we all dip into the same collective subconscious. But it is a construct that has been constrained at every step by external reality. If the world's scientists all got together and decided that elephants are weightless and rise into the air if they are not held down by ropes, it would still be foolish to stand under a cliff when a herd of elephants was leaping off the edge.

In science, there has to be a reality check. That check is carried out by beings who perceive reality through imperfect and biased senses, so it cannot be perfect – but it has to survive some very stringent scrutiny. Far more stringent than any scrutiny of religious beliefs, artistic merit, or a trial in a court of law.

Because our every perception of the world is carried out through the intermediary of the senses, many philosophers have worried that reality may be just a figment of our imagination. This may be true – and there seems to be no way to disprove it – but it requires God to have a rather warped sense of humour, and puts us firmly in omphaloidean territory. A much more important point is that a mind is a figment of reality. Our minds evolved in the real world, and they learned to detect patterns in order to help us survive events outside ourselves. If those patterns bore no genuine relation to the real world outside, they wouldn't help us survive, and such minds would eventually die out. So our figments must correspond, to some extent, to real patterns.

In the same way, mathematics is a construct of the human mind: it is our way of understanding certain features of nature. But we are part of nature, made from the same kind of matter, existing in the same kinds of space and time as the rest of the universe. Therefore the figments in our heads are not arbitrary inventions.

It is then tempting to assume that only one kind of mathematics is possible: 'the' mathematics of the universe. This is an awfully parochial view. Would aliens necessarily come up with the same kind of mathematics as us? Not in fine detail – for example the six-clawed cat-creatures of Apellobetnees Gamma would no doubt use base 24 notation, but they would still agree that twenty-five is a perfect square, even if they write it as 11. However, we're thinking more of the kind of mathematics that might be developed by the plasmoids of Cygnus V, for whom everything is constant flux. They would understand plasma dynamics a lot better than we do, but we wouldn't have the foggiest idea how they do it. They would have nothing like Pythagoras's

Theorem, though, because there are few right angles in plasmas. In fact, it's unlikely that they would have the concept 'triangle'. By the time they had drawn the third vertex of a right triangle, the other two would be long gone, wafted away on the plasma winds.

Mathematics is *invented*. What we have today is only a tiny fraction of the universe of possible mathematical thinking: the particular parochial fraction that the human brain, honed and tuned as it has been, has invented and found congenial. Similarly, human physics is not The Universal Physics, but *a* physics. If we had started from somewhere other than the electron, we would now have a different zoo of particles, different 'fundamental' constants, and these would describe the universe equally well. Just think about creatures for whom gravity is important but not light, and the equally 'true' physics that they would invent.

The non-universality of our kind of mathematics makes Jack&Ian worry about the same effect occurring with tools or weapons. Inventions like swords, throwing sticks that become boomerangs, or bows and arrows, seem obvious to *us*. But they do so because they relate to our brain-structures, not because they are fundamental weapons waiting in the wings for any extelligent creature to find. Our brain-structure found it appropriate, easy, consonant with other mechanisms we perceived, to develop that kind of tool or weapon. So we cannot know in any detail what tools or weapons alien cultures will have. They will be the aliens' own parochial instances of the universal concept 'tool'.

What about language? On Earth, language provided our main entry-route into the world of extelligence, by letting individuals communicate their thoughts to others. Do we have languages, and honed brains, able to grasp concepts-upon-concepts that take us out of our evolutionary limitations, or can we only go so far with the brain we've got, even multiplied as it is in each developing individual by the wonderful Make-a-Human Kit of each culture? There is no doubt that language, the ability to communicate *ideas* to other minds, was the single most significant invention on the road to human extelligence. Language allows the tribe to preserve legends and know-how for the benefit of future generations; it lets individuals build on prior experiences of others, and avoid having to reinvent the wheel in each generation. Once language exists, which is already a symbolic representation of the universe, then the way is open for further symbolic representations – such as writing, which takes extelligence out of human minds and puts it down on clay, stone, papyrus, or vellum. Electronic media are not so far away, then . . . and who knows what comes after those?

Would aliens have language? The universal here is not speech, which relies on a very special collection of organs; probably not even sound, although that would be a common option. It is symbolic communication, the representation of a real thing by a more or less arbitrary label. Labels are simpler to 'carry around' than things – it is much easier to say 'crocodile' than to bring one with you to point to – and they allow an intelligent being to model things, and convey ideas to others, with ease. It seems likely that most aliens will therefore have their own form of language, but it might be based on smell, changing colours, tentacle-waving rhythms, or patterns of polarisation in light.

However, language may be just as parochial as many aspects of mathematics are: it will not be easy to decipher an alien language, because language is deeply rooted in cultural assumptions, and alien culture will be *very* alien. We examined just this point in *Wheelers*: the humans find that making contact with Jovian blimps is a lot simpler than making *meaningful* contact. There we cheated, solving the communication problem with the aid of a human boy, Moses, who possesses amazing empathic powers. We never explained where those powers had come from, or how they could work. We doubt they are really possible, but despite that we are currently planning a prequel that offers some kind of explanation, related to Neanderthals.

There is a basic biological issue here, and for a change it's a rather simple, obvious one. Different organisms have different specialities, and some of these can be seen as tools or weapons. Human beings, paradoxically, are specialised to be unspecialised – to be generalists. But we do have one speciality. Let's start a list, and add ourselves to its end to see what makes sense. The common amoeba is specialised for catching and eating small ciliates and flagellates on the silvery upper surface of the mud of ponds, about 3mm above the anaerobic mud. Fruit flies are specialised, as larvae, to eat decaying fruit; adults are specialised to find decaying fruit and lay their eggs. In this kind of list there is no doubt: humans are specialised to have communication by vocal language.

The question of alien intelligence, extelligence, and language can then be phrased like this. Amoeba can feed on (almost) any prey of about the right size; fruit fly larvae get most nutrition from the yeasts, ubiquitous in decaying fruit pulp, and the adults find the overripe fruits by the smell of alcohol. Can human language have a similar generalising ability, taking our brains into areas we did not originally think about? Because, if such generalising is possible – or if the development of extelligence is constrained, convergent – we may indeed have intellectual overlap with some aliens.

Pepperberg proved, over many years and with impeccably designed experiments, that Alex the parrot, with a very different and relatively tiny brain, can do much of the characteristically language-oriented behaviour of people, with every appearance of putting 'meaning' into it. Alex is the most persuasive argument we have come across for the probability that when we find aliens that have become ('their') extelligent, we and they will indeed overlap enough in our interpretations of the universe that – after twenty years or so – we can begin to understand each other's worlds.

Earlier, we introduced an analogy between the evolution of technology and evolutionary innovation in biology: the cautionary tale of the rocket and the space elevator (with the bolas as halfway house). The point of this analogy is that constraints that appear to be universal may cease to apply (even though no natural laws are broken) if the context changes. Moreover, biology creates new contexts as a matter of course. So apparently strong limitations may not be valid – and it is especially important to recognise this when considering alien life. Reverting to the original province of the analogy, the same goes for alien technology.

The bicycle is a good example of a bolas in human technology, metaphorically speaking, and the steamship is a higher-orbit bolas. Computer technology, which brings cheap computing power to everybody, is a space elevator that takes the PC-user into high orbit at trivial cost compared to the capital investment needed to develop the technology. Rockets to space elevators is our standard example of the later evolutionary product having evolved into a whole new set of rules, and efficiency undreamt of in the initial invention. But that was just one example of how our technology, like biological technology, has evolved. The fluorescent lights illuminating this book as you read it are space elevators, too, aren't they? The enormous technical investment in power stations, the wiring up of all our dwellings and workplaces, represents a capital sum as unthinkable to the electrical pioneers as the space elevator is to our present rocket technicians. The cable linking you to the power station would have been reckoned impossibly efficient at the beginning of the direct-current age of electrical distribution. *Fiat lux*, indeed.

We've argued that extelligent aliens will have technology, but now we've thrown a spanner in the works by pointing out that evolved technology, like biology, employs clever tricks and seems to achieve the impossible. Given this, will we understand alien technology or biology when we come across it?

The old SF scene of taking apart an alien spaceship, or an alien weapon, to see how it works, is unconvincing. Look into a personal telephone: unless you know which parts are resistors, which capacitors, and what the chips do in detail when connected like that, you can't even get started. Furthermore, extelligent aliens themselves will be very puzzling. Not only will they have combined their body parts with a variety of electronic and mechanical and synthesised-biological adjuncts, as we already do (do you wear glasses?), they will have altered their genetics technologically so that their evolutionary lineages may be very hard to relate even to their own planetary lineages.

As well as alien technology, we must consider alien sociology: in an extelligent species, the two are inextricably linked. The major issue that has forced people to ask, 'What are alien societies like?' has been to answer variants of the Fermi Paradox. There has been a spate of suggestions as to why alien societies don't last very long, or are not interested in Contact. They are validated as generalisations by the constant, in human societies, that there are always people who see their society as 'inevitably' going to Hell in a handbasket.

In the 1960s there was a suggestion along these lines, based on the 'fact' that only carnivores evolve high intelligence. Jack's justification of this viewpoint in POLOOP was: 'You don't need much intelligence to creep up on a blade of grass.' The argument was that carnivores compete for food, so that when they became intelligent (we would now say 'extelligent') they start wars over scarce resources, and wipe out life on their planets. This is why we have had no extelligent visitors, and their absence should be a warning to politicians about the cosmic dangers of our imminent atomic warfare. However, it is no longer clear that the allegedly factual starting point about carnivores is genuine. In particular, herbivores *also* compete for food, so why can't that lead to increased intelligence too?

There were many similar suggestions, in which current bogeymen were used to warn politicians: they ranged from pollution to genetic modification, from biological warfare to not enough fresh water, from running out of trace minerals in soils to new viruses, from epidemics of old ones to destroying the Earth's ecosystems. All of these, imputed to aliens because it seemed we humans were about to do that particular lethal trick, were suggested as 'very likely' or even 'inevitable' conse- quences of pursuing a technological path. The general point is that competition for technological resources, or even mere belligerence, must be terminal for cultures.

Proponents of these doomsday scenarios for alien extelligence did occasionally realise that if humanity had not taken that technical path, but had just remained in the Elizabethan Garden, or taken up Taoist or Hindu culture, writing wonderful poetry and theology but dying of smallpox, then *we* wouldn't be going out to visit other solar systems anyway, or even signalling to them, at least by as mundane a means as radio waves. And that explains Fermi's Paradox just as well: they're staying at home tending their gardens and writing poetry.

This reading of Fermi's Paradox might be seen to receive con- firmation from our present situation. Having apparently survived (at least one) threat of extinction by atomic warfare (though some of the other disasters may still be imminent), we are now backing away from space exploration. The space-fever of the 1960s and 1970s has given way to leisurely robotic investigation of the solar system, a not-very-exciting Space Station, and some hi-tech activity listening for radio signals from Out There. You can now contribute to SETI on the web, downloading software to search for alien radio messages, and coordinate your actions with those of other enthusiasts. It's great fun, and it might work, so huge numbers of people have joined the effort, making this probably the largest scientific collaboration ever on the planet.

So, if we *are* indeed typical, we need seek no further, alas, for the answer to Fermi's Paradox. On dozens of nearby planets, all of the aliens are searching for messages and wondering why nobody is sending any – but they're not sending any themselves, because it's so much easier to let the other guys (if they exist) solve all the horrendous technological problems involved in doing that.

Why aren't they here? Because they're waiting for *us* to go *there*.

All of this, fun though it is, is very bad xenoscience. It is sloppy generalisation, from human practices to alien motives. We now discuss three examples of more rigorous considerations of alien extelligence: alien communication, alien transport, and alien societies. Each of these has been illuminated in particular areas by numerous SF stories, but the explicit consideration of the geography of a phase space – the anatomy of the generalisation – has seldom been undertaken in fiction. Instead, authors have picked up certain facets of these areas, particularly those with 'magical' or 'occult' connections – such as telepathy, matter transmission, and spiritual Shangri-Las.

There is a wide generalisation about technology that we should establish before considering the individual areas. It is one that humanity is currently experiencing in western societies and in China, as we are writing. High technology used to mean very finely crafted machinery, very complicated gadgetry that required a specialist to repair it. Now the word 'repair' is becoming obsolete, and nearly every advancing technical field is turning into magic. Most of humanity's previous technology could be fudged, repaired with not-quite-the-right-bit or with an item fabricated on the spot. There were general handymen, 'tinkers', able to turn their hands to the repair or renovation of a variety of technologies, and indeed to societal problems. However, more and more of the technical equipment that we use has the dreaded warning on its case: NO USER-SERVICEABLE PARTS INSIDE. The verb 'repair' is rapidly becoming antiquated.

Clarke's dictum in *Profiles of the Future*, 'Any advanced technology is indistinguishable from magic', is capped by Benford's 'Technology distinguishable from magic is insufficiently advanced'.

Jack was delighted to discover a little piece of magic in his first American car, a Rambler, in 1963: it had a button called COLD START. In England, Jack had been accustomed to putting a paraffin (kerosene) heater under the engine while he had breakfast, then pulling out the choke, retarding the ignition and pressing the starter. Here was a magical spell, indeed: press this button and the car starts in the cold without you doing anything else. All of the translation of verbal instructions into real-world action was undertaken in a secret place under the bonnet (hood). The advance of technology is a downhill bicycle race. In a hundred years, most intentions will be realised by voice instructions, and then only those that are not already taken care of 'under the bonnet' by intelligent robots running society. Aliens that come to us across galactic distances will certainly be much further along that path than we are. Unless, as in the SF cliché, they have degenerated in the generation ship so that what arrives are savages. But then we should contact – or beware of – the *ship's* technology, of course.

Our models of the advanced technology of aliens who come visiting us can come *only* from a prediction of our own future technology, from looking ahead down the bicycle-race hill. Our imaginations cannot conceive of anything *truly* alien. However, even if alien technology is based on a different experience of the universe from ours, and thus on different science, we can still predict certain things about it. Certainly, because advanced technology is genuinely magical, translating wishes into sophisticated and remarkable actions, we won't be able to understand alien magic, even more than we won't be able to under-stand future-human magic. We can be sure that, as in our own imminent future, highly technological aliens will have computer power; the ability to make artefacts out of appropriate materials probably constructed on the spot; instant communication by – at least – electronic means; and replacement of parts, perhaps all, of their biology by improved machinery (not immediately recognisable as machinery).

We guess, and unlike our arguments from Earthly biology we claim that it is no more than a guess, that technology converges, that highly technical races will share technological knowledge, because there is only a limited number of ways to twist the tail of the Universe to make it carry out your wishes. But this may simply be a failure of our imaginations. Perhaps if aliens start from a different place, if they start with gravitics and never think of chemistry, then they will diverge into a technology that we cannot begin to conceive. Even so, there are some generalisations that feel useful in showing that there are common problems, common solutions. We are optimistic that communication can be achieved, at least occasionally – though the wyrdbeasts of Arkansaw–4 may for ever be mysteries to us.

With this background established, we can now consider how aliens might communicate with each other. Our starting-point, as usual, is to think about how Earth's animals communicate with each other – and with plants and bacteria. From these parochials we can derive universals, and invent new instances that aliens might employ. As always, these will serve as placeholders, but they are not predictions: no alien will ever do exactly what we suggest.

Communication on Earth is not confined to sound. There is a lot of chemical communication, such as molecules wafting from a flower and being picked up on special receptor molecules on the sensory cells of a moth. The moth may be a male, more interested in pheromones from a female. The moth can see the flower and the female – by the way, despite its compound eyes, it will *not* see the multiple images that simple-minded filmmakers assume, any more than you see two images of everything because you've got two eyes. There may be a wind, picked up by the bending of little sensory hairs, and there may be sounds, vibrating the membrane of the moth's 'ear'. The moth is colder than it 'wants' to be, so it flies into a patch of sunlight, opening its wings to get more warmth. All this in a brain less than a cubic millimetre! Bees are much brainier, and can carry out much more complicated sensory reception to affect their behaviour; they've got really good internal maps, too.

Sensory perception never seems to be simply a matter of receiving molecules, or of receiving patterned light or sound. Human ears have more nerve fibres going out to the ear from the brain, modulating the sensitivity of different parts of the cochlea (the analytical part of the inner ear), than they have nerves leading from the cochlea to the brain. All animals seem to modulate their sense organs – as well as receiving sensory impressions, their brains interactively control how their sense organs respond. Technology is just beginning to exploit this trick, for example with 'active optics' in astronomical telescopes, where many tiny mirrors are constantly on the move, seeking to produce the clearest image despite blurring caused by Earth's turbulent atmosphere. This finding, in nearly all animals and all sensory modalities, will surely also occur in alien sense organs; they will hear and see what they want to, rather than being subject to overload and then having to sort it out. An extelligent alien will doubtless have evolved a broad set of sensory organs, and is unlikely to use only one for communication with its fellows. We pick up a lot of information by watching 'body language' as people speak, as well as listening; smell plays some part, too. In fact, it is surprising that telephones work so well for us, considering the sensory cues that are missing.

Hospital Station (James White 1962)

In Galactic Sector 12, midway between the rim of the Milky way and the Greater Magellanic Cloud, a huge structure is gradually taking shape: a 'hospital to end all hospitals'. Its wards are engineered to provide hundreds of different environments – any extreme of heat or cold, any level of gravity, any kind of atmosphere. Hundreds of alien races have contributed prefabricated parts, which are now being assembled. Two members of one such race, the Hudlarians, classification FROB, have been caught between the faces of two sections as they are being joined, and have become 'an almost perfect representation of a two-dimensional body'. They are not *so* alien that they have survived the experience. This accident is typical of the day-to-day life of the hospital.

Hospital Station is one of a series of novels – 'fix-ups' obtained by stringing together several short stories into a coherent whole – all in the same medical setting. Aliens are classified by a four-letter scheme. FLGIs are elephantine in proportions; Illensan PVSJs are chlorine-breathers. Humans are DBDG, but so is a beast that stands four feet tall, with seven-fingered hands, and resembles a cuddly teddy bear. There are more alien species than the half a million that can be designated by four letters, so the classification is coarse; nonetheless it gives us an overview of the alien phase space.

A particularly sympathetic character is Dr Prilicla, classification SRTT. The double-T means that he is of variable shape. The R indicates high heat and pressure tolerance. The S, in that context, reveals that among other characteristics, Prilicla is an empath. (A 'telepath' is a being capable of telepathy: similarly, an 'empath' is a being capable of empathy.) A doctor that can instinctively sense his patient's feelings is a major resource for a hospital that may at any moment be faced by an unknown entity with an unknown illness.

Such a doctor can also, of course, be a liability.

What senses will aliens employ? We would not be surprised to see light used, or sound of course. There's more to light than we humans can sense. In 1996, for example, Nadav Shashar and Thomas Cronin discovered that cuttlefish can distinguish objects, and each other, using patterns of polarised light. In fact, cuttlefish seem to employ polarised body patterns in aggressive displays. When two males meet, their polarisation patterns become prominent, controlled by the orientation of skin cells known as iridophores. If neither male backs down, both switch off the patterns and resort to violence. Depolarising filters introduced by experimentalists completely defused the aggression, so it does seem that the cuttlefish were responding to the polarisation, not to some associated pattern of a kind visible to creatures like us.

Alternatively, or in addition, smell could be used. It is by dogs, but they probably don't have to discuss matters as complicated as 'Please can I have some fuel for my spaceship?' Electrical impulses are possible, as are various kinds of electromagnetic radiation. We came to use the electromagnetic spectrum for our civilised communication, recently: mobile telephones. Dr Prilicla, in James White's Sector General stories set in a gigantic space hospital with countless alien species as patients, possesses an innate sense of empathy that functions across species. It's a lovely idea but it makes little scientific sense: empathy is heavily influenced by culture.

There are many myths about communication among animals, and to some extent they have slopped over on to alien stories. The idea that male moths can pick up individual molecules of the female's pheromone – and then fly accurately to where she is – has misled many people about the accuracy and sensitivity of chemical communication. The original experiments, releasing a tiny amount of pheromone from a caged female while your colleague is observing a male five miles downwind, seemed unassailable: when the theoretical local concentration of the pheromone was about one molecule per cubic metre, up flew the observed male and headed off upwind! Only when it turned out that Indian moths didn't 'work' in America, and *vice versa*, did the mechanism become obvious: there were wild moths in the bushes outside the lab. When the pheromone got to them, they released an excitement chemical with much the same effect, and that excited males downwind, and so on. The signal was amplified several thousand times by the time it reached the male that was being observed by the scientists. But he wasn't responding to the female, he was responding to upwind males having a party. It turned out that the level of sensitivity was entirely reasonable, not much better than your own sensitivity to musk, and not as good as trained dogs by a long way.

What about telepathy as a means of communication? This is, for some reason, assumed to be possible for aliens (and mutant humans) much more often than is reasonable. Of course, it depends what you mean by telepathy. The nearest thing to telepathy that has been invented on this planet is probably the electrosensitive sense in sharks and some other fishes. Using organs known as 'ampullae of Lorenzini', they can detect the bursts of nerve impulses (and probably muscle contractions) in worms or fishes buried in the sand on the seabed as they attempt to burrow into it to avoid the predator. It could be the case that some scientifically respectable communication occurs between aliens in ways that we could not detect: we didn't know what the ampullae of Lorenzini in dogfishes did for about a hundred years.

If by 'telepathy' you mean transmission of thought from brain to brain directly, without turning the message into electromagnetic radiation (as in a mobile phone), we doubt that this can ever happen. There is zero credible evidence that telepathy occurs here (many sincere and clever people believe that it does, but their reasoning is flawed). We therefore see no more reason for proposing that it might exist in real aliens than to propose laser-weapon emissions from eyes. The big problem here is that even if there were a possible channel of communication apart from the electromagnetic, the two brains would need to be genuinely congruent – the 'same' nerve cell in both brains would need to be used for the same function, exactly. We know that this is not so for *our* brains, even those of identical twins (who are said, by believers, to experience telepathy more than the rest of us). Twins will have very different experiences programming their brains in the first years of life, so they will have 'teddy bear' spread across different nerve cell associations, with all kinds of different connotations. It is difficult to see how these two different mental architectures could 'resonate', or whatever would be required for messages to pass from one to the other.

Alien extelligence probably requires a Make-an-Alien Kit, communicating to the internal intelligence, as we use our myths and nursery songs for each culture's Make-a-Human Kit. That cannot fail to make aliens as different from each other as we are, and alien telepathy as unlikely as it is for us. The same goes for all other forms of 'Psi', the suite of physically-impossible abilities attributed to fictional aliens, such as being able to pass through walls, to move objects by mind alone, to destroy enemies by emitting laser pulses from the eyes. You've seen lots of these in films. They're nonsense, borrowed from the supernatural zoo: cultural myths like ghosts and fairies.

In fact most of them can be traced rather directly to the spiritualist movement, which got going in the United States in 1848 when two young girls, Katie and Maggie Fox, fooled their mother into believing that their two-room farmhouse in Hydesville was haunted by spirits, and didn't dare admit it when she went public and started to make a lot of money out of being a 'medium', in touch with the spirit world. Strange raps at night had disturbed previous occupiers, and 'a shower of bumps and raps' occurred over several nights – always near the Fox's daughters. Mrs Fox proved to her own satisfaction that the noises were made by a spirit: if she snapped her fingers once, or twice, she elicited the same number of raps in reply. Mrs Fox attributed the phenomena to a spirit, and by 1850 she and her daughters had gone public with their alleged skills as mediums. Around 1890, Maggie Fox confessed that the whole thing had been a childish prank involving an apple tied to a string. Later she withdrew the confession, but in any case it had no discernible effect on the spiritualist movement.

The 'x-ray vision' attributed to other aliens – and to Superman – is also based on a mistaken, ancient belief that eyes 'emit' vision. Ask yourself if Superman's eyes emit x-rays, or receive them. (Then recall that the technology of ultrasound scanning uses a probe that *does* both emit the pulse and receive the echoes.) Aliens will surely communicate in ways that we can't (at least initially) understand; perhaps these will be additional to sound waves, electromagnetic waves and smell. But we think that they will be equally susceptible to scientific analysis.

'Magic' is another story, if by that we mean, as in nursery stories, affecting material objects by symbolic means ('spells' or 'potions'). That does come later in the evolution of extelligence, as we have seen. Aliens that can cross interstellar voids (or humans, some hundreds of years hence) will have many developed technologies that would seem magic to us. But that is no reason to believe in magical aliens, or telepathic ones.

Technological telepathy is another matter. We – and they – might be able to learn to access an internal modem, to send messages to another electronic device, which translates them for the brain it's connected to – not very different from mobile phones, really. But it might look like telepathy from the outside. We recall an SF short story in which the aliens, who cannot imagine sound, decide that humans have telepathy because the flapping of our lips obviously can't carry the complicated messages that seem to be passing between us. The 'Tines' in Vinge's *A Fire Upon the Deep* link their minds by sound-waves, produced by drum-like organs: it 'feels' to them like telepathy, but really it's a form of speech. Advanced aliens – like human beings several hundred/ thousand years hence – will doubtless have effective mechanical aids to communication at vast distances. We should assume that aliens who have crossed from another solar system have communication systems that are immensely effective, and probably impossible to intercept. For all practical purposes, they *will* have telepathy.

People who know little about radio communication assume that it's just a matter, as it apparently is for the *Star Trek* crew meeting an alien vessel, to tune into the right waveband and there they'll be, tentacles waving. Radio communication is much more versatile than that, and therefore it will be more difficult to work out what system aliens are using to communicate. Even in humanity's brief play with radio waves, we have gone through three revolutions, making the present messages very difficult to pick up and recognise as communication, impossible to decode by previous radio operators. The original amplitude-modulated waves, decoded by simple diode rectification, were succeeded by frequency-modulated systems. The AM operator would see the frequency wandering about, and that would suggest that it wasn't a communication signal. Even FM experts would not recognise a modern digital signal as a signal, even if they were familiar with digital signals as they had been used up to, say 1985. They would be completely baffled, for example, by a digital TV channel that has been encrypted or compressed. We are now using communication channels that are specific to special tasks – mobile phone stations have a variety of verifying codes, for example – which guarantee a receiver that doesn't respond with the right code receives no communication. And we're only at the beginning. So what is an alien species that has been using radio for ten thousand years going to be doing? By Lachmann's theorem, any sufficiently advanced communication will be so efficiently coded that it will be indistinguishable from noise. We certainly won't be able to snoop on their communications, that's for sure. And when we go out into the galaxy, and meet that First Contact spaceship, a radio signal will be about as useful as a smoke signal. Communication evolves.

What about 'ansibles', a term introduced by Ursula Le Guin in a series of short stories, detailed in her novel *The Dispossessed*, and adopted widely by the SF community? Ansibles are communicators that can pass messages much faster than light. They are necessary staples of many widescreen baroque SF stories, as is FTL (faster than light) travel. So far as we can see now, the issue concerns the reality of a common 'now' across light-years of distance. From our, perhaps naive, physics the combination view of time and space that Einstein taught us has no place for a universal 'now' of the kind necessary for a human-interest story spread across the stars. Because of this, alien extelligences that have had their Einstein will be reluctant to embark on journeys to other stars . . . this by itself could be the major answer to Fermi's Paradox: the distances are so enormous. Time contraction makes journey times shorter, but communication across different speed frames may be impossible. Astrobiology may show us that the journeys are possible, but we wonder if the biology of real organisms will allow them to cut themselves off from the rest of their species.

This issue carries over into the question of transport, to which we now turn. Is FTL travel feasible? According to astrobiology, it's not, because astrobiology is based on physics *now*. But even today's physics contains hints that the speed of light may not be the ultimate limit. On the basis of Einstein's general relativity, Miguel Alcubierre has shown that while matter cannot move through space faster than light, *space* can move through space faster than light. In principle. So the standard justification of 'warp drives' – put your spaceship inside a small region of space, relative to which it remains stationary, and then move that surrounding space faster than light – might actually *work*. There are huge obstacles to turning this idea into a usable gadget without violating general relativity, but the germ of FTL technology now exists. Mind you, the next major advance in physics could wipe it out again: there's nothing sacred about general relativity.

There could be other obstacles to space travel, too, resulting from alien biology. And what about the psychological effects on living organisms that might, entirely reasonably, be reluctant to leave their sibs, their home, their country, their planet (delete whichever is inapplicable). Even 'generation ships', in which several generations must pass during the trip to reasonable destinations, might be impossible for some alien races to make and crew. But we know that our appreciation of the problems must be far from reality.

We have in mind the 'proof', in about 1860, that if steamships went faster and faster, people would be crossing the Atlantic in seven hours in 1960. That prediction was a puzzle, because it was well known that friction of water against the ship's side increases as the cube of the speed, so the steel hull would melt. No one thought, then, that it would be aeroplanes that made the journey, taking hundreds of people at a time; instead of extrapolating exponentially upwards, be prepared for your variable to turn out of the thickness of the paper, at right angles. That example tells us that we can't know what *our* transport systems will be doing in two hundred years, so it would be silly to come to any firm conclusions about specifics of alien transport.

Generalities are another matter. Lightspeed or not, fast transport has made enormous difference to the Terran societies that have it. This phenomenon is probably a universal for taking all kinds of pests and diseases all over the home planet, and perhaps out to the stars. There are all the little parochials whose provenance for aliens is dubious: the Ford Model-T, afforded by middle-class Americans, gave them unsupervised privacy for the first time, so very many members of the next generation started life in the back seat of a Ford. With trains, where you can get into a 'room' at York station and get out of it at London Euston, transport becomes so easy that even Granny can travel between cities.

As technology, alien or human, improves, the transportation of people and goods will become effectively 'free'. Boundaries and barriers will remain, however; remnants of territorial guarding in the alien or human brain. It is amusing, in contemporary Britain, to compare the older generations, for whom 50 miles an hour is fast and 'doing the ton' (100mph, 160kph) is teeth-clenching, with the generation that has grown up with motorways, doing 70–80mph routinely and not being very bothered by a highly illegal 110mph. The Make-a-Human-Being Kit has obviously moved on in this regard. So have the cars.

Will aliens, and us, have matter transmission? This is another old trope of SF novels – and of course, of the 'Beam me up, Scotty!' generation. On the one hand, we can't see how matter transmission could work on the basis of today's science, but on the other hand, it is just the kind of thing that technology might turn into magic. In Crichton's *Timeline* a combination of matter transmission and time travel is realised using a quantum computer: this is plausible fictional gobbledegook but makes no scientific sense. Instantaneous matter transmission is probably impossible, but technology can so often substitute a process that has the same effect – aeroplanes crossing the Atlantic – that we can imagine some such technology becoming commonplace in some advanced civilisations. Our inability to understand something doesn't stop it from happening. Just as the Victorians couldn't imagine aeroplanes crossing the Atlantic with hundreds of people on board, all taking it pretty calmly, so we can't imagine the transport of the future. Could there be flying saucers? Stranger things have happened. In 1976 the French magazine *Science Vie* published an article by the physicist Jean-Pierre Petit, about a plasma motor for a spacecraft. He pointed out a curious side effect of that technology: the ideal shape for the spacecraft is a disk with rounded edges – the classic 'saucer'.

While we're on the subject of transport, and of understanding, there is the question of wheels. It is quite easy to come up with reasons why wheels have not been 'invented' by organisms, but less easy to decide whether those reasons are correct. The usual suggestions are that it is difficult to build a device to keep the rotating part of the wheel supplied with blood or nutrients, and that legs are generally more effective if travel is not on surfaced roads. It is widely believed that Central American Maya and Aztec peoples did not have wheels, except perhaps on toys, and for the sake of argument let's assume that they didn't, even though the evidence is not totally conclusive. If so, this is a very important ingredient in our consideration of alien technologies, because it prompts two questions. The first is whether there is a large category of such 'obvious' tricks, of which we have only found one; but perhaps aliens have discovered others. The second is to wonder what a Maya or Aztec would respond, if they were shown a wheeled cart; would they say, 'Wow', and invent roller-skates, or would they find it difficult to understand, while their children were skating about? (The modern parallel is programming VCRs.)

Clifford Simak's novella *The Big Front Yard* examines this question: interstellar 'rats' rebuild the house of a Yankee trader handyman so that its front yard is a plain on to which other alien 'front yards' abut, and that serves as a trading station for ideas. The first alien comes riding up on an antigravity saddle, and the trader swaps this idea for the idea of paint. Our picture of human technology, illuminated by that failure of the meso-American civilisations to use wheels, is of a tiny, parochial usage of a few tricks; our engineering science is probably as limited, rooted as it is in this little gadget-garden. Aliens will have their own patches, so it is very educational to wonder whether wheels are parochial or universal.

The word parochial means 'of this parish', and that brings us to the third big topic. Will alien social groups be organised as we are, in families, in villages grouped as parishes, in nations with a hierarchy of allegiances? It seems to us that this arrangement is natural, but to what extent is this because our ancestors lived in family groups, out on the savannah or the seashore? These groups were probably loosely organised in patches of more-or-less friends and more-or-less enemies.

There is now considerable evidence, from the work of the Adapted Brain school of neurophysiologists and psychologists, that our brain 'organs' have built-in preferences of very distinct kinds. Most humans, from Inuit to Zulus, choose a slightly hilly savannah landscape (preferably with a stream) when deciding between pairs of flashed-up scenes; they all prefer this to their current home landscape. English 'parkland', fields with occasional trees, is never found in nature in temperate climes, but has been produced by all of the famous landscape architects – Capability Brown is the best known – and disseminated into innumerable homes by artists like Constable and Turner.

Aliens would presumably have totally different preferences; is 'having preferences' the generalisation to a universal? There have been many explorations of this question in the SF literature, such as the naive 'Their reptilian ancestry determined that their attitude to nearly every question was "Can I swallow that?"' and 'Their feline nature gave them "growl" or "purr" responses.' Here the author carries the terrestrial nature of the animal into the psyche of an alien of the same shape. Usually it is not the real nature, of course, but the iconic nature that we learned in the nursery: owl-like aliens are wise, bear-like aliens eat honey . . . we exaggerate, but not by much. Many of these authors must find it hard to realise that other humans have different labels for the same animal images.

There is a persuasive thread in the SF literature that sees all extelligent species as improving themselves, with better biology and technical adjuncts, as their technology improves. This movement toward the cyborg, the mechanical/biological synthesis – guyed as The Borg in later *Star Trek* series – probably means that when we do meet aliens, they will have designed new sociologies for themselves. Being able to turn yourself off for a long journey, or to avoid present unpleasantness or danger, must make the culture very different. We played with this idea in *Wheelers*, where the Jovians can aestivate (hibernate if you prefer – there is neither winter nor summer under Jupiter's clouds) for tens of thousands of years in their blisterponds. Out in the galaxy we will meet mostly or entirely mechanical/electronic creatures, who have become what they are because of their biological history. That history will not be available to us, nor perhaps to them. By the time we meet them our xenoscience may be more mature, and able to assist with the politics . . .

No, we won't even *start* to discuss alien politics.

13

HAVE ALIENS VISITED US?

'*E*RROR,' *SAYS* *CAIN*. '*Misjudgement. Mistake. Blunder. Bloomer. Bungle, boo-boo, boob, blooper—*'

'*All right, you don't need to squirt it all over. I wish I'd never shown you the thesaurus option, anyway. It was a tiny oversight, that's all.*'

The recycled Galactic Prospectus *had gone to Headquarters, all ten million volumes of it, even the half million on plasma-dwellers. It still needed to be proofread, but the worst was over. Or at least, that's what they'd thought, until Cain noticed that something unusual was happening.*

'*I think that waking the Maracot Deep Complexity into consciousness is more than a mere oversight, Abel.*'

'*Look, it was a simple mistake, anyone could have—*'

'*It's calling you Daddy. Headquarters will not be pleased.*'

'*It's calling you Daddy, too.*'

'*I am no more pleased than Headquarters will be. Tell me how it happened.*'

The glitch had slipped past them when they had first arrived on Earth and looked at a coral reef. 'You remember those two humanoids?' Abel said. 'Well, the male has written a book about centipedes like you, set on some imaginary planet that he calls Mesklin . . . It's sold ever so many copies, the house says. And it came to the attention of the Maracot Deep Complexity, as these things will, and now the weather is behaving strangely and I'm not sure we can calm everything down before the Manifestation Police get a sniff of it, and—'

'*Abel, calm down, it's too late for all that. We can't put the* ⚐ 𝍫 🜸 *back into the bag. We'll have to confess, and take what punish—*'

'*No, no, we may just get away with it,*' says Abel, *struck by a sudden idea. 'We just have to decontaminate the mythology before the authorities notice. Discredit the story, so that the humanoids will think it's fiction.'*

'They think that now.'

'Yes, but it's still too popular. We've got to turn it into an isolated sub-genre.'

Cain is intrigued. 'That might work. But the weather would still be behaving strangely.'

'We can spread rumours blaming the weather on global warming, Cain. All we need is to divert that kind of story into a cultural backwater . . . maybe we could use a gadget like this?' Abel has materialised a curious item of Terran clothing, a cap with a small propeller on the top. 'We can organise meetings at which small groups of humanoids will wear these and talk about aliens.'

Cain thinks about it. 'You're right, Abel. Of course, it will be a bit of a con.'

Have aliens visited Earth? If you believe that democracy is the way to decide issues, and if you take note of how many people are motivated to vote, and if you look at the number of publications, you will not doubt that many kinds of aliens have visited us for a great variety of reasons.

However, if you have followed what we have said so far, and in general agree with our arguments that alien evolutions cannot produce anthropomorphic – humanoid – creatures, then your belief in these 'alien' visitations will be tested very strongly by the pictures that the witnesses have produced. Nearly all of the alleged visiting aliens are slight variations on the human form. Even those that are claimed to be radically different usually have faces, chests, upper and lower limbs with opposite joints (like knees and elbows), and even teeth (erstwhile scales, in a mandible that was clearly related to the front end of a bony gilled filter-feeding ancestor). These characters unrelentingly identify their carriers as being related to Earthly vertebrates, indeed land vertebrates, and nothing else in the universe. This fact cannot have been taken on board by those witnesses, or by the writers who claim that the trivial differences (four digits, for example, not as different from us as cows or horses are) show alien derivation.

Meeting aliens, recognising them, and communicating with them are all very problematic areas. All the 'proofs' of alien visits that we've seen are unconvincing . . . no, *anti*-convincing, because the corpse, photo, film, or recollection is of something humanoid. Therefore they cannot be alien. We address this paradox here. In a sense it is a part of the Fermi Paradox: if intelligent aliens exist, why aren't they here? As a sample of possible answers, we offer nine suggestions, all of which have been proposed both by professional scientists and in SF stories.

- There are no aliens, and there never have been. Humanity is unique in the universe.
- There have been plenty of aliens, but civilisations only moderately more advanced than ours always blow themselves up in nuclear wars.
- The lifespan of an alien civilisation is only a few million years. They visited us ten million years ago, and another bunch will turn up in ten million years' time, but there is nobody around right at the moment.
- Aliens exist, but interstellar travel is impossible because of relativistic limits on the speed of light, or because living creatures cannot survive it, or because they don't enjoy it.
- Aliens exist, but are not interested in interstellar travel.
- Aliens exist and have interstellar travel, but they are not interested in contacting us.
- Aliens exist, but galactic law forbids any contact with us because we are too primitive, or violent, or ✷✲✽✲✽✲✾✲.
- Some aliens see it as their duty to eliminate all other forms of life that come to their attention. Any technological civilisation will develop radio and TV, attract their attention, and be eliminated. They are on their way now.
- They are here already (the preferred answer on the internet's UFO pages).

The evidence for the last assertion, as for the others, is poor. It consists mostly of fuzzy accounts of unusual events, with anthropomorphic creatures assumed to be aliens involved in the action. There are two opposing ways of interpreting these observations, three if we include a self-destroying stance.

The first is to say that these 'visitations' are all illusions; we know that the human mind indulges in such things much of the time. The common patterns result, perhaps, from Jungian archetypes, which we all share, or from copycat behaviour, or from the dredging up of recent media pictures to 'fill in' something either dimly glimpsed or so odd that it cannot be interpreted. This would include the visions experienced by people who wake up while still in 'sleep paralysis' and which are assembled like waking-up dreams from a litter of memories and outside events – like the way we incorporate an alarm-clock call into a dream as if it had been part of the ongoing 'story', even though the call interrupted our thoughts and woke us up. The psychologist Sue Blackmore has found extensive evidence for this explanation of 'alien abduction' experiences and the like. While this would be an interesting branch of psychological pathology, it throws no light whatsoever on what real alien lifeforms are like. All it says is that our brains invent things.

The second is to say that a few of the alleged observations may be 'real', but that most are not good original observations, or are *post hoc* reinterpretations of genuinely alien (but perhaps not extraterrestrial) experiences, which put them into the mundane world because most people have mundane minds. 'Flying saucer' (UFO) observations clearly fall into this category: the serious end of the literature publicly regrets so much of the 'silly' and fake stuff, and concentrates on a few cases, which have irreproachable witnesses or lots of witnesses. There are at least some tens of these, and they must, we think, be taken seriously and not simply dismissed as 'mass hallucinations' or attempts at tourist bait by otherwise unattractive villages. That UFOs are actually to do with alien planets in any way, though, is a weak link. However, a few of the sightings were apparently of living or strangely-mechanical entities, so they deserve to be considered here, in case they really were aliens visiting. Even if only one sighting were real, we ought to know about it and recognise its importance.

The third interpretation is a self-destructive stance which specifically denies that critical approaches are appropriate. Many scientists therefore believe that this is a 'lay' position, one not to waste much time on. This third position on alien visitations is very diffuse, and hovers around the New Age style of assertion: 'My granny's cancer was cured by homeopathy, and here's her picture to prove it.' The act of being critical is greeted by many people as discourteous, so when someone makes this kind of remark, we tend not to point out the obvious holes. For instance: some cancers undergo spontaneous remissions, so how do you know it was the homeopathic remedy that worked? The theoretical basis of homeopathy, that substances become more potent the more they are diluted, is arrant nonsense anyway, and homeopaths are not consistent about it. Jack once saw a shop selling homeopathic remedies for headaches: 'take one for mild headaches, two for severe ones'. He went into the shop to point out that for a devotee of homeopathy, it should have been the other way round. If we require more stringent evidence for events that are very extraordinary, but related by 'ordinary' (not scientifically trained) people, we are believed to be defending our cold scientific unemotional prejudices from the warm, folksy loving woman whose story differs on successive tellings because she has not had the benefit of an education like us. And why should consistency be important anyway, they ask? We should not be so rude and brutal as to get her to tell her story several times, or so discourteous as to disbelieve her in the first place.

The fact that scientists are critical of their own positions, and not just those of others – and hope to be more critical, and better at it – is regarded by proponents of the warm-emotions lobby as evidence that scientists' lives are 'ruled by cold rationality', and that they need to be cured of that by transcendental meditation, the love of a good woman, exorcism, or scientology. But these same people show their lack of useful causal thinking when they are addressing other subjects, such as the cure of illness, and are against 'chemicals', without realising that so much of our natural world is chemical. A recent news article had a shock headline about a 'horrific chemical spillage'. The chemical was (liquid) carbon dioxide, and the 'spillage' merely increased its concentration in the atmosphere by an insignificant fraction. Another news story pointed out that a fast-food strawberry milkshake contains over forty chemicals, as if this fact alone is damning. How many chemicals do they think a *real* strawberry milkshake contains? It must be hundreds of thousands. There are plenty of good reasons to criticise fast-food outlets, and 'synthetic' food, but simple-minded chemical counting isn't one of them.

What *kind* of chemical, what effect it has on children . . . now you're talking.

People of this persuasion approve of 'natural' herbs without wanting to know that many herbs (or even well-known natural foods like cabbage or pineapple) have carcinogens in much higher quantities than the processed foods they despise and fear, whose quality is regulated by law – mostly an effective sanction in these cases. Their fear of radioactivity cannot be ameliorated by the discovery that coal-fired power-stations (and of course domestic hearths, *pro rata*) release much more radioactivity into the air, in the form of carbon-14 and tritium, than nuclear power stations do. When there was a 'scare' about some birth control pills being very slightly more likely to cause circulatory problems, many women stopped taking them, and many other kinds of contraceptive pill, but they didn't replace the pills by other methods of birth control. The number of pregnancies that resulted was fifty times more dangerous than all the women in England transferring to the suspect pills for a year. But that was the coldly rational view, not taking into account how the women felt, of course.

This kind of causal picture of the universe is so loosely put together that it can admit endless stories of the kind 'The aliens took my baby' without asking, 'Were you pregnant?' As we mentioned, Jack asked just that, and considerably embarrassed a woman on live radio, because she hadn't been. 'They're magic, you know,' she said. And that's the self-defeating point of these loosely woven attitudes to evidence. There's always another 'warm' explanation, ready for the opposite evidence – or ready to cover any lack of evidence.

Not only that. If you take such thinking seriously, it leads into very murky waters. You mustn't be *too* imaginative. In about 1965, Ivan Robinson, a devotee of ESP, was very keen to show the closed-mindedness of scientists, and tackled Jack with an ESP story. ESP, you'll recall, stands for 'extra-sensory perception' – supernatural phenomena like telepathy. Robinson's tale was characteristic of a particular genre of these stories, in that it was very well attested by the local judge, Chief of Police and other notables of an American Midwest town. The gist of the story is that a local man was very proud of his dog's responsive intelligence, and one day he asked the dog, in the presence of several notables, to bring him a particular newspaper from the local drugstore. To everyone's surprise, the dog came back with the right paper. This was repeated, for a larger and larger collection of local notables, for a different newspaper every day for a couple of weeks, with everyone commenting that the dog had to be able to read, as well as being able to understand the request, and so forth. Ivan had several newspaper clippings, all extolling the dog's abilities. He was convinced, as were the newspaper reporters, that this must be a very special dog.

Stop reading: spend two minutes thinking how this result could be achieved by simple trickery. We bet you can find at least half a dozen explanations, all entirely straightforward. 'Yes, but it *could* have been ESP . . .' Sure. So could any of the tens of thousands of other entirely prosaic things that happen to you every day. Maybe the dog was really teleporting the paper into its mouth. Maybe the law of gravity gets suspended every so often, but intelligent rain still manages to make itself fall downwards by the power of thought.

Jack, though, had an entirely different objection, and it's far harder to answer. The problem with this kind of credulity is not that there are prosaic explanations. It is that so many different *supernatural* explanations are possible. 'Nope,' said Jack to Ivan. 'Not a telepathic dog – he's just a dumb pooch who likes picking up newspapers. But his *owner*, there's someone with astonishing precognition!'

The view that we should believe whatever anyone tells us as long as they seem to be sincere, good-hearted people, is absolutely fine for a philosophy of life, but rotten as an exercise in editing our beliefs towards what the universe really does. Even if the event had happened exactly as related, as with the dog-story above, there is no guidance as to how to change your worldview. If the dog could read, and the performance was because of the dog's abilities, we would have to reject much of what we think we know about animal brains, about the nature of being taught to read, about linguistics . . . Or do you throw out all your beliefs about the electromagnetic spectrum, or about the dissimilarity of brains, to allow telepathy? Do you throw out your beliefs about the nature of time and the indeterminate nature of the future, chaos theory or quantum theory, to allow precognition?

The wonderful abilities shown by Pepperberg's African Grey parrot Alex are completely different. Because she constantly took trouble to exclude other explanations than the mundane but exciting one that parrots can learn our language much better than we had thought, we have an unambiguous message. This enables us to change our minds piecemeal, in a disciplined way.

Compare this with Yuri Geller bending spoons. Leave aside all those stage magicians who say that it's a standard conjuring trick; over the centuries conjurers have developed ways to simulate almost every supernatural phenomenon, so any 'real' supernatural effect would be dismissed out of hand if we allowed that kind of evidence. No, the real problem is that if we are convinced that Geller really does have the power to bend spoons using his mind alone, then we don't know whether we must change our minds about metallurgy, about what 'mind' is, or about the relationship of mind-mechanism to metals.

We know very well that the Magic Circle can puzzle us with 'magic' tricks, of course. When a stage magician cuts a woman in half on the stage in front of us, nobody in the audience has to change their mind about anatomy, about how the woman can possibly live with her halves separated. We know that it didn't happen, but that all the evidence presented to us made us think that it did. However, if it really *did* happen (and the ability of conjurers to fake it does not prevent such an occurrence, only our current understanding of nature) then we wouldn't know what assumptions to change. We could start a scientific programme to find out – but then, the first thing we would do is repeat the experiment, trying to find the trickery that is overwhelmingly likely to be there. Only when we had excluded that would we start the more difficult process of rewriting our entire theory of the universe.

Either we think scientifically – or at least rationally and critically – about this kind of event, or we end up unable to think sensibly about anything. You choose.

The same goes for alien-sightings, and disbelieving the people who tell us about them – especially when all of our experience says that humanoid aliens are impossible. We believe these people to have been mistaken, like so many of us are when a conjurer fools us. Maybe the universe has a set of phenomena that do this to everybody, making us see ghosts or sea serpents or UFOs. This is why the Roswell film of alien dissection is not persuasive: it shows exactly what wouldn't be there, namely a small humanoid creature with humanoid organs like muscles and liver. It's no more credible than the sawn-in-half woman. On the imaginary world of Epona, devised by people with imaginations, muscles were originally water-storage tissues in plants, expanding and contracting as they took up water or expelled it. Then they were used to adjust the angle of the leaf to the sun, then to strangle other plants. Then some plants evolved into animals. As a placeholder for an alien parochial, it's a far more convincing scenario.

There is a film marketed as *Miracle of Life*, about human reproduction, which cheated in a similar way. It portrayed what people expect to see. One shot of spermatozoa, for example, consists of rabbit (or perhaps bull) sperms – much more like tiny tadpoles than human sperms are. But the clincher is that the fluids in the tubes, purporting to be in a living body, have a surface, like water in a transparent hosepipe or, come to that, in a cup. Liquid in your reproductive tubes is not accompanied by air. The liquids fill the tube, and if there's not much liquid, then the tube collapses and no longer looks like a tube. Because people *expect* to see a tube, and therefore a liquid surface, this is what the film showed. Similarly with Roswell: because people expect to see a little human-like corpse being dissected, that's what the mock-up shows. And that's why it was always virtually certain to be a fake.

We would be absolutely delighted to meet real aliens. But there is no good evidence that they're here, or ever have been here. At least, not here in *The Hitchhiker's Guide to the Galaxy* fashion.

There are many well-attested, but apparently very puzzling, phenomena – coincidences, strange occurrences. Some of these have resonances with alleged alien visitations. Even though these reports often seem baffling, we are not *obliged* to believe them just because we can't think of a way to explain or disprove them. Science is the best protection we have against believing things because we want to. A bit of critical scientific thinking can focus our minds on reasonable explanations instead of unreasonable ones. At least, then, we appreciate the range of possibilities, instead of falling for the first vaguely plausible suggestion that we are offered.

We have a general point of view that sheds light on all such reports. It is based on a classification of these strange observations into two categories.

The first category consists of 'weird phenomena', such as sightings of large rotating fans of light below the sea surface, observed by sailors and passengers on ships. Many strange, apparently metallic objects have been found which, when 'analysed by a local laboratory' belong to 'no known groups of materials'. This surely says more about the range of expertise of the analysers than about the materials. One classic example, a large, shiny sphere, had the mysterious ability to roll to the centre of a table, and was declared to be made from an unknown metal. It turned out to be part of the valve gear of a hydroelectric power station, and it rolled to the middle of the table because it was *heavy*, causing the tabletop to sag.

There is clearly a spectrum here, up to 'double-decker bus seen on Moon's surface', but there is certainly a plethora of mundane events that do not lend themselves to the kind of explanation that we are used to giving. 'Oh, that's caused by a . . .' leaves us with a line of dots, mouth open, at a loss. One of the most famous of these phenomena is crop circles: areas where standing corn or wheat has been flattened into some curious shape, like a circle or a Mandelbrot set. Crop circles certainly do appear, and there is no doubt whatsoever that some of them are hoaxes, made by people. Most likely they all are, but there are claims that some of them have appeared overnight to the accompaniment of floating lights and without any disturbance of the corn stalks of a kind that people must inevitably make. There are films of these oddities. It is also claimed that satellite photos have proved that very complicated patterns have appeared in the two hours between observations.

It is entirely possible to dismiss these by claiming that the films have been faked (some probably have: there are 'transparent' humanoid figures carrying the lights, showing double exposure) and that the satellite photos are not successive ones (as judged by shadow length).

These things are not like ghosts, telepathy, or clairvoyance, for which most of us are happy to seek explanations in ethereal realms like misty air, statistics or gullibility of some human minds. There actually is a material basis, which people can observe in the same way that we deal with the mundane world of the kitchen, the public-health laboratory, or items found on country walks. People have sectioned the bent-over stalks of the corn for microscopy, and have found – so they tell us –

differences in the pattern of the vessels from those formed when people have deliberately bent corn stalks. On the other hand, people have successfully constructed corn circles, but not very complicated ones, under observation. It would be possible to imagine, or to duplicate, a mechanism for producing large rotating fans of light below the sea surface. In many ways, the problem with these oddities is not that they exist: it is how they come to be when and where they are.

The second category of 'supernatural' events was a phenomenon in the quasi-non-fiction book industry some decades ago: the 'Ancient Astronauts' phase, of which Erich von Däniken was the doyen. Vast tracks on South American plateaux, the technology of the pyramids, the origins of cuneiform writing, and some Biblical descriptions of visitors with odd headgear, were all attributed to (humanoid) visitors from other planets (or at least other places: sometimes it was Other Dimensions). This was clearly a popular way of associating these puzzling aspects of our world, of our reported history, with one over-arching explanation. It had more excitement, more exotic detail, more enjoyable associations with other puzzles (is *that* why the Aztecs and the Egyptians had similar pyramids?) than the previous over-arching explanation: God did it. And there was the link to everyday life: you could imagine yourself going on holiday to South America and being taken by bus to those vast tracks, in principle judging for yourself whether they could be seen as a cross only by landing spaceships, or whether they were simply places where two llama-cart tracks crossed at right angles.

The Sirens of Titan (Kurt Vonnegut 1959)

Multimillionaire Winston Niles Rumfoord and his dog have sailed their spaceship into a chronosynclastic infundibulum – a place where all the different world-views about the universe fit together perfectly. Every fifty-nine days Rumfoord (and his dog) manifest themselves. On one such occasion he tells his former associate Malachi Constant what his future will be, and it's not pleasant.

Constant's father Noel has made a mega-fortune investing in the stock-market, using a scheme based on the letters of the Bible. Everything he bets on is a big winner. The son takes over where the father had left off, and for five years the scheme continues to bring in big bucks. Then, suddenly, the family business lies in ruins. Constant and Rumfoord's wife Beatrice flee Earth in a spaceship, incidentally fulfilling the first part of Rumfoord's prophecy.

Constant ends up on Mars, in the army, with no memory, bearing the name Unk. Later, he and another army man, Boaz, spend several years in a cave a hundred miles beneath the surface of Mercury. It is inhabited by *harmoniums*: creatures that plaster themselves to the cave walls like leaves, and give off a yellow light. They like to arrange themselves in patterns, a common one being the word THINK. It is an intelligence test, and Unk and Boaz fail it, until one day the harmoniums give Constant the clue he needs to get back out of the cave: 'UNK, TURN SHIP UPSIDE DOWN.'

He, Beatrice, and their son arrive on Titan, where Rumfoord is materialised all the time. There they find Salo, an 11 million-year-old Tralfamadorian. In 483,441 BC Salo left Tralfamadore on an important mission across the galaxy, bearing a sealed message, which he must on no account open. In 203,117 BC his spaceship broke down on Titan, and ever since, he has been waiting for the spare part to arrive. Now it turns up, as a battered strip of metal kept by Constant's son as a lucky charm. The Tralfamadorians, who are intelligent machines, have been manipulating the whole of Earth's history for hundreds of thousands of years to get that metal strip to Titan – which is why the Constant family's investment scheme worked so well, and why it then failed.

The shame of this discovery provokes Salo to open the message. It is a single dot: Tralfamadorian for 'greetings'.

The above two categories justify our explanation for all such phenomena, whether attributed to alien intervention or not. The explanation is that humans have a very poor estimation both of how clever they are, and of how clever the supposed aliens must have been.

Most people today expect to be able to comprehend both the existence and the provenance of the objects they come across: 'Ah, that exotic method of ground transport can be located precisely, it has a Ford medallion on its radiator grille.' Actually, we have very little idea how anything in our *own* world works – so how can we possibly judge a corn circle to be impossible without alien intervention? And most people today fail to grant even quite rudimentary technical ability to their forbears: 'What, they brought those enormous stones all the way from Wales? They must have had a truck. Or (wow) done it by magic.'

The mute testimony of the stones themselves, which are *there*, so must have been transported there, cannot compete with the eagerness to embrace the transcendental.

As the Clarke–Benford dictum indicates, most of us today are surrounded by magic. Real magic, magic that works, magic that we cannot comprehend, cannot explain, but can use without even thinking. From turning on lights to starting cars, from television to the removal of excreta, we live in a world in which other people have expertly taken the effort, especially the intellectual effort, away from us. This means that we assess our powers of understanding as being much higher than they really are. On the level of 'What turns the light on?' we offer 'The switch on the wall.' We all seem to have our surroundings pretty well under our volitional control. That means that when we see something which lies outside the provenance of our extelligence, but which looks like the kinds of thing which we expect to find inside, like large rotating fans of light below the sea surface or corn circles, we fail to explain them. There isn't a switch-on-the-wall there. We expect to find a mechanism, and we fail; we look for a physical manifestation of the 'spell' and there isn't one. So we attribute the extelligent-looking mechanism to an *outside* extelligence: ancient astronauts, aliens.

Hollywood films of pretend-pyramid construction up to the 1980s showed the stones being moved either by giant levers, with the stones on rollers, or by gangs of slaves with poles slung under the stones, and two or three men at each end of a pole. Archaeologists examining the huge stones of Solomon's Temple in Jerusalem have located exactly the scratches that would be made by the first kind of technology, but only when the stones were being slid down a prepared slope to their final location – this applies to four out of some tens of stones. The second technology, carrying them on poles, has never been validated, for the very good reason that it works fine for large expanded-polystyrene blocks, but it is *really* bad, awkward, injurious even to experienced slaves, when rocks of some tonnage are to be transported. People could see this, and that was part of the attraction of von Däniken's attribution of the technology to aliens.

However, just because *we* can't do it doesn't mean that our ancestors couldn't. We now know a simple way that it could have been done, and it's probably what was used. If not, there's probably one (or more) methods that we haven't thought of. That method is to support the rocks using rope nets instead of poles; then the weight is distributed much better, and the slaves can manage about a hundred pounds each without much problem. Further, unlike poles, one person can let go of a net, and his contribution doesn't fall solely on his neighbour (making him drop the pole because it's too heavy, then think dominos), but on everyone. The use of rope nets doesn't make it easy to move huge rocks, but it does make it possible with the technologies we know the ancients had. The technology of carrying large rocks on nets *works*, and it doesn't need alien help. The old stupid technology of carrying on poles, doubtless invented by film directors who would not ask quarrymen, doesn't.

There is a third class of observations, which are puzzling but in a rather different way. Their explanation is different, too: we 'see' these things because we use a special bit of our brains for thinking about living creatures. People often see strange, frequently 'impossible' animals, ranging from mammoths and dinosaurs to giant rats and insects to dragons and Siberian death snakes. Two American sightings claimed to be of mammoths were certainly elephants from a local visiting circus, giant rats observed in East Anglia in England were probably coypus (genuinely giant rats), and various giant insect reports have been exaggerated sightings of mantises in Southern England, of *Extasoma* stick insects (about six inches long, and heavily built) in school corridors, or frequently of the caterpillars of the elephant hawk moth. Most English police forces are familiar with the 'Aliens have landed, they're about a foot long with eyes on stalks at their back ends, and they spray with a hose at the front!' variants; elephant hawk moth caterpillars are indeed very odd, quite common on rosebay willow-herb patches, about two-and-a-half inches long. Their relatives the poplar hawk moth and the alder hawk moth feed, respectively, on poplar and alder trees; elephant hawk moths would be even odder if their name was based on a similar dietary pattern.

What about sea serpents? We think that these are a mixture of genuine observation of ribbonfish, lines of dolphins or seals, basking sharks, perhaps giant squid writhing at the surface, with a lot of bad observation, exaggeration and simple making up of stories by mariners in search of a free drink. Are there sea serpents? Well, there may be. Like the Loch Ness Monster, the evidence for their existence is *just* sufficient that if it were any less, we'd say, 'Nonsense!' But at this point, we wouldn't be very surprised if one were caught, though not in Loch Ness, which can't provide anything like an adequate food supply. We'd certainly be less surprised than everybody was in 1938 when a coelacanth was caught, because in many ways (the length of time during which it had been supposed extinct, for example) that was more surprising than Professor Challenger's Lost World, where dinosaurs never became extinct, would have been – though not as impossible as Jurassic Park.

It also has to be said that not all strange sightings are accurately reported. Indeed, most *familiar* sightings, like car accidents, are not accurately reported. There are three possibilities: the reports may be lies; they may relate to genuine (but misinterpreted) experiences like dreams; they may be accurate reports of actual events.

If the reports of puzzling phenomena are genuine – which many probably are, from sea serpents to crop circles – they are often dismissed by newspaper scientists because they don't fit into the normal scheme of causes. Newspaper scientists expect everything to have a neat, tidy scientific explanation (educators' simplifications of the kind we have called 'lies-to-children'). In contrast, the composite entity Jack&Ian expects most things, on close examination, to have many aspects that are difficult to comprehend, and expects many things to be pretty well totally incomprehensible. So it enjoys corn circles, UFOs, and the Loch Ness Monster, despite which it is certain that nearly all of them (probably all) are fakes or misperceptions; either way, it sees no reason to associate any of them with aliens.

There is no doubt that some people imagine things that have no physical existence. From Disney animations to SF films to mundane novels to Elm Street horror movies, we are adept at making up alternative realities. We always have been. Cave paintings show things that don't (and almost certainly didn't) exist, like dog-headed men. Medieval belief in succubi (devil-women) and incubi (ditto-men) was associated with sexual dream-experiences of (mostly) the other sex when asleep, and so can be tied down to actual events. But witches, elves, fairies, and (especially) ghosts, soar progressively higher into the cultural imaginative stratosphere. There is sometimes a complicity with nursery stories, in which this supernatural zoo is over-represented; but because it is a complicity, with each component feeding the other, we can't attribute the imagination of these creatures to an archetypal fairy story, or blame their appearance in the stories on adult observations. All human cultures seem to have this overlay of imaginary creatures, with ghosts, as representatives of people now dead, constituting a major component.

We do not believe that any of these creatures exist in physical reality, but we are very interested that people see them, sometimes feel them or hear them – or, more often, claim to have a friend, or an aunt, or a grandmother who did. We appreciate these constructions as part of our cultural life, from the Ghost of Hamlet's father to the Cottingley Fairies that so convinced Sir Arthur Conan Doyle, and indeed to 'Lady Angelica Cottington's' gruesome (spoof) *Pressed Fairy Book*. But we have not seen any ourselves, and neither has anyone we know. We do have friends and colleagues who suggest, or at least agree, that they have interpreted an observation in one of these frames. And of course we ourselves have observed phenomena that are immediately interpretable as poltergeists, or misty human shapes, or malevolent invisible spirits, especially when we have been required to explain breakages of particularly sentimental crockery to our spouses. And we, like all of you, have misidentified many scenes at first exposure: the grotesque man wound up in the gnarled tree becomes a gnarled tree, the three approaching assassins with upheld knives become three students with rolled papers. Humans are not reliable witnesses, and they have a predilection for interpretations that fit cultural models.

You see where we are going. Recall that Aldiss presented this view in *Nature*'s Astrobiology supplement. Aliens, he said, are the pseudo- or quasi-scientific stand-ins for the old zoo of fairies, elves, ghosts and ghoulies. We agree with him completely. The Greys, the genitally-interested Abductors, the Cow Eviscerators, the Little Green Men, and the Wise Martians are all place-holders for previous cultural icons, now dismissed as 'unscientific' and therefore not worthy of today's serious attention. By dressing them up in 'science', these new models have attained a kind of respectability not now possessed by the previous Child Ghosts, Body-Snatchers, vampires, elves, and Spirits of the Woods. However, Aldiss goes too far in his argument, extending it to deduce the non-existence of all other lifeforms on all other planets, anywhere. As we've already argued, the aliens (if they exist) don't *care* about human myths, and they won't vanish from the universe just because their counterparts in our culture are mythical. However, Aldiss has done well to point out that when people 'see' alien abductors now, they are behaving just like our ancestors 'seeing' leprechauns.

A very acute test of whether what is seen is 'outside' in the reality, or inside the human mind and therefore subject to cultural interpretation, is to ask what people from non-western cultures see. The icons of the supernatural, played through the imaginative poverty of film companies forced to use actors, have played a large role in forming the alien myths of the western world. Do Chinese UFO-watchers, saucer-seers, see little green men, like George Adamski did? Or grey 'alien' ghostomorphs? Or do they see creatures like demons and dragons, lifted wholesale from the central areas of *their* supernatural zoo? We have no idea, and would like to know what evidence exists. Oriental films are not much help, of course, because they are made for western audiences, where the money is. Indian films use the icons of the *Jungle Book*, for example: it's all giant cobras and metamorphosis, sex and western sin. We haven't seen any SF films from the East, because SF itself has no roots in that culture. Japanese *manga* is close, but again uses fashionable western icons.

The Hitchhiker's Guide to the Galaxy (Douglas Adams 1979)

Arthur Dent is upset, because his house is about to be demolished so that the council can build a bypass. He meets a young man named Ford Prefect. The demolition of his house is interrupted by the appearance of a fleet of gigantic starships. Prostetnic Vogon Jeltz of the Galactic Hyperspace Council announces that 'plans for development of the outlying regions of the galaxy require the building of a hyperspatial express route though your star system, and regrettably your planet is one of those scheduled for demolition. The process will take slightly less than two of your Earth minutes. Thank you.' Observing the ensuing panic, the Vogons point out that the relevant application has been on display at the planning department on Alpha Centauri for the last fifty years.

Prefect stows away on board a Vogon ship, taking Dent with him. A Babel fish placed in Dent's ear allows him to understand the Vogon language, and the first thing that he hears is that he and Prefect have been discovered, and will be thrown off the ship after being read some excruciating Vogon poetry. They survive only because they are picked up by the two-headed Zaphod Beeblebrox in his Infinite Improbability Drive ship.

On the planet of Magrathea they meet Slartibartfast, who designs planets, and did the crinkly bits for the Norwegian fjords on Earth. They hear the tale of Deep Thought, the supercomputer that was asked the Answer to the Great Question of Life, the Universe, and Everything. After seven and a half million years of computation, it replied 'forty-two'. At which point the assembled philosophers realised that perhaps they needed to understand the question better, first.

Dent and Prefect are taken before the mice, who for ten million years have been using the Earth and its inhabitants as a giant computer, including running innumerable experiments on humans, in order to find the Ultimate Question. Like all mice, they are really the visible three-dimensional extremities of pan-dimensional hyper-intelligent beings. They think that the Ultimate Question may *just* happen to be encoded in the structure of Arthur Dent's brain. So they want to buy it from him.

His brain, that is.

So far as we can tell, then, all of the standard, well-publicised claims about aliens visiting this planet are nonsense. Despite this, real aliens, non-iconic ones, could have come visiting; they could even be here now, today. Would we know? The disguises available to us, now, would fool many non-technological peoples. 'Stealth' techniques that would trick us would be no problem whatsoever for an alien extelligence that can cross the void between stars. We simply would not know they were here unless they wanted us to, as Douglas Adams made clear in *The Hitchhiker's Guide to the Galaxy*. The standard evidence that they have, from von Däniken's collection of lucrative 'Our ancestors couldn't have done that without help' stories to the dissection video from the Roswell event, is totally unconvincing, indeed anti-convincing, as we've argued above.

On the other hand, people who have 'experienced' an alien abduction are sometimes very convincing – until you turn the story round and look at it from the point of view of the putative aliens: there is no reason for them to be concerned about our genital areas. But if it had been a witchcraft/succubus/ghost story we would have no doubt of the intra-personal provenance of the tale; Aldiss's explanation sorts those out, and takes away any evidence that they might have provided about real aliens.

We've got to get away from all those comfortable ideas that aliens will be like us, except for a few minor differences that don't challenge our imaginations. Real aliens will be very alien indeed. Life is a universal, so it will evolve in any habitat that supports the required complexity of organisation.

We cannot as yet define, in sufficient generality, the properties of habitats that are necessary to support life, but it is likely that our familiar water/oxygen planet is only one of many possibilities. SF has explored many others, including the surfaces of other planets and asteroids, the atmospheres of gas giants, stellar interiors, interstellar space or molecular clouds, and even the surfaces of neutron stars (*Dragon's Egg*). Some of these locations, conventionally regarded as passive environments, have been regarded as lifeforms in their own right, an example being the sentient ocean of *Solaris*. In fact, it is difficult to imagine a habitat that could not support a suitable form of life. Anywhere that physical matter can exist, and that offers a rich enough energy substrate, can in principle harbour highly organised processes carried out using matter and energy of the same kind. As far as we are concerned, that would be alien life.

A balloon-like creature floating in the atmosphere of Jupiter (as in our novel *Wheelers*) would probably regard the terrestrial environment as being lethally unattractive. Most aliens would not wish to visit Earth at all, any more than we would care for a ramble across the surface of a neutron star – or care to live, as do some extremophiles even here on Earth, in boiling water. We might suppose that the aliens least disinclined to visit us are those who have evolved in a habitat like present-day Earth, and such habitats might comprise an unknowably small subset of all possible life-supporting habitats. The chance that such aliens exist within 10,000 light years of us at the present time is small. There are plenty of other places to visit: why Earth?

However, non-humanoid aliens might be keeping a cold,

unsympathetic eye on us for their own scientific purposes, writing yet another small footnote in their xenoscience texts (such as *What Does a Human Look Like?*). As discussed above, they are unlikely to do anything as obvious as abducting gullible readers of supermarket tabloids – and if they did, they would be even less likely to make the mistake of returning them afterwards. They will possess technology that to us would appear incomprehensible, in accordance with the Clarke–Benford dictum. So if they were here, we would never recognise them, probably never even encounter them or their technology in any meaningful way, because they would be masters of disguise. They would be able to change their bodies (if they still had bodies) at will. When they booked their tour of Quainte Olde Earthe at the cosmic outlet of Going Places, they would get the necessary phenotypic reconstruction cubelet to plug in during the short ✴✦✤✦✤- powered excursion to our region of the galaxy. They would leave home as silicon carbide pyramids higher than a human house, and by the time they arrived they would look like Demi Moore.

Certainly, aliens would not look like the canonical Little Green Men . . . unless they wanted to (which brings us into *Hitchhiker* land again). They might look exactly like people. Or cats. Or houseflies. Or they could be invisible, or lurking just outside our spacetime continuum along a fifth dimension, observing our insides like The Sphere in Edwin A. Abbott's *Flatland* observing A. Square. Or they could be concealed inside atoms. Or they could exist only in the gaps when human perceptual systems are in their refractory phase and cannot register their existence.

Most likely, though, they are not here at all – for reasons of alien extelligence rather than non-existence. Why run the risk of travelling to exotic places when you can put on a headset and walk through Virtual Venice or Artificial Africa? When VR becomes as real as RealR, an actual visit might seem bothersome, expensive, unsafe, even boring.

We can see the germ of this introspective trend within humanity, so far the only extelligent species we know. More than thirty years ago we landed on the Moon. Our last visit was in 1972, and we no longer have a ready capability to land there. A low-Earth-orbit space station is laboriously taking shape, amid little real enthusiasm. We talk of future manned expeditions to Mars, but as we write, NASA is probably about to cancel an unmanned probe to Europa (which might well possess some living chemistry at least) in favour of reinstating a previously cancelled one to Pluto (no chance we'd recognise that kind of life). The question is not about whether aliens have visited us, and if so, why they aren't here. The important question is why we have not ventured further into space. It would be sad indeed if it turns out that the inability (or reluctance) of an extelligent species to leave home turns out to be a universal.

Even if most aliens are like us and stay at home, they could still make themselves known to us. So should we be watching and listening for messages from the stars?

We've seen that in 1960 there was a systematic attempt to listen for radio signals from Outer Space and since then the SETI project has grown to include hundreds of listening sites, hundreds of scientists – and computer people and SF fans and authors, science writers, journalists . . . It has recently achieved a metamorphosis in its computer capacity by invading the screen memories of thousands of computers all over the world. This vast increase of processing power is devoted, however, to listening for a signal that we cannot imagine would be the method of choice for any alien civilisation. The sociology of the project, though, has achieved many good things; in particular it has changed many minds towards seeing humanity in a galactic context, and it has fostered many friendships, community activities and scientific projects. Almost by itself, it has changed the outlook of most people on the planet on the question of the existence of aliens. Even those senators who disapprove of spending money on SETI do so with an approach that is like that towards other kinds of foreign aid, but a bit more so. No one today criticises SETI because there aren't any aliens.

However, it is probably pointless to search the heavens for radio signals from other worlds (as the SETI project does). It would be equally sensible to look for smoke signals. Radio did not exist on this planet a hundred years ago, and it will probably be obsolete a hundred years from now. If the aliens are out there, and if they communicate at all, they probably do so using gravitons, squarks (supersymmetric partners of quarks), telepathy, or the power of ✶✦✧✦✧. And even if they *are* using radio, they will have encoded their transmissions for optimal efficiency. Lachmann's theorem shows that if so, their messages will be indistinguishable from black-body radiation. (Imagine a Second World War radio operator picking up one of today's encrypted satellite TV channels. It would sound like static.) Is this the true meaning of the cosmic background radiation? Is it the buzz of an indecipherable, perfectly encoded, intergalactic mobile phone system? Vogonfone?

In the SF and technothriller literature there is a persistent belief that a radio message can become an alien organism – usually humanoid, female, and deadly. This scenario occurs, for instance, in Fred Hoyle and John Elliot's *A For Andromeda* and in the very silly movie *Species*. Fuel has been added to the flames by the metaphor of DNA as both message and blueprint. The error here is the same as the most fundamental of the errors in Jurassic Park. Namely, once you are in possession of the DNA 'information' needed to make a dinosaur, then you can make a dinosaur. Not so: we've seen that one further vital item of equipment is needed – mother dinosaur. As with *all* information, an appropriate context is needed to access the raw information and turn it into something with meaning. The binary information on a CD can be turned into Bach or Bon Jovi only with the aid of a CD player. You can study the sequence of 0's and 1's indefinitely, and not be able to tell whether the CD is a recording of a string quartet, rock music, or the mating cries of the howler monkey.

The idea of 'information' being something real is seductive. It appeals especially to physicists of a fundamentalist turn of mind. John Archibald Wheeler's felicitous phrase 'It from Bit' captures the attitude: things from information. In this view, which is an approach to quantum theory, the basic building-blocks of the universe are not matter, not even quanta: they are information ('bits', binary digits), and the rules of physics (laws of nature) turn the bits into 'its' – things. The idea behind Wheeler's phrase is that the quantum state of any particle, or system of particles, is analogous to a binary choice between 0 and 1, or between some sequence of bits and some other sequence or sequences. Is this electron spinning in its *up* state or its *down* state? The answers to enough such questions determine the quantum state of the universe: now throw away the electrons and keep only the questions and answers.

Greg Bear made clever use of this idea in *Anvil of Stars*, whose prequel *The Forge of God* ended with the total destruction of the Earth. The perpetrators of this outrage must be tracked down by a few survivors, to wreak vengeance; it turns out that these destructive aliens are in possession of technology that can flip the bits of fundamental particles – so that, for instance, one of the human characters, and her ship, can be turned into antimatter, leading to a slow death as she is annihilated step by step by particles of ordinary matter in the not-so-perfect vacuum of space.

'It from Bit' *could* be right. It is certainly a neat and original way to formulate quantum mechanics mathematically.

However, it looks more like the standard physicists' trick of reifying mathematical concepts: taking ingredients from mathematical descriptions and interpreting them as real things. Examples include such 'basic' concepts as energy, momentum, force, entropy . . . even velocity. And, of course, information. However, it is not clear that any of these concepts is real. Not even velocity. In the real world, objects move, and their motion takes time. Mean velocity is a measure of how far the object moves in a given period of time; velocity is an 'instantaneous' version of this idea, the limiting value of mean velocity over a time interval that becomes as small as we please. No physicist has ever measured a true velocity, because there is a lower limit to the time intervals that can be studied experimentally; therefore 'velocity' is a mathematical abstraction and may not correspond to any real thing. That doesn't stop the concept being useful – indeed it is the main reason *why* it is useful, because if velocity were a thing then we wouldn't need the abstraction – but it does stop it *being* physically real. Yes, the concept has a sensible *interpretation* in the physical world, but it is not the same as that interpretation.

The same goes for energy, which is a unifying mathematical construct. It was devised when mathematicians pursued Newton's laws of motion and gravitation, and noticed that the rules become very simple if a cunning trade-off between velocity and position in a force field was formalised into two quantities: 'kinetic energy', proportional to the square of the velocity, and 'potential energy', the gradient of the force field's landscape. This discovery led to a unifying concept, *energy*, defined to be the sum of kinetic and potential energy. Its great simplifying feature is that it is conserved – stays the same – in any

Newtonian mathematical model of a mechanical process. As mechanics broadened, it always turned out that the energy concept could be carried over in to the new interpretation, while retaining the conservation property. Thus heat can be identified with the kinetic energy of molecules. The concept even extends into the quantum realm.

There is no question that energy is a concept of huge significance for physics. It is a major tool for dealing with the mathematics, and it represents how material objects behave with considerable accuracy. It is certainly possible that the reason why the formal concept of energy extended so readily as physicists improved their models of nature is that energy genuinely is real, that the universe runs on an energy economy, trading it like hard currency. There is no question that when doing physics, it makes life far simpler if you assume that energy is real and don't think too hard about what that would entail. But we can question whether energy has a genuine existence beyond that of a mathematical construct. Is the universe made from a thing, energy, along with perhaps other things? That's not so clear. Maybe energy is merely an epiphenomenon – a concept derived from reality, but not part of it.

Theoretical physicists may be aghast at the suggestion that energy is not real; but then, theoretical physicists are the ultimate Platonists. And they're very happy nowadays to tell us that space, time, and matter are not real, so why stop there? Consider an analogous concept, that of a negative number. If a chicken lays three eggs and the farmer removes two, how many are left? Clearly $3-2 = 1$, one egg. Mathematicians learned, long ago, that it is convenient and useful to enlarge the number system to include negative numbers, such as -2. By exploiting negative numbers, subtraction can be reduced to addition: $3-2 = 3+(-2)$. No logical inconsistency can be introduced by extending the number system in this way: that's been proved, it's a theorem. And the extended system correctly describes transactions with eggs. However, although -2 eggs occur in one version of the arithmetic, -2 eggs do not exist. You can't start with one egg and remove three. You can *interpret* -2 eggs as a debt of two eggs, but that sets up a new context, and in any case a debt is a concept, a social construct, not a thing.

Energy may be no more real than minus two eggs.

The same goes for information. There is no good reason to suppose that it is anything beyond a mathematical epiphenomenon. Useful, yes; suggestive, yes; real – no.

In order to turn a message into a beautiful, deadly, alien female, you need a context in which that message will be interpreted correctly. Making an aircraft is not merely a matter of e-mailing someone the blueprint: something has to turn the blueprint into a flying machine of aluminium and plastic, and the only something that we know will work is an aircraft factory. Similarly, the only context for turning 'information' into an organism that we know of is another organism, or perhaps one of its cells – if it has any. As every IVF clinic knows, it takes far more than a printout of a DNA sequence to make a baby. Even the DNA sequence is no use on its own: its 'message' must be impressed upon a fertilised egg. Moreover, to make a human baby, it has to be a human egg, and the DNA has to be a human genome. So *Species* doesn't work, and the plot could be rescued only with some very clever, sophisticated, and carefully engineered ultra-high technology – test-tube aliens. So you don't need to worry that if you take part in SETI's internet search for alien intelligence, an incoming message will turn your PC into a slinky, death-dealing, humanoid beauty.

More's the pity.

There is one way for an alien message to wreak havoc, though. Aliens could transmit a computer virus into the internet, and disable it. They would have to observe internet traffic very closely, and learn how Windows™ works, but that's child's play to a superintelligent being. *Independence Day* plays the same trick in reverse, with humans hacking into the computer system that controls the aliens' invasion fleet. The invading commanders would do well to heed the advice displayed on one of the 'pages' of the World Wide Web, which offers instruction to alien programmers on how to write secure code that can counter this disturbing threat to their plans for dominating the universe.

14

GALACTIC EMPIRES

*D*EAR *HEADQUARTERS,*
   Warning of Imminent Rise to Technical Competence of the Humanoid Terran Species. (Confidential/Urgent.)

We have just discovered something that we inadvertently omitted from the Earth *Prospectus. Earth humanoids, primate type DBDG, have attained technical competence. Barely. They should now be considered as properly extelligent (for evidence consult Appendix 1).*

However, a second group of species on the same planet, in an ecosystem almost totally disconnected from the humanoids' habitat, has made contact with us. This species-complex is a biologically engineered set of tool-organisms of type JZZX*, complicit with a bacterial cloud of type UUDD that regards Abel as its progenitor (Appendix 2).

To anticipate your question: yes, we have checked that there has been no genetic contamination.

These are both very primitive extelligences. Neither species has yet experimented with discorporation or interstellar spaceflight, and each still believes that the universe was designed for them alone.

That there are two of them should soon disabuse them of such alienthropic philosophies.

However, Rule 4932 states that covert tourism must be discontinued on any world that harbours genuine extelligence. We request guidance in this matter, which threatens to disrupt your interesting – and profitable – tourist agency on Earth.

Perhaps a Galactic Emissary could be sent to make proper diplomatic

contact? Embodied in a form that the indigenes will recognise *as a Galactic Emissary, we humbly suggest? Please give this matter your earliest attention.*
   Sincerely yours,
      Cain and Abel

Let's start to draw the threads together and come to some conclusions. Several major themes have emerged from our discussion. The most important one is the appreciation of just how alien a real alien would be. Seriously alien creatures would be beyond our understanding; indeed, we probably would not recognise them as creatures and would not realise that any understanding was necessary. This leads into a second big theme: the need to distinguish between astrobiology and xenoscience.

Astrobiology is comfortably based in known science, and for a few purposes that solidity may offer advantages. If we are trying to detect signs of life on distant planets by making observations from the vicinity of our own, then it is reasonable to concentrate on our kind of life. That way we at least know what to look for. But when it comes to anything more ambitious, the astrobiological mind-set can be seriously misleading, especially as regards alien biology. Most of what we know about biology concerns terrestrial parochials; DNA is a case in point. Possibly carbon-based aliens on worlds very like ours will use DNA for their genetics, but even then we would expect many differences in detail, such as different bases and different genetic codes. The necessary chemical research has not yet been done to determine whether there are alternatives for carbon-based life, but there are good reasons to think that there should be. For more exotic lifeforms, the sky's the limit. They might not even have genetics at all, though they probably will have, in some form. But it could be encoded in the topology of magnetic vortices or the frequency-levels of a radio signal.

Xenoscience has to consider precisely these exotic possibilities, so that for the moment it must remain largely a theoretical science. It does have some observational evidence – for example, the experimental confirmation that it is possible to add new bases to DNA, or to introduce new amino acids within the scope of the genetic code, or to create new transfer RNA that changes the genetic code altogether. Mostly, though, it has to be rooted in theory and computer simulations: artificial life, exotic chemistry, autocatalytic systems ... However, other perfectly respectable sciences suffer from similar problems, among them cosmology and palaeontology. Palaeontology is mostly historical, and it is difficult to make predictions and even harder to test them; the closest we can get is 'postdictions', finding old evidence for new theories. Cosmology has even bigger problems: you can't go back to the Big Bang, introduce some changes, and re-run the birth of the universe to see whether it does what you expect. Until we meet some aliens, either within the solar system or beyond it, we can't do much observational or experimental xenoscience. Indeed, we can't be sure that the objects of that science actually exist at all – though here astrobiology will help, with its search for aqueous planets round distant stars, because those obviously offer excellent prospects for alien life, albeit of a relatively familiar kind.

This paucity of observational support is likely to change. Unless something drastic happens – and with humans, this is entirely plausible – people much like us will visit Europa soon, say within the next couple of hundred years, very likely sooner. When they get there it is likely that they will find, at the very least, complex pre-life chemistry. When people used to ask, in the 1980s, whether mankind would establish a base on Mars, visit Jupiter's moons, push habitations out to Saturn's orbit, and colonise Titan, the smart answer was 'Yes, we think so, people more or less like us'. It now looks as though they will probably be Chinese or Japanese though, not westerners. We seem to have lost our desire for frontiers; we've taken to pouring all of our money into the holy buckets of Defence, Health, and Welfare. Such spending can in principle take all of our wealth, however large that may be, and it won't return it or multiply it. In contrast, all pioneering multiplies, or at least returns, the initial investment, allowing more to be spent on Defence, Health, and Welfare in the longer term.

Perhaps the western world is beginning to understand this again, though. The Space Station up there now is a traditional start, but we would be happier to see a bolas or two on the drawing board, because there is a schema available for a set of space-bolases and a space elevator, and because we are nearly at the point where our technology could consider building them. We want humanity to start thinking seriously about colonising the solar system, and we want it to start developing the necessary kit.

There's an entire universe out there to explore and exploit: Earth cannot continue to support an ever-growing population for much longer. We can wallow in growing misery in a resource-limited, over-populated environment – or we can head out to the stars. The most obvious alternative – effective population control and a focus on renewable resources – *could* happen, but the way George W. Bush recently sabotaged the Kyoto agreement on the reduction of carbon dioxide emissions for short-term gain, and the Roman Catholic Church's dangerous attitude to birth control, do not augur well.

And even if staying down here worked, we'd still be stuck in one tiny corner of a huge, wonderful universe. Humanity degenerates without new frontiers.

One thing we know about the future: it never happens as planned. As Niels Bohr is reputed to have said: 'prediction is very hard, especially about the future.' In the 1870s, when people began to calculate the future speeds of transatlantic steamships, they could not imagine what the future would bring. It brought aeroplanes, not super-fast ships. The idea that thousands of times more people would fly across the Atlantic than had ever gone by steamship could not be comprehended then, never mind foreseen. Similarly, we cannot know what changes will happen to people, or to our technology, even in a couple of hundred years. The important lesson from steamships and aeroplanes is that the predicted seven-hour journey time was spot on – and customs and airport delays take only another seven hours. So having the concepts of the space bolas and space elevator means that people will indeed do the things that those technologies would have made possible, but they will do them in other ways. So in this book we will continue to write as if bolases and space elevators will be the way human technology will develop. Just as none of the origins-of-life scenarios that we've portrayed are actually the way that life on this planet appeared, but are placeholders for the real one, so our hypothetical future of humans in space will be a placeholder, employing space elevator technology as a metaphor for whatever actually gets used.

The great thing about that technology, or whatever will supplant it, replace it, or make it redundant, is that it will make getting down on to planetary surfaces easy. And back up. Until now everyone has been concerned about getting enough people and hardware *up*. Up and out. But once we have them up, with the kind of robotic technology that is now being developed, they will be able to park in Clarke (synchronous) orbits, and dangle an elevator down to the surface. We like the idea of rolling the elevator cable up and using it again for the next planetary descent, instead of throwing lots of rocket-fuel away to go *down*. Perhaps planetary exploration really is easier with rockets, though, or of course ✻✲✽✾✿, which we haven't invented yet. Perhaps, though, it will be water and moon rock that's thrown away, and solar powered rockets. Sustainable rockets, as they aren't at the moment. Whatever we use, we'll certainly visit all the local sights, if enough of us want to.

And it is xenoscience that will inform the thinking that takes us there, not astrobiology. Astrobiology is a science of restrictions. It starts with the whole universe, and successively narrows it down until all that is left is the surface of one unrepresentative planet. Its starts with the rich potential of self-organisation systems and cuts it down to orthodox DNA chemistry. It starts with what might be if things are different, and cuts it down to dull repetition of what has already happened (and only a rather orthodox part of that). It starts with universals, and cuts them down to parochials.

Xenoscience – if it existed as a fully fledged subject – would be very different. It would be inclusive, not exclusive. What kind of process is life? Can we think of *new* ways to realise that kind of process? New materials, new interactions, new environments? Where could those environments occur? How could those interactions arise? *What have we forgotten?*

Planet Earth is wonderful, rich, apparently inexhaustible. It is our home. It is, near enough, all we know.

It is irrelevant.

Planet Earth is one tiny, unrepresentative lump of rock in an exceedingly ordinary part of a vast, incomprehensible universe. We can either follow Ward and Brownlee, and rule out entire *galaxies* on the basis of our limited understanding of life, or we can ask what kind of life might arise in a galaxy *unlike* our own. If we assume Earth is typical, we are lost. We are like the Easter Islander who sees nothing broader than coconuts and clams. We look at the diversity and resilience of life on our own planet, and see nothing but narrow escapes and fragility. Does *Rare Earth* envisage *anything* that has not already happened on

338

this planet? No. For most sciences, that might be acceptable, might even make sense. But for a science of the *alien*?

Beyond The Blue Event Horizon (Frederick Pohl 1980)

Pohl's *Gateway* introduced the Heechee, a mysterious race that disappeared from the universe while leaving behind incomprehensible hi-tech gadgetry. Most of which works, if only you can find out what it does. One of them is Gateway, a structure whose elliptical orbit overlaps those of Mercury and Venus, which contains nearly a thousand small spacecraft shaped like fat mushrooms. Risk-taking humans can volunteer to get inside, and head off in excess of lightspeed to an unknown destination, taking an unknown time, about forty-five days on average. Usually the ships return. If the passengers are still alive by then, they are rewarded for anything they find. Sometimes the reward is enormous, billions of dollars.

There is a jokey book they sell on Earth, called *Everything That We Know About the Heechee*. It has 128 pages, and every one of them is blank. By the time described in *Beyond The Blue Event Horizon*, however, those pages could be filled many times over. The Heechee are curious, they have their own form of science – and in some areas it closely resembles our own. They expanded throughout the universe, and studied it, until they noticed something that terrified them.

They knew that the physical constants on which the universe was based were 'optimal': changing them would make the universe less suitable for life. But the Heechee found evidence that someone – they didn't know who – was slowly destroying the universe, in order to rebuild it with new constants: less suited to the Heechee, but presumably *more* suited to Themselves . . .

So the Heechee set some promising lifeforms on the track to intelligence, and they left clues like Gateway all over the universe. And then they found themselves a convenient region of space, a few light years across, filled it with stars, entered it in their ships . . . and watched from the inside as it collapsed around them to form a Black Hole. From that moment onwards, they were sealed off from the remainder of the universe, safe from whoever was doing the remodelling.

Time, for the Heechee, flowed very slowly. Now, all they had to do was wait. When the creatures that were rebuilding the universe came back to occupy it, they'd have to get past an awful lot of other races before they could get to the Heechee.

Admittedly, xenoscience is in its infancy. It has none of the comfortable certainties of astrobiology. Instead of 'knowing' exactly what limitations the universe places on living creatures, it admits to ignorance of as yet unknown alternatives. It addresses possibilities, not facts. But it is the possibilities that matter. Nature doesn't care about the limited mind-set of moderately intelligent apes on planet three of an undistinguished star. Whatever is possible will be realised, somewhere out there in the vastness of a trillion galaxies.

We cannot, in a single book, lay the foundations for predicting what will be realised and where it will occur. But we can lay the scientific foundations to think sensible thoughts about what it might be. Here are a few of the problems that xenoscience must grapple with, in increasing order of scope.

How common are aqueous planets? How long do they stay aqueous? How often does DNA arise on them? What are the alternatives to DNA? New proteins? New amino acids? New bases? New molecular architectures for genes? New control systems for managing networks of genes? New body plans? New directions for evolution? New kinds of planet? New habitats that might not be planets at all? New organisms for new environments? New metabolic pathways? New reproductive methods? New sources of energy? New routes to self-organisation and self-complication? New materials, new structures, new habitats? New kinds of life? New processes that go beyond life? New societies, new technologies, new histories, new geographies? New novelties?

How can we forge such a science?

Not by being narrow-minded. Not by being unimaginative. Not by promoting parochials to universals. Not by adopting the Goldilocks ploy that there is only one 'right' way to do things.

Xenoscience is the science of deducing universal principles from parochial instances. 'Let $2=n$,' as mathematicians say. It is irreducibly interdisciplinary; it will grow from the gaps *between* conventional sciences like astronomy and biology, not be encompassed by merely adding them together. Its general principles will be interdisciplinary commonalities.

What we need is not Rare Earth, but Common Ground.

The potential rewards are boundless. If we can put this interdisciplinary science together, it will help to take us to the stars, open up the galaxy to exploration, and – subject to who else is out there – colonisation. Depending on what is politically correct at the time, we might even build an empire, assuming no one else has got there first. SF contains many galactic empires. Asimov's *Foundation* trilogy begins with the predicted fall of just such an empire, whose severity must be reduced so that civilisation can rise from the ashes. Hari Seldon, inventor of 'psychohistory', which can predict the behaviour of people in large masses, saves the day. But not easily. Asimov's capital planet of Trantor, the universe's largest urban development, covered entirely in buildings except for the odd park, is guyed by Harrison in *Bill the Galactic Hero*. (How does the heat escape?) The evil empire of the *Star Wars* movies closely resembles Asimov's, with Coruscant in place of Trantor.

Often these are alien empires; more often still, they have been founded by an expansion of humans away from their home world. Suppose that we do develop xenoscience, go to Europa, and start to colonise the solar system. Is that really such a good idea? Some people think it will be a terrible mistake to let these dreadful *human* creatures loose in Outer Space. They've made a big mess of their home planet: all they'll do is make an even bigger mess of all the others.

However, it's absurd to think of our solar system as a substantial or important part of the universe. Even Pluto is still on our doorstep, not even out of the porch, compared to the nearest star. Out in the galaxy, entire stars are exploding; colliding neutron stars are filling huge regions of space with bursts of gamma rays so energetic that if one of them went off within a hundred light years it would sterilise the Earth completely. So any damage that we did within our own system wouldn't even be noticed by the rest of that vast universe. None of this means that we should cheerfully vandalise our own system, of course: *we* would notice, and not like what we saw. But we don't need to worry about becoming a galactic plague.

There's no question that we can *get* to the nearest star. We could almost put together a suitable means of transport now. The large distances mean that it would have to be a generation ship, supporting successive generations of people during centuries-long voyages. We don't know for sure how humans would survive in an interstellar environment with only centrifugal force substituting for gravity as the ship is spun on its axis, but that's a technical issue which should have some technical solution. But if we started planning and building now, and invested enough money, then we could get to the nearest star in a few hundred years. As more than one SF story points out, however, that approach is probably not worth the effort, because such a ship could be overtaken (literally) by new technology. Before the ship arrives at Proxima Centauri, we may well have developed FTL travel, though not by any of the routes that we envisage today. If not, then a few hundred years more should do it. By one method or another, our descendants will go out to the nearer stars.

By then, however, they will not be humankind. We wouldn't recognise them. They will have little in common with us, except for a few of their genes and a predilection for Shakespeare or Confucius. They will probably look more like machines, or like nothing even comprehensible to us. We should not be denying those very different creatures, who will be wiser than us in far more ways than we are wiser than the ancient Romans, their chance to open out their lives beyond our solar system. We can't, anyway: it's coming one way or another. And there is another reason for getting out there. If we can, in all probability someone else already has. And that opens up very many scenarios, and closes few, as our boxed synopses of SF stories demonstrate.

Even if our descendants do drift out like thistledown to the nearer stars, we won't have corrupted the universe. Even if we get out of our spiral arm and reach the central stars of our galaxy (a bit close to the central star-gobbling Black Hole – see Frederick Pohl's *Beyond The Blue Event Horizon*); even if we corrupt this entire galaxy . . . the universe won't notice. Our whole galaxy is but one dust mote in a hurricane. So no one should worry that we'll spoil anything. At most, we'll be as local a plague as the ants in your kitchen; and of as much concern to other inhabitants of the galaxy as Texas fire ants are to Mrs Chen in Beijing.

We now know that there is considerable traffic between the inner planets, as asteroids and meteorites bombard us, bounce our rocks off into solar orbits, and splash our oceans into deep space to freeze-dry into salty crispy bits of plankton. We have found tens of pieces of Mars rock lying about on the ancient snow of Antarctica. We have found intriguing shapes and chemistry inside which, if they had been found in Earth rock of that age, would immediately have been accepted as evidence of early life here. We are much more sceptical when the same shapes are advanced as evidence of extraterrestrial life, and that is only right and proper: 'extraordinary claims require extraordinary evidence'. The best way to settle the question of life on Mars once and for all is to explore that planet, extensively, and see what we find.

However, it is worth remembering that those pieces of rock came to us in very short journey times, by solar system timescales: only sixty million years. We must assume, therefore, that at least some of our junk – at least a few million-ton bits of freeze-dried ocean with all its viruses, bacteria, microplankton and the like – has been bounced up into solar orbit by collisions with asteroids or comets, and distributed around the solar system. So we should not regret that our space ventures have already infected the Moon, and probably Mars and Venus, with Earthly bacteria. The bacteria make the trips anyway, which is why the discovery of life just like ours elsewhere in the solar system would *not* be proof of two origins and convergence, but only of natural infection. Only if the life we find is very different from that here would it be material evidence for the central proposition of this book: that life turns up anywhere wet-and-warm, and without doubt it turns up elsewhere too, in unexpected environments and forms.

Great Sky River (Gregory Benford 1987)

Once, the humans on Snowglade had dwelt in huge citadels, and even vaster arcologies. Then came Calamity, and the arrival of the mechs. Now, straggling bands of humans are constantly on the run through the ruins of their world, pursued by relentless death-dealing machines.

The humans had known about the mechs before coming to Snowglade, but they had to flee *somewhere*. A mechanical civilisation had arisen in the vastness of the galaxy, and it had dedicated itself to wiping out organic entities. They knew the mechs would come eventually, but they were hoping to buy time. In the event, it made no difference.

Machines do not function well in the warm, wet environments that suit organisms. So when the mechs arrived, a hundred years ago, they set out to change the ecology of the planet. They mostly ignored the humans; they just kept on with the business of lengthening and deepening the winters and cutting down on the rain. Slowly the humans realised that the mechs were bringing clouds of gas and dust into Snowglade's orbital path, creating a dust-fall that smothered the equatorial regions. And the mechs were evaporating the icepack at the poles. It also became horribly clear that the mechs had done all this before, that they were masters of the difficult science of planetary engineering. And then they stopped ignoring humans.

The Family, on the run, pursued by killing-mechs, comes across a manhunter of a kind they have never seen before, the Mantis. The Mantis is a collector of human minds and a weaver of murderous illusions. Killeen, one of the Family, wonders why the machine is using such elaborate forms of slaughter when it could easily destroy all remaining humans in a single explosion. To find out, he needs to know what it is like inside the Mantis's sensorium, where a myriad purloined human personalities are still preserved . . .

When we visit our local astronomical backyard, *we* will be the advanced lifeform. We will not find Bradbury's Martians on Mars, or indeed anywhere but in our own heads; nor Burroughs's Barsoom – no Red Princesses, human except that they lay eggs. Venus certainly doesn't have moist jungles as Burroughs envisaged, and it's not an ocean world like Lewis's *Perelandra*. Perhaps, though, we will terraform Mars, the Moon, even Venus in the next millennium, becoming their conquering aliens, taking over their living space, and turning them into the fictional worlds that they currently are not. There have been many fictional portrayals of that process, the most ambitious being Robinson's trilogy *Red Mars, Green Mars, Blue Mars*. With space elevators, or placeholders for them, it really does not cost much to get off-planet, or down to your eventual destination.

We see our species, with luck and a following (solar?) wind, at least visiting, and – considering our multiplication rate now that we have our extelligence to keep us warm – probably colonising all the local lumps of rock. We may well change our selves so that our descendants are very different, very alien compared to what we are now, as imagined so charmingly in Blish's *The Seedling Stars*, and less charmingly in *Childhood's End*. Then we might colonise Jupiter's atmosphere (Blish again: see *They Shall Have Stars*, the prelude to his Okie Cities trilogy), or turn ourselves into aliens that can live there (see Poul Anderson's 'Call Me Joe'). During the next millennium we will find ourselves, our descendants, on that path.

However, the interesting travel will be between the stars. Perhaps there genuinely is no FTL travel, but there certainly is subjective time contraction for those approaching the speed of light, the basis for Poul Anderson's *Tau Zero*. So the subjective time taken to reach a star can remain within reasonable bounds. When you come home, everyone you knew will have aged a lot more than you: with current lifespans they'd all be dead, but in a thousand years' time, who knows how long we will live?

With the energy available to the affluent even in 200 years' time, and reasonable assumptions about future technology, it seems clear that we will soon be able to cross the gaps between stars. What will happen then is beyond speculation, but some placeholders can be found in the SF works of Gregory Benford, Greg Bear, David Brin, Paul MacAuley, Dan Simmons, Iain M. Banks, and many others.

Will we be the first? The first locally, in our spiral arm of the galaxy? Will we find alien life? Or is there a supervening reason for us to be the only ones, ever, in the Great Adventure?

Surely not.

In less than 200 years, we will know.

Popular Xenoscience Reading List

C. Adami, *Introduction to Artificial Life*, Telos, New York 1998.

P. Aldhous, Angry cuttlefish see one another in a different light, *New Scientist* (30 March 1996) 16.

J.F. Atkins and R.F. Gesteland, The twenty-first amino acid, *Nature* **407** (2000) 463–464.

E. Ashpole, *The Search for Extraterrestrial Intelligence*, Blandford, London 1990.

S. Blackmore, *The Meme Machine*, Oxford University Press, Oxford 1999.

D.F. Blake and P. Jennieskens, The ice of life, *Scientific American* **286** No. 2 (2001) 44–51.

A. Böck, Invading the genetic code, *Science* **292** (2001) 453–454.

M. Chown, Made to measure, *New Scientist* (30 June 2001) 12.

A.C. Clarke, *Profiles of the Future*, Heinemann, London 1962.

J. Cohen and I. Stewart, *The Collapse of Chaos*, Viking, New York 1994.

P. Cohen, Let there be life, *New Scientist* (6 July 1996) 22–27.

B. J. F Cowie, Detecting extraterrestrial life, *The Science Fact and Fiction Concatenation* **8** (1994) 7–12. (More articles on exotic science as well as science fiction book reviews on www.concatenation.org)

I. Crawford, Where are they? *Scientific American* **283** No. 1 (2000) 38–43.

R.J. Davenport, Making copies in the RNA world, *Science* **292** (2001) 1278.

P. Davies, *Other Worlds*, Penguin, Harmondsworth 1990.

A. Delsemme, *Our Cosmic Origins*, Cambridge University Press, Cambridge 1998.

S.J. Dick, *The Biological Universe*, Cambridge University Press, New York 1996.

L.R. Doyle, H.-J. Deeg, and T. M. Brown, Searching for shadows of other Earths, *Scientific American* **284** No. 3 (2000) 58–65.

F. Dyson, *Disturbing the Universe*, Basic Books, New York 1979.

F. Dyson, *Infinite in All Directions*, Harper & Row, New York 1988.

G. Ferry, Beetles show why it pays to have sex, *New Scientist* (28 October 1995) 20.

H. Gee, *In Search of Deep Time*, Free Press, New York 1999.

R. Irion, The science of astrobiology takes shape, *Science* **288** (2000) 603–605.

R. Irion, Giant planets 'on the loose' in Orion? *Science* **290** (2000) 26.

B. Jakosky, *The Search for Life on Other Planets*, Cambridge University Press, Cambridge 1998.

R.A. Kerr, Beating up on a young Earth, and possibly life, *Science* **290** (2000) 1677.

R.A. Kerr, Wet stellar system like ours found, *Science* **293** (2001) 407–408.

V. Kiernan, J. Hecht, P. Cohen and D. Concar, Did Martians land in Antarctica? *New Scientist* (17 August 1996) 4–5.

J. Knight, The immortals, *New Scientist* (28 April 2001) 36–39.

R. Lorenz, Mars attracts, *New Scientist* (19 May 2001) 38–39.

J. Marchant, Life in the clouds, *New Scientist* (26 August 2000) 4.

L. Orgel, A simpler nucleic acid, *Science* **290** (2000) 1306–1307.

R.T. Pappalardo, J.W. Head, and R. Greeley, The hidden ocean of Europa, *Scientific American* **281** No. 4 (1999) 54–63.

R.J. Parkes, A case of bacterial immortality? *Nature* **47** (2000) 844–845.

T. Pratchett, I. Stewart, and J. Cohen, *The Science of Discworld*, Ebury Press, London 1999.

T. Pratchett, I. Stewart, and J. Cohen, *The Science of Discworld II: The Globe*, Ebury Press, London 2002.

A. Rogers, Come in, Mars, *Newsweek* **128** No.8 (1996) 41–45.

E. Samuel, Young suns go for it, *New Scientist* (15 September 2001) 10.

G. Schilling, Weird new planets leave theory behind, Science **291** (2001) 409–412.

S. Schmidt, *Aliens and Alien Societies*, Writer's Digest Books, Cincinnati OH 1995.

J. Schnabel, *Dark White: Aliens, Abductions and the UFO Obsession*, Hamish Hamilton, London 1994.

M. Schrope, Expanding life's alphabet, *New Scientist* (8 April 2000) 12.

S. Shostak, *Sharing the Universe*, Berkeley Hills Books, Berkeley CA 1998.

M. Sincell, Shadow and shine offer glimpses of otherworldly Jupiters, *Science* **286** (1999) 1822–1823.

D. Stevenson, Europa's ocean – the case strengthens, *Science* **289** (2000) 1305–1306.

I. Stewart and J. Cohen, *Figments of Reality*, Cambridge University Press, Cambridge 1997.

C. Tudge, *The Variety of Life*, Oxford University Press, Oxford 2000.

G. Vogel, Expanding the habitable zone, *Science* **286** (1999) 70–71.

G. Wächtershäuser, Life as we don't know it, *Science* **289** (2000) 1307–1308.

M. Walter, *In Search of Life on Mars*, Perseus, Cambridge 1999.

P.D. Ward and D. Brownlee, *Rare Earth: why Complex Life is Uncommon in the Universe*, Copernicus Books, New York 2000.

C. Zimmer, 'Inconceivable' bugs eat methane on the ocean floor, *Science* **293** (2001) 418–419.

B. Zuckerman and M.H. Hart (eds.), *Extraterrestrials, Where Are They?* Cambridge University Press, Cambridge 1995.

TECHNICAL XENOSCIENCE READING LIST

B.W. Aldiss, Desperately seeking aliens, Insight Astrobiology, *Nature* **409** (2001) 1080–1082.

R.T. Anderson, F.H. Chapelle, and D.R. Lovley, Evidence against hydrogen-based microbial systems in basalt aquifers, *Science* **281** (1998) 976–977.

S.Y. Auyang, *Foundations of Complex-System Theories*, Cambridge University Press, Cambridge 1998.

M.Balter, Evolution on life's fringes, *Science* **289** (2000) 1866–1867.

J.P. Bradley, R.P. Harvey, and H.Y. McSween Jr., No 'nanofossils' in Martian meteorite, *Nature* **390** (1997) 454.

M.H. Carr, Evidence for a subsurface ocean on Europa, *Nature* **391** (1998) 363–365.

S.B. Carroll, Chance and necessity; the evolution of morphological complexity and diversity, Insight Astrobiology, *Nature* **409** (2001) 1102–1109.

C.R. Chapman, Probing Europa's third dimension, *Science* **283** (1999) 338–339.

C.F. Chyba, Energy for microbial life on Europa, *Nature* **403** (2000) 381–382.

G.D. Cody, N.Z. Boctor, T.R. Filley, R.M. Hazen, J.H. Scott, A. Sharma, and H.S. Yoder Jr, Primordial carbonylated iron-sulfur compounds and the synthesis of pyruvate, *Science* **289** (2000) 1337–1340.

B.A. Cohen, T.D. Swindle, and D.A. Kring, Support for the lunar cataclysm hypothesis for lunar meteorite impact melt ages, *Science* **290** (2000) 1754–1756.

J. Cohen, Speculations on the evolutionary history of the tribble, *The Spang Blah* **18** (1979) 3–7.

J. Cohen and I. Stewart, Where are the dolphins? Insight Astrobiology, *Nature* **409** (2001) 1119–1122.

A.R. Collar and J.W. Flower, A (relatively) low altitude 24-hour satellite, *J. Brit. Interplanetary Soc.* **22** (1969) 442–457.

M.J. Crowe, *The Extraterrestrial Life Debate, 1750–1900*, Dover Publications, Mineola NY 1999.

V. Döring, H.D. Mootz, L.A. Nangle, T.L. Henrickson, V. de Crécy-Lagard, P. Schimmel, and P. Marlière, Enlarging the amino acid set of *Escherischia coli* by infiltration of the valine coding pathway, *Science* **292** (2001) 501–504.

M.J. Fogg, Temporal aspects of the interaction among the first galactic civilisations: the Interdict Hypothesis, *Icarus* **69** (1987) 370–384.

E.I. Friedmann, Endolithic microorganisms in the Antarctic cold desert, *Science* **215** (1982) 1045–1053.

T. Gold, The deep, hot biosphere, *Proceedings of the National Academy of Sciences of the USA* **89** (1992) 6045–6049.

C. Huber and G. Wächtershäuser, Peptides by activation of amino acids with CO on (Ni,Fe)S surfaces: implications for the origin of life, *Science* **281** (1998) 670–672.

J.D. Isaacs, H. Bradner, and G.E. Backus, Satellite elongation into a true 'sky-hook', *Science* **151** (1966) 682–683.

S.A. Kauffman, *The Origins of Order*, Oxford University Press, Oxford 1993.

S.A. Kauffman, *Investigations*, Oxford University Press, Oxford 2000.

E.F. Keller, *The Century of the Gene*, Harvard University Press, Harvard 2000.

J.P. Kennett, K.G. Cannariato, I.L. Hendy, and R.J. Behl, Carbon isotopic evidence for methane hydrate instability during quaternary interstadials, *Science* **288** (2000) 128–133.

R.A. Kerr, Putative Martian microbes called microscopy artifacts, *Science* **278** (1997) 1706–1707.

M.G. Kivelson, K.K. Khurana, C.T. Russell, M. Volwerk, R.J. Walker, and C. Zimmer, Galileo magnetometer measurements: a strong case for a subsurface ocean on Europa, *Science* **289** (2000) 1340–1343.

A. Labeyrie, Snaphots of alien worlds – the future of interferometry,

Science **285** (1999) 1864–1865.

M. Lachmann, M.E.J. Newman, and C. Moore, The physical limits of communication, Working paper **99–07–054**, Santa Fe Institute 2000.

J. Laskar and P. Robutel, The chaotic obliquity of planets, *Nature* **361** (1993) 608–614.

J. Laskar, F. Joutel, and P. Robutel, Stabilization of the Earth's obliquity by the Moon, *Nature* **361** (1993) 615–617.

L. Li and S. Lindquist, Creating a protein-based element of inheritance, *Science* **287** (2000) 661–664.

H. Lipson and J.B. Pollack, Automatic design and manufacture of robotic lifeforms, *Nature* **406** (2000) 974–978.

J.J. Lissauer, Three planets for upsilon Andromedae, *Nature* **398** (1999) 659–660.

D. Mange and M. Sipper, Von Neumann's quintessential message: genotype + ribotype = phenotype, *Artificial Life* **4** (1998) 225–227.

G.W. Marcy and R.P. Butler, Detection of extrasolar giant planets, *Ann. Rev. Astron. Astrophys.* **36** (1998) 57–97.

D.S. McKay, E.K. Gibson Jr., K. L. Thomas-Keprta, H. Vali, C.S. Romanek, S.J. Clemett, X.D.F. Chillier, C.R. Maechling, and R.N. Zare, Search for past life on Mars: possible relic biogenic activity in Martian meteorite ALH84001, *Science* **273** (1996) 924–930.

S. Messenger, Identification of molecular-cloud material in interplanetary dust particles, *Nature* **404** (2000) 968–971.

J. von Neumann, *Theory of Self-Reproducing Automata*, University of Illinois Press, Urbana IL 1966.

D.E. Nilsson and S. Pelger, A pessimistic estimate of the time required for an eye to evolve, *Proceedings of the Royal Society of London B* **256** (1994) 53–58.

E.G. Nisbett and N.H. Sleep, The habitat and nature of early life, Insight Astrobiology, *Nature* **409** (2001) 1083–1091.

V.J. Orphan, C.H. House, K.-U. Hinrichs, K.D. McKeegan, and E.F. DeLong, Methane-consuming archaea revealed by directly coupled isotopic and phylogenetic analysis, *Science* **293** (2001) 484–487.

I.M. Pepperberg, *The Alex Studies: Cognitive and Communicative Abilities of Grey Parrots*, Harvard University Press, Cambridge 1999.

J.-P. Petit, Les OVNIs: s'ils existent, voici comment ils peuvent voler et plonger, *Science Vie* **129** No. 702 (1976) 42–49.

J.A. Reggia, J.D. Lohn, and H.-H. Chou, Self-replicating structures: evolution, emergence, and computation, *Artificial Life* **4** (1998) 283–302.

M. Roos-Serote, A.R. Vasavada, K. Kamp, P. Drossart, P. Irwin, C. Nixon, and R.W. Carlson, Proximate humid and dry regions in Jupiter's atmosphere indicate complex local meteorology, *Nature* **405** (2000) 158–160.

L.J. Rothschild and R.L. Mancinelli, Life in extreme environments, Insight Astrobiology, *Nature* **409** (2001) 1092–1101.

J. Ruiz, The stability against freezing of an internal liquid-water ocean in Callisto, *Nature* **412** (2001) 409–411.

M.J. Russell, D.E. Daia, and A.J. Hall, The emergence of life from FeS bubbles at alkaline hot springs in an acid ocean, in: *Thermophiles, the keys to molecular evolution and the origin of life* (eds. M.W.W. Adams, L.G. Ljungdahl, and J. Wiegel) Taylor and Francis, Washington DC 1998, 77–126.

K. Scherer, H. Fichtner, J.D. Anderson, and E.L. Lau, A pulsar, the heliosphere, and Pioneer 10: probable mimicking of a planet of PSR B1257+12 by solar rotation, *Science* **278** (1997) 1919–1921.

K.-U. Schöning, P. Scholz, S. Guntha, X. Wu, R. Krishnamurthy, and A. Eschenmoser, Chemical etiology of nucleic acid structure: the a-threofuranosyl-($3' \rightarrow 2'$) oligonucleotide system, *Science* **290** (2000) 1347–1351.

M. Sipper, Fifty years of research on self-replication: an overview, *Artificial Life* **4** (1998) 237–257.

K.O. Stetter, Hyperthermophiles: isolation, classification, and properties, in: *Extremophiles: microbial life in extreme environments* (eds. K. Horikoshi and W.D. Grant) Wiley-Liss, New York 1998, 1–24.

A. Tough, Fresh SETI strategies, *Journal of the British Interplanetary Society* **52** (1999) 286–289.

H.L. True and S.L. Lindquist, A yeast prion provides a mechanism for genetic variation and phenotypic diversity, *Nature* **407** (2000) 477–483.

M. Vargas, K. Kashefi, E.L. Blunt-Harris, and D.R. Lovley, Microbiological evidence for Fe(III) reduction on early Earth,

Nature **395** (1998) 65–67.

D. Vokrouhlicky and P. Farinelle, Efficient delivery of meteorites to the Earth from a wide range of asteroid parent bodies, *Nature* **407** (2000) 606–608.

R.H. Vreeland, W.D. Rozenzweig, and D.W. Powers, Isolation of a 250 million year old halotolerant bacterium from a primary salt crystal, *Nature* **407** (2000) 897–900.

L. Wang, A. Brock, B. Harberich, and P.G. Schultz, Expanding the genetic code of *Escherischia coli*, *Science* **292** (2001) 498–500.

B.P. Weiss, J.L. Kirschvink, F.J. Baudenbacher, H. Vali, N.T. Peters, F.A. Macdonald, and J.P. Wikswo, A low temperature transfer of ALH84001 from Mars to Earth, *Science* **290** (2000) 791–795.

R.H. White, Hydrolytic stability of biomolecules at high temperatures and its implication for life at 250°C, *Nature* **310** (1984) 430–432.

R.J. White and M. Averner, Humans in space, Insight Astrobiology, *Nature* **409** (2001) 1115–1118.

T.L. Wilson, The search for extraterrestrial intelligence, Insight Astrobiology, *Nature* **409** (2001) 1110–1114.

M.R. Zapatero Osorio, V.J.S. Béjar, E.L. Martín, R. Rebolo, D. Barrado y Navascués, C.A.L. Bailer-Jones, and R. Mundt, Discovery of young, isolated planetary mass objects in the σ Orionis star cluster, *Science* **290** (2000) 103–107.

INDEX